The Nature of Explanation

THE NATURE OF EXPLANATION

PETER ACHINSTEIN

New York   Oxford
OXFORD UNIVERSITY PRESS

Oxford University Press

Oxford New York Toronto
Delhi Bombay Calcutta Madras Karachi
Petaling Jaya Singapore Hong Kong Tokyo
Nairobi Dar es Salaam Cape Town
Melbourne Auckland

and associated companies in
Beirut Berlin Ibadan Nicosia

Copyright © 1983 by Oxford University Press, Inc.

First published in 1983 by Oxford University Press,
Inc., 200 Madison Avenue, New York, New York 10016
First issued as an Oxford University Press paperback, 1985

Oxford is a registered trademark of Oxford University Press.

All rights reserved. No part of this publication may be reproduced,
stored in a retrieval system, or transmitted, in any form or by any
means, electronic, mechanical, photocopying, recording, or otherwise,
without the prior permission of Oxford University Press.

Library of Congress Cataloging in Publication Data
Achinstein, Peter.
The nature of explanation.
Includes index.
1. Explanation (Philosophy)   Addresses, essays,
lectures.   I. Title.
BD237.A25   1983   121   82-22571
ISBN 0-19-503215-2
        0-19-503743-X (pbk.)

Printing (last digit): 9 8 7 6 5 4 3 2
Printed in the United States of America

For my son Jonathan

Preface

For more than a third of a century philosophical discussions of explanation have been dominated by various formal "models." Included among these are the deductive and statistical models of Hempel and variants proposed by others, and the statistical model of Salmon. Yet there are well-known counterexamples to these models, and good reasons for supposing that they fail to provide even necessary conditions. The typical response is to add more conditions to the model or to modify others. But each new variation on the old theme is simply an invitation to philosophical sharpshooters to hit the mark with fresh counterexamples.

What this book presents is not another variation but a different theme. Unlike the standard theories, it focuses, to begin with, on the explaining act itself—the act in which by uttering or writing words someone explains something. From that act a "product" emerges: an explanation. To characterize what kind of entity that product is, as well as how it can be evaluated, essential reference must be made to the concept of an explaining act. Otherwise, I argue, we will be unable to distinguish explanations from products of non-explaining acts; and we will be unable to say why various explanations, particularly in the sciences, deserve praise or blame.

A theory of the explaining act, of the product and its ontological status, and of the evaluation of explanations, is presented in the second, third, and fourth chapters. Following this the theory is brought to bear on a number of issues: why have the standard models of explanation been unsuccessful (are they seeking the impossible)? What is a causal explanation, and must explanations in the sciences be causal? What sort of explanation is one that appeals to the function of something (is it, e.g., causal)? Are some things too fundamental to be explained, and what does that mean? What is the relationship between explanation and evidence (can the latter concept be defined by reference to the former)? These issues, and others, will be approached by making essential use of the theory of explanation developed in the earlier chapters.

One trend in recent philosophy of science is to reject the formalism and precise schemas of logical positivism, from which the early models of explanation emerged. Although I too will not accept the tenets of positivism and its attendant picture of explanation, I will try to develop the theory of explanation in a precise way. This will necessitate some "formal" definitions. (Those who want to take their philosophy at a gallop are forewarned.) However, numerous examples are presented in order to make these definitions clear. Where appropriate, the examples are simple, everyday ones. In some cases—particularly in the discussion of the evaluation of scientific explanations, of causal explanations, and of the limits of explanation in science—I have chosen examples from physics.

The theory of explanation presented here had a forerunner in Chapter 4 of my book *Law and Explanation,* published in 1971. That account I no longer regard as adequate. However, in some of the chapters that follow I have made use of revised and expanded material from recent articles published in the *American Philosophical Quarterly* (vol. 14), *Midwest Studies in Philosophy* (vol. 4), *Mind* (vol. 76), and *Philosophy of Science* (vol. 44).

I am indebted to the National Science Foundation for research grants during the tenure of which this book was written. In the spring of 1976 when I served as Lady Davis Visiting Professor at the Hebrew University of Jerusalem I offered a graduate seminar on explanation, and began to think particularly about the problem of functional explanation discussed in Chapter 8. This and many other issues were also aired in an N.E.H. seminar for college teachers I directed in the summer of 1978 and in graduate seminars at Johns Hopkins. Participants in these seminars as well as persons from various universities in the U.S. and Israel where material from several chapters was read bestowed vigorous criticisms on my proposals. They are to be credited for my numerous attempts in what follows to counter objections.

Stimulating discussions with Jaegwon Kim, who read the entire manuscript, sharpened my thoughts, as did perceptive comments by George Wilson, Dale Gottlieb, and Barbara von Eckardt. These persons made important suggestions for improving the quality of the work and I am deeply indebted to them. Luckily for me some of their suggestions conflicted; otherwise I would still be revising.

Alan Berger raised questions about Chapter 2 that I had not previously considered. Robert Causey and Bas van Fraassen helped me better understand views of theirs which I treat. David Zaret, with whom

I taught a graduate seminar on evidence in the fall of 1981, offered criticisms which improved the last chapter. Carl G. Hempel noted a problem in the technical formulation of the NES condition in the footnote on p. 159; in this paperback edition I offer a revised formulation. Robert Cummins and Daniel Garber, both lively philosophical conversationalists, gave me the opportunity to explore ideas over lunch at the Hopkins faculty club. And David Sachs furnished wise counsel when it was needed.

I want to express my thanks to Nancy Thompson, the Philosophy Department's chief secretary, for her generous help on many matters, to Cecelia Hrdlick and Molly Mitchell for their excellent typing, and to Vera Herst and Fred Kronz for assistance with the proofs.

To my wife Merle Ann, who shared with me the exhilarating as well as the difficult moments in the enterprise of explaining explaining, and who provided moral and physical support, I am forever grateful.

Baltimore   P.A.
November 1982

Contents

1. Introduction, 3
2. Explaining, 15
3. What Is an Explanation?, 74
4. The Evaluation of Explanations, 103
5. Can There Be a Model of Scientific Explanation?, 157
6. The Causal Relation, 193
7. Causal Explanation, 218
8. Functional Explanation, 263
9. The Limits of Explanation, 291
10. Evidence and Explanation, 322
11. Evidence: Additional Topics, 351

Index, 383

The Nature of Explanation

CHAPTER 1
Introduction

1. THE PROJECT

Suppose that a speaker S explains something q by uttering (or writing) a sentence u. For example, S explains why that metal expanded, by uttering

   That metal expanded because it was heated.

An *act* of explaining has occurred, which took some time, however short. We might also speak of the *product* of this act—viz. the explanation given by S—which did not take any time, but was produced in or by the act of explaining. ("The explanation given by S" can be used to refer either to the act or to the product; in what follows it will be used for the product.) If S repeats the sentence above on different occasions when he explains this phenomenon, he has engaged in several explaining acts (he has explained several times), even though the product—his explanation—is the same on each occasion.

Various questions can be raised about explaining acts and products. The ones of concern to me in this chapter and the three that follow are quite general:

a. What is an explaining act? More particularly, can necessary and sufficient conditions be supplied for sentences of the following form?

   (1) S explains q by uttering u.

b. What is the product of an explaining act? (What is an explanation?) Can necessary and sufficient conditions be supplied for sentences of the form

   (2) E is an explanation of q.

c. How should explanations (i.e., products) be evaluated? Can necessary and sufficient conditions be supplied for sentences of the form

(3) E is a good explanation of q.

(Perhaps we will want to consider other terms as well for positive or negative evaluations.)

My aim is to develop a theory of explanation that will answer these questions, and in particular, propose informative conditions for sentences of forms (1), (2), and (3).

The terms "explain" and "explanation" can be used broadly to refer to explaining acts and products that may or may not be good (adequate, successful, "scientific"). They can also be used more narrowly to refer only to acts and products that are (regarded as) good. In accordance with the broader, but not the narrower, use an atheist could admit that his religious friends are explaining the origin of man when they assert that man was created by God. And he could refer to the product of such acts as an explanation. In what follows, the terms "explain" and "explanation" in (1) and (2) will be used in this broader way. The narrower, evaluative use will be considered when we turn to sentences of form (3).

The theory of explanation I propose begins with explaining acts, and in particular with sentences of form (1). These will be my concern in Chapter 2. Questions (b) and (c) will be addressed in Chapters 3 and 4, respectively. My thesis is that the concept of an explaining act is fundamental, and that the concept of an explanation (as product), and that of a good explanation, must be understood, in important ways, by reference to the former. Thus, the answers I will propose to questions (b) and (c) will depend on a prior consideration of question (a).

This is very different from the usual strategy of explanation theorists, which is to concentrate on question (c); to offer some views, or at least clues, about (b) but not dwell on this (since the concern is not to characterize explanations generally, but only good ones); and largely to ignore (a) by focusing on what its proponents call the "logic" of explanation rather than its "pragmatics." In the remainder of this chapter I shall describe some leading theories of this sort. My aim is not to try capturing all the details of these theories, but only their main outlines. In later chapters I will show that by developing views about the product of explanation, and about the evaluation of such products, that are independent of the concept of an explaining act these theories, and others like them, provide inadequate accounts.

2. ARISTOTLE'S DOCTRINE OF THE FOUR CAUSES

There are, Aristotle believed, four causes or determining factors which correspond to the meanings of the question "why": the material, formal, efficient, and final causes.[1] Aristotle is concerned with why-questions of the form

(1) Why does X have (property) P?

In *Metaphysics* Z he writes:

> The "why" is always sought in this form—"why does one thing attach to some other?" . . . We are inquiring, then, why something is predicable of something. (1041a)

And in the *Physics*, where the doctrine is most fully developed, his examples of why-questions include "Why is he walking about?" and "Why did they go to war?"

Typical of the form of the answer to the why-question—i.e., the form of the explanation—Aristotle considers is "because r," or more fully,

(2) X has P because r.

("They went to war because they were raided.") It is Aristotle's view that the r-position will be filled by reference to various causes: the matter or constituents of which X is composed in virtue of which it has P (material cause), the form or structure of X (formal cause), an external source of motion or change by which X comes to have P (efficient cause), and "that for the sake of which" X has P (final cause).

Aristotle makes little attempt to define his four causes. He seems to treat them as primitives, that can be illustrated but not further analyzed. A complete explanation in physics, and perhaps in science generally, invokes causes of all four types:

> Now, the causes being four, it is the business of the physicist to know about them all, and if he refers his problems back to all of them, he will assign the "why" in the way proper to his science—the matter, the form, the mover, "that for the sake of which." [*Physics*, 198a]

I follow Moravcsik in construing Aristotle's doctrine of the four causes as a doctrine about explanation.[2] (Perhaps, then, "explanatory factor" is a better term to use than "cause" to express Aristotle's thought.) However, I do not agree with Moravcsik's view that Aristotle selects the four basic causes he does solely because of his onto-

---

1. This is Aristotle's view in the *Physics* and *Metaphysics*. A more elaborate view in the *Posterior Analytics* will be noted in Section 5.
2. Julius M. E. Moravcsik, "Aristotle on Adequate Explanations," *Synthese* 28 (1974), pp. 3–17.

logical doctrine of substance (which, according to Moravcsik, commits Aristotle to there being these four kinds). Rather, I take Aristotle's view to be, in part, a semantical one concerning the meaning of the word "cause" (or "explanatory factor" or "explanation"). In the *Physics*, after introducing the doctrine, he writes:

> This then perhaps exhausts the number of ways in which the term "cause" is used. (195a)

And in *Metaphysics* D (1013a):

> Cause means (1) that from which, as immanent material, a thing comes into being. . . . [Aristotle continues with all the causes.]

Aristotle's view is that by an explanation (i.e., by something which gives a cause) we mean something which gives one or more of the causes he mentions. However, he does go on to make the ontological claim that the four types of causes exist in nature. His doctrine, then, seems to combine both semantical and ontological features, as follows. Each substance has both matter and form (ontological doctrine). But these are two senses of the word "cause" (semantical doctrine). So each substance has both a material and a formal cause. Furthermore, substances exhibit behavior for the sake of an end (ontological fact). But one sense of "cause" is teleological, i.e., for the sake of an end. Therefore, there exists behavior that has a teleological cause. Finally, substances do exhibit behavior produced by external sources. But one sense of "cause" is external, i.e., efficient, cause. So efficient causes exist. In each case, the conclusion that there is in nature a cause of the sort Aristotle cites is reached by means of two premises, one ontological and one semantical.

Turning now to a simple example, consider the question

Why did that metal expand?

Assume that someone offers the explanation

(3) That metal expanded because it was heated.

The explanation here (i.e., the product), Aristotle seems to be saying, is a proposition, viz. the proposition expressed by (3). This is an explanation because it is a proposition that purports to describe a cause—in this case, an efficient cause—of the metal's expansion.

More generally, if we consider questions of the form

(1) Why does X have (property) P?,

Aristotle's answer to question (b) of the previous section (What is the product of an explaining act, i.e., what is an explanation?) can perhaps be formulated as follows. Where q is an indirect question whose direct form is given by (1),

(4) E is an explanation of q if and only if E is a proposition of the form "X has P because r," in which r purports to give one or more of Aristotle's four causes of X's having P.

Aristotle's answer to question (c) of the previous section (How should explanations be evaluated?) might then be given as follows. Where q is an indirect question whose direct form is (1),

(5) E is a good explanation of q if and only if E is a proposition of the form "X has P because r," in which r correctly gives (and does not merely purport to give) one or more of Aristotle's four causes of X's having P. (Or perhaps Aristotle's view is the stronger one that E must give causes of all four types, if they exist.)

Aristotle's doctrine of the four causes does not concern itself with explaining acts. It does not address question (a) (What is an explaining act?). However, taking a cue from what has been said above, Aristotle might have wished to espouse the view that explaining acts consist in the uttering of sentences that express propositions that are explanations. That is, where q is an indirect question whose direct form is (1),

(6) S explains q by uttering u if and only if S utters u, and u expresses a proposition of the form "X has P because r" in which r purports to give one or more of Aristotle's four causes of X's having P.

For example, S explains why that metal expanded by uttering "that metal expanded because it was heated," since S utters a sentence that expresses a proposition of the form "X has P because r" in which r is replaced by a sentence purporting to give an efficient cause of the metal's expansion.

However, it should be emphasized, I am not here attributing (6) to Aristotle himself. The main point I want to stress is that Aristotle seems to be defending a view which divorces an account of explanations from that of explaining acts. On this view, one can understand the concept of an explanation independently of that of an explaining act. Neither (4) nor (5) invokes the latter concept. Indeed, if (6) were Aristotle's position on explaining acts, then his theory of such acts would depend on an independent account of explanations.

3. HEMPEL'S DEDUCTIVE-NOMOLOGICAL MODEL

The view that an explanation of a phenomenon brings it under a law is implicit in many writings, but its fullest and most influential expression is due to Hempel.[3] According to this view, an explanation is an argument whose premises include laws and whose conclusion is a description of the phenomenon to be explained. Hempel distinguishes explanations of two types: deductive-nomological (D-N), and inductive-statistical. In the former, the premises entail the conclusion; in the latter, the premises make the conclusion probable without entailing it. Hempel is concerned with what he calls explanation-seeking why-questions.[4] These have the form

Why is it the case that p?

in which "p" is replaced by a sentence. Let Q(p) be a question of this form which presupposes that p is true, and let q(p) be the indirect form of this question. Hempel's view can now be put as follows:

(1) E is a potential D-N explanation of q(p) if and only if E is a valid deductive argument whose premises contain lawlike sentences that are actually used in the deduction and whose conclusion is p.[5]

(2) E is a good (correct, scientific) D-N explanation of q(p) if and only if E is a potential D-N explanation of q(p) all of whose premises are true.

For example, let Q(p) be

Why is it the case that this metal expanded?

Consider the argument

(3) This metal was heated
All metals expand when heated
Therefore,
This metal expanded.

By (1), (3) is a potential D-N explanation of why this metal expanded, since (3) is a valid deductive argument whose premises contain a lawlike sentence (the second premise) that is actually used in the deduction. If the premises of (3) are both true, then, by (2), (3) is also a good (correct, scientific) explanation of why this metal expanded.

Let us now turn to questions (a), (b), and (c) in Section 1 (What is an explaining act? What is an explanation? How should explanations be evaluated?). I take (1) to be an answer to (b), and (2) to be an answer to (c). An explanation, Hempel is urging, is an argument of a sort described in (1). (Or else it is a certain type of inductive argument

---

3. Carl G. Hempel, *Aspects of Scientific Explanation* (New York, 1965).
4. *Ibid.*, p. 334.
5. This is a simplification of Hempel's model; some additional conditions will be discussed in Chapter 5.

which also contains lawlike sentences; but I shall ignore the inductive cases in what follows.) A *good* explanation is an argument of the sort described in (1) in which all of the premises are true.

By contrast, Hempel fails to devote much attention to acts of explaining. He does recognize what he calls a "pragmatic" dimension of explanation. The use of the term "explanation" and its cognates "requires reference to the *persons involved in the process of explaining*."[6] And in this pragmatic sense, what explains something for someone might not explain it for someone else, because of differences in beliefs, puzzlements, or intelligence. Although he does not explicitly say so, perhaps Hempel would suggest a condition such as the following for explaining acts:

(4) S explains q by uttering u if and only if S utters u, and u expresses a deductive argument of the sort described in (1) (or a comparable inductive argument of the sort prescribed by the inductive-statistical model).

In accordance with this condition, if S utters (3) then S is explaining why this metal expanded, since (3) expresses a deductive argument of the sort described in (1). However, it should be stressed, Hempel does not devote himself to the project of providing conditions for explaining acts. Rather, he identifies his task as one of

> constructing a nonpragmatic concept of scientific explanation—a concept which is abstracted, as it were, from the pragmatic one, and which does not require relativization with respect to questioning individuals any more than does the concept of mathematical proof. It is this nonpragmatic conception of explanation which the covering-law models are meant to explicate.[7]

Hempel believes that there is a concept of explanation which can be understood without reference to the idea of an explaining act. (1) and (2) above invoke no such idea. Indeed, if (4)—or something like it—were Hempel's account of explaining acts, then an understanding of explaining acts would depend on a prior understanding of explanations.

4. SALMON'S STATISTICAL RELEVANCE (S-R) MODEL

Like Hempel, Salmon stresses the need for laws in explanations.[8] Unlike Hempel, he construes an explanation not as an argument but simply as a set of sentences. These sentences provide a basis for an

---

6. *Ibid.*, p. 425, my emphasis.
7. *Ibid.*, p. 426.
8. Wesley C. Salmon, *Statistical Explanation and Statistical Relevance* (Pittsburgh, 1971).

inference concerning the event to be explained; but Salmon, by contrast with Hempel, does not require that this inference be to the conclusion that the event definitely, or even probably, occurred. The explanation need provide only a basis for inferring with what probability (however small) it was to be expected to occur. Salmon writes:

> An explanation does not show that the event was to be expected; it shows what sorts of expectations would have been reasonable and under what circumstances it was to be expected.[9]

Salmon's statistical relevance model embodies this idea. It is concerned with explanations which answer questions of the form

Why is X, which is a member of class A, a member of class B?

An explanation consists of a set of empirical probability laws relating classes A and B, together with a class inclusion sentence for X, as follows:

(1) $p(B, A \& C_1) = p_1$
$p(B, A \& C_2) = p_2$
.
.
.
$p(B, A \& C_n) = p_n$
$X \in C_k \ (1 \leq k \leq n)$

Salmon imposes two conditions on the explanation. One is that the probability values $p_1, \ldots, p_n$ all be different. The other is

> *The homogeneity condition:* $A \& C_1, A \& C_2, \ldots, A \& C_n$ is a partition of A, and each $A \& C_i$ is homogeneous with respect to B.

$A \& C_1, \ldots, A \& C_n$ is a partition of A if and only if these sets comprise a set of mutually exclusive and exhaustive subsets of A. Set A is homogeneous with respect to B if and only if there is no way, even in principle, to effect a partition of A that is statistically relevant to B without already knowing which members of A are also members of B. (C is statistically relevant to B within A if and only if $p(B, A \& C) \neq p(B, A)$.) Intuitively, if A is homogeneous with respect to B, then A is a random class with respect to B.

Consider a simple example in which the explanatory question is

(2) Why is this substance, which is a member of the class of metals, a member of the class of things that melt at 1083°C? (That is, why does this piece of metal melt at 1083°C?)

---

9. *Ibid.*, p. 79.

Letting

>A = the class of metals
>B = the class of things that melt at 1083°C
>C_1 = the class of things that are copper
>C_2 = the class of things that are not copper,

we can construct the following explanation for (2):

> (3) $p(B, A \& C_1) = 1$
> $p(B, A \& C_2) = 0$
> $X \in C_1$.

What this explanation tells us is that this substance, which is a metal, melts at 1083°C, because

> the probability that something melts at 1083°C, given that it is copper, is 1; and
>
> the probability that something melts at 1083°C, given that it is not copper, is 0; and
>
> this substance is copper.

Salmon's two conditions are satisfied: the probability values in (3) are different; $A \& C_1$ and $A \& C_2$ form a partition of A and both these subclasses are homogeneous with respect to B.

Salmon's statistical relevance model is, I suggest, best construed as an attempt to answer question (c) of Section 1. It is an attempt to provide conditions for being a *good* (adequate, scientific) explanation. Perhaps Salmon would say that his model gives at least this much of an answer to question (b): An explanation is a set of sentences of form (1) containing probability laws and a class-inclusion sentence. However, the only necessary and sufficient conditions he proposes are ones for being a good explanation. Nor does Salmon address the question of what conditions are necessary and sufficient for an explaining act. Clearly, the conditions for explanations that he does formulate are independent of the concept of such an act.

5. BRODY'S TWO MODELS

A theory of explanation developed by Baruch Brody combines certain features of Hempel's D-N model with some ideas from Aristotle. Brody formulates two models, in both of which an explanation is a deductive argument satisfying the conditions of Hempel's D-N model. According to the first, the causal model, Hempel's D-N conditions must be supplemented by the

12   The Nature of Explanation

Causal condition: The premises of the argument must contain essentially a description of the event which is the cause of the event described in the conclusion.[10]

So, e.g.,

(1) This metal was heated
All metals expand when heated
    Therefore,
This metal expanded

is an explanation of the expansion, on Brody's causal model, since it is a deductive argument satisfying Hempel's D-N conditions plus the causal condition above. Its premises contain the sentence "this metal was heated," which is a description of the event that caused the event described in the conclusion.

Brody's second model, the essential property model, requires that Hempel's D-N conditions be supplemented by the

Essential property condition: The premises of the argument contain "a statement attributing to a certain class of objects a property had essentially by that class of objects (even if the statement does not say that they have it essentially) and . . . at least one object involved in the event described in the [conclusion] is a member of that class of objects."[11]

In a later work Brody attempts to explicate the concept of an essential property by developing the idea that a has property P essentially if and only if there is no possible future in which a continues to exist but does not have P.[12] His examples include those in which (he claims) atomic numbers of substances are essential properties of those substances. Thus consider

(2) This substance is copper
Copper has the atomic number 29
Whatever has the atomic number 29 conducts electricity
    Therefore,
This substance conducts electricity.

This, Brody would say, satisfies not only Hempel's D-N model but the essential property condition as well. Its premises contain a statement—the second one—that attributes to copper an essential property; and the object described in the conclusion (this substance) is, indeed, copper.

10. B. A. Brody, "Towards an Aristotelean Theory of Scientific Explanation," *Philosophy of Science* 39 (1972), pp. 20–31; see p. 23.
11. *Ibid.*, p. 26.
12. *Identity and Essence* (Princeton, 1980), pp. 115ff.

Brody's models are, I suggest, proposed as answers to question (c) of Section 1. They are meant to provide necessary and sufficient conditions for being a good explanation. Brody, unlike Hempel, does not introduce the concept of a potential explanation; and it is not clear to me whether what he says can be used to formulate necessary and sufficient conditions for being an explanation (where goodness is not implied). Nor does he characterize explaining acts. But it is evident that the conditions for explanations that he does supply do not invoke the concept of an explaining act.

Brody claims to derive his view from Aristotle's doctrine of scientific knowledge in the *Posterior Analytics*. In this work Aristotle does not develop separate models, as Brody has done, but only one. According to it, an explanation (or what Aristotle calls a "demonstration") is "a syllogism productive of scientific knowledge" (bk. 1, ch. 2). The premises in such a syllogism must show not simply that the phenomenon to be explained did occur, but that it had to occur. To do this, according to Aristotle, they must ascribe to things properties that are essential to them, the having of which caused the phenomenon to be explained. On Aristotle's view, then, an explanation is a deductive argument whose premises cite both an essential property of a substance involved in the phenomenon to be explained and the cause of that phenomenon, which together permit the inference that the phenomenon had to occur.

6. CONCLUSIONS

There are other models of explanation in the literature, a number of which will be discussed in Chapter 5. But, I suggest, the ones briefly outlined here are typical in the following respects. They are concerned primarily with providing conditions for E's being a good explanation of q. In general, the accounts have this form:

> E is a good explanation of q if and only if E is a proposition (or argument, which is a certain kind of compound proposition) satisfying conditions C.

Conditions C impose requirements on the kinds of propositions which comprise the explanation and on their relationships to one another. But they make no mention of persons or types of persons who are explaining, or of particular or general types of explaining acts, or of audiences for whom explanations are given or intended. Nor do the modelists in question even attempt to characterize explaining acts. One might speculate that if they were to do so, their view would be that an explaining act—one in which a speaker explains by uttering something—can be understood quite simply as an act in which the sentence or sentences uttered express a proposition or argument satisfying their conditions for being a (good) explanation. Views of this sort would hold that a sentence of the form "S explains q by uttering u" is true if and only if S utters u, and u expresses an explanation. Thus, the concept of an explanation is to be understood independently of that of an explaining act, while that of an explaining act is to be understood by reference to the concept of an explanation.

I turn now to the development of a very different type of theory. It begins with explaining acts, which are the subject of the chapter that follows. In Chapters 3 and 4, respectively, the theory will characterize the concept of an explanation, and that of a good explanation, by reference to such acts. In the course of the latter discussion I will argue that serious problems in the characterization of these concepts will arise if explaining acts are ignored or de-emphasized.

CHAPTER 2
Explaining

1. CONDITIONS FOR AN ACT OF EXPLAINING

The verb "to explain" is, to borrow a classification from Zeno Vendler, an accomplishment term.[1] It has a continuous present, "is explaining," that indicates that an act is occurring that occupies some stretch of time. But unlike some other verbs which also have a continuous present, such as "to run" and "to push" (which Vendler calls activity terms), it has a past tense which indicates not simply a stop to the act but a conclusion or completion. If John was running, then no matter for how long he was running, he ran. But if the doctor was explaining Bill's stomach ache, then it is not necessarily true that he explained it, since his act may have been interrupted before completion.

Sylvain Bromberger suggests that although the accomplishment use of "explain" is the most fundamental there is also a non-accomplishment use, illustrated by saying that Newton explained the tides. Here

> one need not mean that some explaining episode took place in which Newton was the tutor. One may mean that Newton solved the problem, found the answer to the question.[2]

Using Vendler's terminology, Bromberger classifies this as an "achievement" use of "explains." Achievement terms (e.g., "winning a race"), unlike accomplishment terms, describe something that occurs at a single moment rather than over a stretch of time.

I believe that Bromberger is mistaken. If Newton has simply solved the problem or found the answer, although he may be in a position

---

1. Zeno Vendler, *Linguistics in Philosophy* (Ithaca, 1967), p. 102.
2. Sylvain Bromberger, "An Approach to Explanation," in R. J. Butler, ed., *Analytical Philosophy* 2 (Oxford, 1965), pp. 72–105.

to explain the tides (we say that he *would* or *can* explain them as follows), he has not yet done so until he has said, or written, or at least communicated, something. One does not explain simply by believing, or even by solving a problem or finding an answer, unless that belief, solution, or answer is expressed in some act of uttering or writing. (We sometimes explain rather simple things by non-verbal acts such as gesturing, but such cases will not be of concern to me here.) This does not mean that we must construe "Newton explained the tides" as describing a particular explaining episode. Following Davidson,[3] this can be treated as an existentially general sentence, i.e., as saying that there was at least one act which was an explaining of the tides by Newton.

Explaining is what Austin calls an illocutionary act.[4] Like warning and promising, it is typically performed by uttering words in certain contexts with appropriate intentions. It is to be distinguished from what Austin calls perlocutionary acts, such as enlightening someone, or getting someone to understand, or removing someone's puzzlement, which are the effects one's act of explaining can have upon the thoughts and beliefs of others.

The illocutionary character of explaining can be exposed by formulating a set of conditions for performing such an act. To do so I shall consider sentences of the form "S explains q by uttering u," in which S denotes some person, q expresses an indirect question, and u is a sentence. (I will assume that any sentence of this form in which q is not an indirect question is transformable into one that is.)[5]

The first condition expresses what I take to be a fundamental relationship between explaining and understanding. It is that S explains q by uttering u only if

(1) S utters u with the intention that his utterance of u render q understandable.

This expresses the central point of S's act. It is the most important feature which distinguishes explaining from other illocutionary acts, even ones that can have indirect questions as objects. If by uttering u I am asking you, or agreeing with you about, why the tides occur, by contrast to explaining it, I will not be doing so with the intention that my utterance render why the tides occur understandable. (I shall return to the concept of understanding in Section 3 after formulating the remaining conditions.)

---

3. Donald Davidson, "The Logical Form of Action Sentences," in N. Rescher, ed., *The Logic of Decision and Action* (Pittsburgh, 1967).
4. J. L. Austin, *How To Do Things with Words* (Oxford, 1962). Austin includes "explain" on his list of "expositives," pp. 160–61.
5. In Section 9, the field will be broadened to include cases in which u is not a complete sentence.

To explain q is not to utter just anything with the intention that the utterance render q understandable. Suppose I believe that the words "truth is beauty" are so causally efficacious with you that the mere uttering of them will cause you to understand anything, including why the tides occur. By uttering these words I have not thereby explained why the tides occur, even if I have satisfied (1). The reason is that I do not believe that "truth is beauty" expresses a correct answer to the question "Why do the tides occur?" More generally, assuming that answers to questions are propositions (see Section 3), we may say that S explains q by uttering u only if

    (2) S believes that u expresses a proposition that is a correct answer to Q. (Q is the direct form of the question whose indirect form is q.)

Often people will present hints, clues, or instructions which do not themselves answer the question but enable an answer to be found by others. To the question "Why do the tides occur?" I might respond: "Look it up in Chapter 10 of your physics text," or "Newton's *Principia* has the answer," or "Think of gravity." Some hints, no doubt, border on being answers to the question. But in those cases where they do not, it is not completely appropriate to speak of explaining. By uttering "Look it up in Chapter 10 of your physics text" I am not explaining why the tides occur, though I am uttering something which, I believe, will put you in a position to explain this.

These conditions are not yet sufficient. Suppose that S intends that his utterance of u render q understandable not by producing the knowledge that u expresses a correct answer to Q but by causing people to come to think of some non-equivalent sentence u' which, like u, S believes expresses a proposition that is a correct answer to Q. In such a case, although S utters something which he believes will cause others to be able to explain q, S does not himself explain q by uttering u. For example, to an audience that I believe already knows that the tides occur because of gravitational attraction, I say

    u: The tides occur because of gravitational attraction of the sort described by Newton.

Although I believe that u does express a correct answer to Q (Why do the tides occur?), suppose that I utter u with the following intention: that this utterance will render q understandable not by producing the knowledge of the proposition expressed by u that it is a correct answer to Q, but by causing my audience to look up the more detailed and precise answer actually supplied by Newton, which I don't present. This is like the situation in which I give the audience a hint that in this case is a correct answer, but is not the answer in virtue of which I intend q to be understandable to that audience.

18   The Nature of Explanation

To preclude such cases we can say that S explains q by uttering u only if

(3) S utters u with the intention that his utterance of u render q understandable by producing the knowledge, of the proposition expressed by u, that it is a correct answer to Q.

In the case of the tides mentioned above, I do not intend that my utterance of u render q understandable by producing the knowledge *of the proposition expressed by u* that it is a correct answer to Q, but by producing such knowledge with respect to another proposition. So, according to condition (3), in such a case by uttering u I am not explaining why the tides occur.

Suppose, by contrast, I know that my audience is familiar with the answer supplied by Newton, but its members have no idea whether this answer is correct. Since the audience knows what sort of gravitational attraction Newton describes, I might explain why the tides occur, simply by uttering u. In this case I intend to render q understandable by producing the knowledge, of the proposition expressed by u, that it is a correct answer to Q. It is possible for me to have this intention with respect to u since I know that the audience is aware of the sort of gravitational attraction described by Newton.

Let us change the example once more. Suppose I believe that the audience does not know that the tides are due to gravitational attraction. I now proceed to utter u above with the intention that my utterance of u will render q understandable by the following combination of means (which I regard as jointly but not separately sufficient for rendering q understandable): (i) producing the knowledge, of the proposition expressed by u, that it is a correct answer to Q; and (ii) causing others to look up some different, more detailed, proposition (supplied by Newton) which is also a correct answer to Q. By uttering u am I explaining why the tides occur?

One might be inclined to say that I am *both* explaining q by uttering u *and* giving a clue about where to find another answer to Q. If this is correct, then (3) should be understood in a way that allows S to intend to render q understandable by a combination of means that includes producing the knowledge, of the proposition expressed by u, that it is a correct answer to Q. On the other hand, in the case just envisaged one might be tempted to say that I am doing something that falls between explaining and giving clues but is not exactly either. If this is correct, then (3) should be understood in a way that requires S to intend to render q understandable *solely* by producing the knowledge, of the proposition expressed by u, that it is a correct answer to Q. I am inclined to regard the latter interpretation of (3) as preferable, but I will not press the point. (This, of course, does not preclude

S from explaining q by formulating a number of different propositions whose conjunction constitutes an answer to Q, or from engaging in several acts in which different, though not necessarily competing, answers to Q are provided.)

In Section 9 some further conditions (involving restrictions on q and u) will be suggested whose formulation requires concepts to be introduced later. For the present I shall treat these three conditions as not only necessary but jointly sufficient. If so, then the same honor can be accorded to (3) by itself, since (3) entails both (1) and (2).

Although "explain" may be used in describing an act governed by these conditions, it can also be employed in a more restricted way to cover only correct explainings. We can say that Galileo explained why the tides occur, even though he did so incorrectly, or that he failed to explain this, even though he tried. When one has correctly explained q by uttering u one has performed the illocutionary act of explaining q and in doing so one has provided a correct answer to Q. In what follows, however, when reference is made to acts of explaining I shall mean acts for which this is not a requirement. (The concept of a correct explanation, and, more generally, the question of the evaluation of explanations, will be taken up in Chapter 4.)

2. SOME PRELIMINARY OBJECTIONS

It may be objected at this point that, to explain, what is required is not simply that S utters u with the intention to produce a certain effect, but that S intends to do so *by means of the recognition of this intention*.[6] On this proposal, to (3) of Section 1 we must add that S also intends that q be rendered understandable by means of the audience's coming to recognize that S has the intention to render q understandable.

This addition, I suggest, should not be made. For one thing, S's audience may be tired of explanations; at the moment its members may not want anything rendered understandable to them. In such a case S could explain q to this audience with the intention that its members not recognize that S has the intention to render q understandable to them. It is possible for S to explain q to an audience while concealing from it his explanatory intentions. More importantly, the means by which S intends to render q understandable is already included in (3)—viz. producing the knowledge, of the proposition expressed by u, that it is a correct answer to Q. S does not intend that q be rendered understandable by means of the audience's recognition

---

6. See H. P. Grice, "Meaning," *Philosophical Review* 66 (1957), pp. 377–88.

of his intention. The latter is neither necessary nor sufficient for S to produce a state of understanding in the audience.[7]

A different objection has been raised by Robert J. Matthews.[8] In a criticism of (3) he points out that in an explaining episode S may not only be supplying a ("direct") answer to Q; he may also be providing needed background information, correcting certain assumptions of the audience, showing the audience how the answer is compatible with its beliefs, and so forth. As a result, what S utters (a conjunction $u_1 \& u_2 \& \ldots \& u_n$, let us suppose) can be construed as an "answer" to Q only in a very broad sense (one not intended by (3)). Indeed, Matthews urges, S can explain q even if S does not believe he is providing any ("direct") answer to Q.

His own account, which reflects these ideas, is this:

> S explains q to (audience) A by citing (proposition) E if and only if S cites E with the intention of (i) presenting to A such information as A must learn in order to bring A to understand q, and (ii) having A come to understand q as a result of having recognized that E provides this information. (p. 75)

The proposition E may contain information in addition to an answer to Q. Indeed, according to Matthews, it need not even contain (what S believes to be) an answer to Q.

In response, let me acknowledge that during the course of uttering $u_1 \& u_2 \& \ldots \& u_n$, S can be explaining q while performing other illocutionary acts as well, such as providing background information and correcting assumptions. These other acts may help set the stage for, clarify, defend, or otherwise buttress, the act of explaining q. This is true for illocutionary acts generally. Suppose I say

> Three weeks ago I bought a watch at Cartiers. It has four diamonds, and it cost $10,000. No, it is not the one I am now wearing. I promise to give the watch to you in three months. I always keep my promises to you: remember when I promised you a car.

During the course of uttering these sentences I have performed the illocutionary act of promising to give you the watch in three months. I have also buttressed my promise by describing the watch, correcting a mistake you may be making about its identity, and defending my promising record. However, it is by uttering the fourth sentence ("I promise to give the watch to you in three months") that I have performed the act of promising. Without uttering this sentence there is

---

7. For general arguments against this Gricean condition, see Stephen R. Schiffer, *Meaning* (Oxford, 1972).
8. Robert J. Matthews, "Explaining and Explanation," *American Philosophical Quarterly* 18 (1981), pp. 71–77.

no such act; with it there is, even if the other sentences are not uttered.

Similarly, suppose I say

> Three weeks ago I bought a watch at Cartiers. It has four diamonds, and it cost $10,000. No, it is not the one I am now wearing. The reason that I keep the watch in my safe deposit box is that I don't want to pay the high insurance premiums. I know the premiums are high because I checked with my insurance broker.

During the course of uttering these sentences I have explained why I keep the watch in my safe deposit box. I have also set the stage for this act by describing the watch, I have corrected a mistake you may be making about its identity, and I have defended a claim I make in my explanation. However, it is by uttering the fourth sentence that I have performed the act of explaining why I keep the watch in my safe deposit box. Without uttering this sentence there is no such act; with it there is, even if the other sentences are not uttered. This is not to impugn the importance of these other sentences for my explaining act. Uttering such sentences may be necessary to make the explaining act effective in this case (e.g., by allowing you to correctly identify the watch in question).

Turning to Matthews's own definition, note that it does not require that S provide what S believes to be a ("direct") answer to Q. Because of the absence of this requirement it is subject to a difficulty mentioned earlier. Suppose S believes that citing the proposition "truth is beauty" is so causally efficacious with his audience that the mere citing of it will cause the audience to understand anything, including why the tides occur. Suppose further that S believes that his audience must learn that truth is beauty in order to be brought to understand (to be caused to understand) why the tides occur. Suppose, finally, that S cites the proposition "truth is beauty" with the intention of (i) presenting to A such information as (he believes) A must learn in order to bring A to understand why the tides occur, and (ii) having A come to understand why the tides occur as a result of having recognized that the proposition "truth is beauty" provides this information. Matthews's conditions are now satisfied. But if S does not believe that the proposition that truth is beauty is a correct answer to the question Q (Why do the tides occur?), then he is not explaining q.

In the remainder of this section I shall briefly note some other uses of "explain" and some concepts that remain to be discussed.

The verb "to explain" is used not only in cases in which a particular explaining act is being described. As already noted, we may say that Newton explained the tides even if we are not describing a particular explaining episode. The truth of our claim, however, depends on the existence of such an episode: Newton explained the tides only if some such act occurred. We may also make claims such as "Newton explains the tides by saying that they are due to gravitational attraction," in which, again, no particular explaining act is being described. A claim of this type will be true if there was such an act and if there is or was a disposition on the part of Newton to produce such acts. We may even say something like "Newton explained why Skylab fell," when no explaining act occurred. But this, I suggest, is elliptical for a claim about what Newton could, or would, have explained.

There are also claims of the form

Proposition p explains q,

The fact that F explains q,

What you did explains q,

in which the emphasized words do not denote something that performs an illocutionary act of explaining q. It is possible to restrict an illocutionary account to sentences in which *speakers* are described as explaining, and to suppose that there are other senses of explaining as well. However, I seek to avoid multiplying such senses. My claim is that the illocutionary concept of explaining is fundamental and that other explaining concepts are explicable by reference to this. Thus sentences of the forms above can be understood as equivalent to

One can explain q by uttering something which expresses (or describes) _____ ,

in which the blank is filled by the emphasized words.[9] This, of course, does not suffice to establish the primacy of the illocutionary concept. In Chapter 3, I will argue in favor of such a view by showing that it can handle difficulties which alternatives cannot.

Condition (3) for explaining, given in Section 1, introduces several concepts which will need attention in developing the illocutionary theory. One is that of understanding (this will be discussed in the present chapter). Another is that of being a correct answer to an (explanatory) question (which will be considered in Chapter 4, Section 1). A third concept employed in condition (3) is a causal one. Explaining, like a number of illocutionary acts, involves an intention to *produce* a certain effect. In Chapters 6 and 7 certain features of causation will be examined. But still it might be wondered whether any account of explaining acts (or of explanations) that utilizes a causal concept can be illuminating. In response, several points deserve notice.

---

9. These are the kinds of paraphrases that would be offered by what, in Chapter 3, I call the no-product view. A different theory—the product view—would construe the first sentence above as "An explanation of q is that p," whose truth-conditions, as supplied in Chapter 3, make reference to the concept of an explaining act.

(a) If the use of any causal concepts renders an account of explaining acts or products unilluminating, then several theories mentioned in Chapter 1 are vulnerable as well. Aristotle and Brody both invoke causal concepts, indeed unanalyzed ones, in their accounts. On these views, in essence, a proposition or argument is an explanation of p's being the case if (and in Aristotle's case only if) it gives a cause (of a certain type) of p's being the case. Although Aristotle distinguishes types of causes, there is virtually no attempt to define these types, only to illustrate them. Brody, in his causal model, appeals to a notion of causation which is not further explicated.

(b) In the illocutionary account a notion of causation is invoked, but unlike the above accounts, it is not stated that one explains by uttering "p because r" (or that the latter expresses an explanation) *if and only if r gives a cause of p's being the case.* The illocutionary theory does not in this way reduce explaining to giving causes. The causation involved in the illocutionary account does not relate p and r. It is invoked rather in connection with the intention with which a speaker utters "p because r": the speaker by uttering this intends to cause a certain state in the audience.

(c) The illocutionary account is not committed to the idea that causation is undefinable. Condition (3) is compatible with various definitions, though, to be sure, the illocutionary program would be circular if one is wedded to a definition of causation in terms of explanation.[10]

3. UNDERSTANDING AND KNOWLEDGE-STATES

Explaining q has been defined as uttering something with the intention of rendering q understandable (in a certain way). A theory of explaining, however, that invokes an undefined notion of understanding does not take us far enough. Much of the discussion that follows in this chapter is devoted to formulating an account of understanding that will complement the conditions for explaining and allow us to develop the concept of an explanation (product) and that of a good explanation in later chapters.

Since explaining involves an intention to render q understandable—where q is an indirect question—I shall be concerned with cases of understanding where q is also an indirect question. Such understanding I take to be a form of knowledge.[11] One understands q only if one knows a correct answer to Q which one knows to be correct.

---

10. As is Michael Scriven in "Causation as Explanation," *Noûs* 9 (1975), pp. 3–16.
11. Not all forms of understanding are amenable to the "understand q" context or are forms of knowledge—e.g., "understand that" in the sense roughly of "take it as a fact that."

Using quantificational notation, we can say that a necessary condition for the truth of sentences of the form "A understands q" is

(1) (∃x)(A knows of x that it is a correct answer to Q).

If A satisfies this condition, I shall say that A is in a *knowledge-state* with respect to Q.

This condition involves a *de re* sense of knowing. (In the *de re*, by contrast with the *de dicto*, sense, from "A knows of x that it is P," and "x = y," we may infer "A knows of y that it is P.") Now I am construing an answer to a question as a proposition, and *de re* knowledge of a proposition as involving an "acquaintance" with it (a knowledge of its content). Although no definition of "acquaintance with a proposition" will be offered, I do want to say something about it.

The expression "knowledge by acquaintance" was first introduced by Bertrand Russell.[12] We may, says Russell, know that the candidate who gets the most votes will be elected without knowing who that candidate is; if so, we do not have knowledge by acquaintance of that candidate. He writes:

> I say that I am *acquainted* with an object when I have a direct cognitive relation to that object, i.e., when I am directly aware of the object itself. When I speak of a cognitive relation here, I do not mean the sort of relation which constitutes judgement, but the sort which constitutes presentation.[13]

Russell does not define "direct awareness," although he claims that we can have it of sense data as well as of abstract objects.

According to one subsequent theory, acquaintance involves some type of interaction with its object. For example, Carl Ginet says that "one knows R (in the acquaintance sense) if and only if one has had an experience of R that is of the right sort [this varies from one type of R to another] and knows that one has had this experience by remembering it from the experience itself."[14] But unless such a view tells us what counts as "experiencing" a proposition it will not be very helpful for present purposes. (An account invoking a causal relation between the proposition and the "experiencer" does not seem possible, since propositions are dubious causal entities.) Moreover, if there are several ways of "experiencing" a proposition, such a view must tell us which are of the "right sort." Finally, unless strictures are imposed on "remembering an experience" the account is unsatisfactory for any R. Suppose that ten years ago I had an "experience" of the "right

---

12. Bertrand Russell, "Knowledge by Acquaintance and Knowledge by Description," *Aristotelian Society Proceedings* 11 (1910–11), pp. 108–28.
13. *Ibid.*, p. 108.
14. *Knowledge, Perception, and Memory* (Dordrecht, 1975), p. 5.

sort" of a certain candidate for office. (I confronted him, shook his hand, and talked with him.) Now I remember my experience, though, of course, not all aspects of it (e.g., I remember shaking hands with the candidate, talking with him, and what he said). I could do so without remembering *him,* or who he was. Similarly, I might remember my first "experience" of Heisenberg's uncertainty principle (by remembering my feelings of excitement and bewilderment in confronting this proposition and my attempting to apply it to particular cases), without remembering *it.*

Another account, which may seem preferable, is dispositional. To be "acquainted with a proposition p" means to be able to produce a sentence that expresses p, of which one knows the meaning and knows that it expresses p. I do not find this approach completely attractive, since I want to allow the possibility that young children, and others who are not particularly verbal, may be acquainted with a proposition which they are unable to express in language. Moreover, if A does have the disposition or ability in question, we can explain why he does by appeal to the fact that he is acquainted with that proposition (just as we might explain why A is able to give the candidate's name and description by saying that A is acquainted with the candidate). But if to be "acquainted with a proposition p" *means* to have such abilities, then these explanations are precluded. Accordingly, I am inclined to support a "functional" account: to be acquainted with a proposition p is to be in a certain mental state that can be characterized by typical causes and effects. Typically (though not universally) such a state is caused by having seen or heard certain sentences that express p whose meaning one knows, and it results in the ability to produce the same or equivalent sentences, of which one knows the meaning and knows that they express p. But there are other causes and effects as well, and the identification of the mental state in question by reference to these is intended as an empirical characterization, not a definition.[15]

Returning now to (1)—the definition of a knowledge-state with respect to a question Q—there are several points to mention in addition to the fact that it involves *de re* knowledge of propositions. First, being in such a state is not simply knowing that there is some proposition that is a correct answer to Q. It is not defined simply as

A knows that $(\exists x)(x$ is a correct answer to Q),

for this could be true even if A fails to know any correct answer to Q.

Second, being in a knowledge-state with respect to Q entails that Q has a correct answer; therefore, so does understanding q. If there is

---

15. Additional empirical claims might be made by combining such functional characterizations with a "computational" theory of mental states along the lines suggested by Jerry Fodor. See his *The Language of Thought* (New York, 1975), p. 77.

no correct answer to Q, if, e.g., Q is based on a false presupposition, then understanding q is an impossibility. One cannot understand why helium is the lightest element, since it isn't.

Finally, it is not sufficient that A know of (i.e., be acquainted with) an answer to Q which happens to be correct. A speaker who claims that he does not understand how a certain accident occurred might ask: Was it caused by the slippery road, did the driver fall asleep, was there a blow-out, or what? Suppose that, unbeknownst to the speaker, the slippery road did cause the accident. The mere fact that he knows of an answer that happens to be correct is not sufficient to say that he understands. What he does not know of this answer is that *it is correct;* and this is necessary for understanding.

Instead of taking answers to questions to be propositions, could we choose sentences instead? Although I have not provided an ontological theory of propositions, I have assumed that they are abstract entities which are different from, but are expressible by means of, sentences (any sentences? see Section 9); that they can have truth-values; and that it is possible to have *de re* knowledge by acquaintance with them. By contrast, sentences—in particular, sentence-tokens—are not abstract: they are physical marks or sounds. The purported advantage of the latter would be that *de re* knowledge becomes a less mysterious affair than such knowledge in the case of propositions.

Construing an answer to a question as a sentence (token) would require an important change in condition (1) for understanding. A might know of some sentence, say (a token of)

a: Strangeness is conserved in strong interactions,

that it is a correct answer to some question Q. (He knows this, let us say, because a physicist informed him of it.) But unless he knows what (a) means he does not understand q. Accordingly, with sentences as answers to questions, (1) must be revised to read

(2) (∃x)(A knows of (sentence) x that it is a correct answer to Q, and A knows the meaning of x).

The second conjunct in (2) is unnecessary in the case of propositions, since "knowledge of the meaning of a proposition" I take to be equivalent to, or at least involved in, the idea of (*de re*) "knowledge of a proposition." By contrast, one might be acquainted with a sentence without knowing its meaning.

Does (2) really avoid *de re* knowledge of abstract entities? That depends on how its second conjunct is to be construed. "A knows the meaning of x" might be understood as

(3) (∃y)(A knows of y that y is the meaning of x).

Here the y which is the meaning of (sentence) x cannot itself be a sentence, since A could know of one sentence, whose meaning he does not know, that it has the same meaning as that of x. For (3) to work in an account of understanding, we shall have to assume that there are "meanings" (which are not the same as sentences or other linguistic expressions) of which A has knowledge. And this, I take it, involves acquaintance with entities no less abstract than propositions.

Such entities could be avoided in (2) by using substitutional quantification and understanding "A knows the meaning of x" as

(4) $(\exists y)_s$(A knows of x that x means that y).

This requires knowledge only of sentences. But (4) does not adequately capture the idea behind the second conjunct of (2). The novice may know of the sentence (a) above that it means that strangeness is conserved in strong interactions. Yet he may not know the meaning of the sentence (a). (Or, to take a less trivial example, an English speaker fluent in Russian but ignorant of physics may know, of a Russian equivalent of (a), that it means that strangeness is conserved in strong interactions—without knowing the meaning of the Russian equivalent of (a).) Alternatively, by treating "knows-the-meaning-of" as a primitive relation we could understand the second conjunct of (2) as expressing a relationship between A and a sentence rather than between A and a meaning. But this would introduce a semantical complication. Construed in this way, (2) would saddle us with "knows" as a relation between a person and a sentence (in the first conjunct), as well as "knows" as an inseparable part of a semantically unrelated expression "knows-the-meaning-of" (in the second conjunct).[16]

For these reasons I do not regard (2) as an improvement on (1), where the latter is construed as quantifying over propositions. In what follows, then, I will continue to take answers to questions to be propositions.

Condition (1) is a strong one for understanding, and it might be denied. If A knows of no answers to Q at all, or only incorrect ones, then we might be willing to follow (1) and deny that A understands q. But (1) makes the stronger claim that A knows of some answer to Q that it is correct. Suppose that the police believe, but do not know, that the accident was caused by the slippery road. And suppose that their belief is correct. If (1) is a necessary condition for understanding, we cannot say that the police understand what caused the accident. This seems odd, since they can produce what they believe to be, and is, a correct answer to Q.

16. For objections to such semantical complications, see Donald Davidson, "Truth and Meaning," in J. Rosenberg and C. Travis, eds., *Readings in the Philosophy of Language* (Englewood Cliffs, N.J., 1971).

Nevertheless, I suggest that this refusal is justified. If the police suspect, but do not yet know, that the accident was caused by the slippery road, then, although they are well on the way toward understanding what caused the accident, they are not yet in the position of someone who does understand it. Their failure to be in this position derives not from the fact that they have yet to find a correct answer, but from the fact that they have yet to achieve a required epistemic state with respect to that answer, viz. one of knowing that it is correct. To be sure, if the police believe correctly, but do not know, that the accident was caused by the slippery road, then it could be misleading to deny that they understand what caused the accident. This is because we would not be giving the whole truth, only part of it, regarding their epistemic situation. To avoid misleading the audience what might be said is that the police do not yet understand what caused the accident, although they believe it was the slippery road.[17]

On the other hand, our definition of a knowledge-state may be accused of being too weak for understanding. Suppose that A once heard a correct answer to Q which he can no longer remember. However, he knows that the answer he heard is correct. Therefore, he is in a knowledge-state with respect to Q, despite the fact that he cannot remember a correct answer to Q. But surely if A is to be said to understand q he cannot be in this position.

My reply is that in the case envisaged A is not in a knowledge-state with respect to Q, although he once was. At one time, but not now, he was acquainted with a correct answer to Q. He does now know that the answer he once knew is correct. His situation can be described by saying that

($\exists$x)(A knew of x that it is a correct answer to Q), and A knows that ($\exists$x)(x is a correct answer to Q).

But it is not the case that

($\exists$x)(A knows of x that it is a correct answer to Q).

4. COMPLETE PRESUPPOSITION OF A QUESTION

I shall suppose, then, that to understand q one must be in a knowledge-state with respect to Q. For any Q, can we say that if one is in such a state then one understands q? No, since there are questions whose indirect form seems not to be amenable to the context "A un-

---

17. The police could be in (what in Section 12 I call) an alternation state: they might know that the accident was caused by the slippery road or by the driver's falling asleep, but not know which. If so, then (I will claim) they have too much knowledge to say that they do not understand q but not enough to say they do.

derstands q" (or "A does not understand q"); yet one might be in a knowledge-state with respect to such questions.

Quantity questions such as "How high is x?" or "How hot is y?" or "How heavy is z?" may be placed in this class.[18] Do I understand how high the Matterhorn is simply in virtue of the fact that I know of the answer, "The Matterhorn is 14,700 feet high," that it is correct? It sounds odd to say that I do. A similar oddness results from whether-questions. Although I may be (or fail to be) in a knowledge-state with respect to the question "Is the Matterhorn over 14,000 feet high?," can it be concluded that I understand (or fail to understand) whether the Matterhorn is over 14,000 feet high?

A theory of understanding should be able to deal with these cases and others besides. Thus, e.g., Bobby might be said to know that this instrument to which I am pointing is a bubble chamber and still fail to understand what this instrument is. Yet, he is in a knowledge-state with respect to the question "What is this instrument?" Why is this so? To adequately extend the theory of understanding some new concepts will need to be introduced in this section and the two that follow.

A question such as

(1) Why did Nero fiddle?

will be said to *presuppose* a number of propositions,[19] e.g.,

(2) Nero fiddled for some reason;
Nero fiddled;
Nero did something;
Someone did something.

Any proposition entailed by a proposition presupposed by a question will also be said to be presupposed by that question. A *complete presupposition* of a question is a proposition that entails all and only the presuppositions of that question. Of the propositions in (2) only the first is a complete presupposition of (1).

Let us consider only wh-questions (those expressed using interrogative pronouns such as "why," "what," "how," "who," "when," and "where"—but not "whether" questions, which call for a "yes" or "no" answer). To obtain a sentence expressing a complete presupposition of a (wh-)question, it will be necessary to use some existential term or phrase which corresponds to the interrogative pronoun in a sentence

---

18. See Sylvain Bromberger, "An Approach to Explanation."
19. See Nuel D. Belnap and Thomas B. Steel, *The Logic of Questions and Answers* (New Haven, 1976), p. 5. These authors suggest that a question Q presupposes a proposition p if and only if the truth of p is a logically necessary condition for there being some correct answer to Q.

expressing that question. Thus, "for some reason" in (2) corresponds to "why" in (1). (In the case of other questions, "in some manner" could correspond to "how," "at some time" to "when," and so forth.) "For some reason" will be called a reason-existential term ("at some time" a time-existential term, etc.); and, in general, I shall speak of a ϕ-existential term. Now a sentence expressing a complete presupposition of Q can be transformed into what I shall call a *complete answer form* for Q. This is done by dropping the ϕ-existential term, putting an expression of the form "the (or a) ϕ that (or in, or by, which)" at the beginning, and a form of the verb "to be" followed by a blank at the end. The result is an incomplete sentence which is obtained from a sentence expressing a complete presupposition of Q without adding or deleting any information from that presupposition, and which when completed can express an answer to Q.

For example,

(3) The reason that Nero fiddled is _____

is a complete answer form for (1). It is obtained from (1) in the following way. Formulate the sentence "Nero fiddled for some reason," which expresses a complete presupposition of (1). Drop the reason-existential term in this sentence ("for some reason"), and put the expression "the reason that" at the beginning and "is" followed by a blank at the end, yielding (3).

In the case of what- or which-questions the procedure is slightly different. Consider the question

(4) What force caused the acceleration?

and the following sentence which expresses a complete presupposition of this:

(5) Some force caused the acceleration.

Here the ϕ-existential term "some force" corresponds to the interrogative *phrase* "what force." (In this case we must consider the phrase and not simply the interrogative pronoun "what.") To obtain a complete answer form, we drop the "force-existential" term and put the expression "the force that" at the beginning and "is" followed by a blank at the end, obtaining

(6) The force that caused the acceleration is _____ .

In the case of the question

What (cause) caused the acceleration?

a complete presupposition can be taken to be

Some cause caused the acceleration.

To obtain a complete answer form we drop the "cause-existential" term "some cause" and put the expression "a cause that" at the beginning and "is" followed by a blank at the end:

A cause that caused the acceleration is _____ ;

or equivalently

A cause of the acceleration is _____ .

The non-equivalent questions

(7) What university does Mary intend to aid?

(8) What intention does Mary have to aid a university?

both have

Mary has an (some) intention to aid a (some) university

as a complete presupposition. Despite this fact their complete answer forms are different. The ϕ-existential term for (7)—the one that corresponds to the interrogative phrase "what university"—is "some university"; that for (8) is "some intention." Accordingly, a complete answer form for (7) but not for (8) is

A university Mary intends to aid is _____ .

A complete answer form for (8) but not for (7) is

An intention that Mary has to aid a university is _____ .

Finally, since there are alternative formulations of questions, we should not be surprised to find various complete answer forms for the same question. Thus (4) above might be said to be the same question as

What caused the acceleration that is a force?

If so, then not only is (6) a complete answer form for this question, but so is

A cause of the acceleration that is a force is _____ .

5. CONTENT-NOUNS

Let us return to the question

(1) Why did Nero fiddle?

and the complete answer form

(2) The reason that Nero fiddled is _____ .

Compare the following sentences, each of which is obtained from the complete answer form (2) by filling in the blank:

(3) The reason that Nero fiddled is that he was happy.

(4) The reason that Nero fiddled is difficult to grasp.

Although (3) constitutes an answer to (1)—though not necessarily a correct one—(4) does not. (3), we might say, gives the content of the reason; (4) says something about the reason but does not give its content. Some complete answer forms contain ϕ-terms which I shall call content-nouns, and blanks that can be filled with expressions giving the content associated with such nouns. "Reason" is a content-noun, and (3), but not (4), contains an expression ("that he was happy") which gives the content of the reason. Here are various content-nouns:

explanation	penalty	purpose
excuse	consequence	function
complaint	implication	event
danger	meaning	means
difficulty	reason	method
effect	fact	manner
importance	cause	process
question	rule	puzzle

How can such nouns be characterized? Broadly speaking, they are abstract nouns whose content can be given by means of nominalization. In what follows I shall speak of a sentence such as (3) as a *content-giving sentence for the noun "reason,"* and then define a content-noun by reference to content-giving sentences. Before supplying the details, we might note that the noun "reason" is abstract and that in (3) the content of the reason is given by the nominal "that he was happy."

More generally, the nouns on the list above are not nouns for physical objects or substances (e.g., "tree," "copper"), or for physical properties or dispositions ("color," "hardness," "inertia"), or for physical events or processes ("earthquake"; "event" and "process," however, are content-nouns), or for physical places or times ("desert," "noon"). Nor are they nouns for agents ("man," "dog," "God")—whether physical or otherwise. Such nouns are, as I shall say, *abstract from a physical and agent point of view.* To be sure, there can be physical causes, facts, events, and processes; but there may be non-physical ones as well (e.g., mental ones). Not all nouns that are abstract from a physical and agent point of view are content-nouns. Some (e.g., "desire") will turn out to be; others (e.g., "number," "existence") will not.

Second, the nouns in this class can be used to construct sentences of the form

(i) $\left\{\begin{matrix}\text{the}\\ \text{a}\end{matrix}\right\}$ + noun N + $\left\{\begin{matrix}\text{prepositional phrase}\\ \text{that-phrase}\end{matrix}\right\}$ + form of verb *to be* + (preposition) + nominal.

A nominal (which occurs as such within a sentence) is a noun phrase with a verb or verb derivative that may or may not have a subject, object, or other complement. Among nominals are that-clauses (I know *that he is kind*); phrases containing a verb plus the suffix *-ing* (*his singing* was interminable, he pleased us by *arriving on time*); infinitive phrases (*to give* is better than *to receive*); wh-phrases (*why he likes her* is anyone's guess, the question is *what he found*); and a host of other constructions.[20] Most of the nouns above can be used to construct sentences of form (i) in which the nominal is a that-clause. For example,

The explanation of his behavior is that he is greedy.

The reason that Nero fiddled is that he was happy.

The danger in defusing the bomb is that the bomb will explode.

The penalty for creating a disturbance is that the person convicted will be fined $1000.

The fact about Jupiter discovered by Galileo is that Jupiter has moons.

The cause of the Moon's acceleration is that the earth exerts a gravitational force on the Moon.

The purpose of putting the flag there is that it will warn drivers of danger.

The event that is now occurring is that Skylab is falling.

To be sure, the nouns in question in these examples need not be complemented by that-clauses in order to give content. Thus we might say

The explanation of his behavior is greed.

The penalty for creating a disturbance is a $1000 fine.

But such sentences are plausibly construed as equivalent in meaning to ones of the former sort.[21]

Some of the nouns on the list are not normally used to construct

---

20. For discussions of various types of nominals and their grammar, see Robert B. Lees, *The Grammar of English Nominalizations* (Bloomington, Ind., 1960); Zeno Vendler, *Adjectives and Nominalizations* (The Hague, 1968), chapters 2–5; Vendler, *Linguistics in Philosophy* (Ithaca, 1967), chapter 5.

21. That-clauses can complement "meaning" in the case of sentences or paragraphs, and also "meaning" in the sense of "significance," but not "meaning" in the case of words. Nevertheless, I shall count a sentence such as "The meaning of the German word 'Weg' is way" as content-giving.

sentences of form (i) in which the nominal is a that-clause. They, as well as many of the others, can be complemented by other nominals (emphasized below), e.g.,

The function of the heart is *pumping the blood.*
The function of the heart is *to pump the blood.*
The method by which he achieves success is by *working hard.*
The means by which he scaled the wall is by *using a ladder.*
The manner in which he escaped is by *dressing up as a guard.*
The activity that is most healthy is *walking.*
The question is *why Nero fiddled.*

Although, as in the last example, phrases that are nominals can be introduced by (wh)-interrogative pronouns, not all such phrases are nominals.[22] Thus in

The state of happiness is what Tom is seeking

the phrase "what Tom is seeking" is equivalent to "the thing (or that) which Tom is seeking"; the which-clause is a relative clause, not a nominal. Compare this with

The question is what Tom is seeking.

Here the phrase "what Tom is seeking" is not equivalent to "the thing which Tom is seeking." It is a nominal, not a phrase containing an implicit relative clause.

A restriction on (i) will be imposed. Consider sentences such as

The reason that Nero fiddled is that he was either happy or bored,
The function of that machine is (either) pumping water into the system or wastes out of it,

in which the abstract noun is complemented by a disjunction. I shall say that such disjunctive sentences are *distributive* if they are equivalent in meaning to

Either the reason that Nero fiddled is that he was happy, or the reason that Nero fiddled is that he was bored,
Either the function of that machine is pumping water into the system, or the function of that machine is pumping wastes out of the system.

More generally, a sentence of form (i) will be said to be distributive if the nominal contains a disjunction which when distributed in the

22. See Lees, *The Grammar of English Nominalizations,* pp. 60–61; Vendler, *Adjectives and Nominalizations,* pp. 37–39.

manner above yields a sentence that is equivalent in meaning; a non-disjunctive sentence of form (i) will also be said to be distributive if it is equivalent in meaning to a distributive one. And I shall restrict content-giving sentences of form (i) to those that are *non-distributive*. (Intuitively, "the reason that Nero fiddled is that he was either happy or bored" does not give the content of the reason; it gives possible contents.)

This does not mean that all sentences of form (i) with disjunctive complements are precluded. ("The warning on the bottle is that this pill will cause headaches or stomach cramps" is non-distributive; it gives the content of the warning.) However, to avoid truth-functional trivialization, let us restrict the disjunctions in the nominal to those that are not truth-functionally equivalent to one of the disjuncts or to an element in a disjunct. Thus, we will not say that

The reason that Nero fiddled is that he was happy

is distributive on the grounds that it is equivalent in meaning to

The reason that Nero fiddled is that he was happy or happy (or: that he was happy and bored or happy and not bored).

Even if there is equivalence in meaning (which is dubious), the disjunctions are truth-functionally equivalent to one of the disjuncts or to an element therein.[23]

What has been said about disjunctions can be applied to other compounds. Thus, extending "distributive" in an obvious way,

The reason that Nero fiddled is that he was happy, if this textbook is to be believed

is distributive, since it is equivalent to the conditional

If this textbook is to be believed, then the reason that Nero fiddled is that he was happy.

By contrast,

The purpose of putting the flag there is that it will warn drivers of danger, if they should start to cross the bridge

is not distributive. It is not equivalent to the conditional

23. Conversely, one cannot trivially turn distributive sentences into non-distributive ones by coining special predicates. Thus, replacing "is either happy or bored" with "is in an alpha-state," we might write "The reason that Nero fiddled is that he was in an alpha-state." But the latter is distributive, since it is equivalent in meaning to "The reason that Nero fiddled is that he was happy or bored," which is a distributive sentence of form (i).

If drivers should start to cross the bridge, then the purpose of putting the flag there is that it will warn drivers of danger.[24]

We can now introduce the notion of a *content-giving sentence for a noun N* by saying that among the conditions for S to be such a sentence is that S be, or be equivalent in meaning to, a non-distributive sentence of form (i).[25]

Turning to another condition, if S is a content-giving sentence for noun N then the (equivalent) sentence of form (i) is itself equivalent to one of the form

(ii) This + is + $\left\{ \begin{array}{c} \text{the} \\ \text{a} \end{array} \right\}$ + noun N + phrase: (preposition) + nominal,

in which noun N, phrase, and nominal are those in (i). (When the nominal in (i) is a that-clause the word "that" is dropped; when it is an indirect interrogative the direct form is used.) If the present condition is satisfied I shall say that the sentence in question has a *this-is equivalent*. Here are some examples:

This is the reason that Nero fiddled: he was happy (a this-is equivalent of "the reason that Nero fiddled is that he was happy").

This is the function of the heart: pumping the blood (a this-is equivalent of "the function of the heart is pumping the blood").

This is the manner in which he escaped: by dressing up as a guard (a this-is equivalent of "the manner in which he escaped is by dressing up as a guard").

This is the question: why did Nero fiddle? (a this-is equivalent of "the question is why Nero fiddled").

Finally, a certain type of reversibility is possible in the case of content-giving sentences of form (i):

The function of the heart is pumping the blood

---

24. An exception will need to be made for distributive conjunctions. The sentence "A danger of this pill is that it will cause anyone who takes it to get stomach cramps and headaches" is distributive. It is equivalent to the conjunction "A danger of this pill is that it will cause anyone who takes it to get stomach cramps, and a danger of this pill is that it will cause anyone who takes it to get headaches." Yet it seems intuitive to suppose that the former sentence gives the content of the danger. Assuming that the latter sentence is a conjunction of content-giving sentences, we can say that a distributive conjunction which is equivalent to a conjunction of content-giving sentences is itself content-giving.

25. Obviously, I am restricting S's to meaningful sentences. I don't count "The existence of the Moon is that Galileo observed it" as a content-giving sentence for the noun "existence." However, as will be noted in Section 9, there are sentences of form (i) which, though syntactically or semantically deviant in certain respects, are not without meaning. ("The reason the tides occur is that truth is beauty" may be one such example.) I do not want to exclude such sentences.

is equivalent to

Pumping the blood is the function of the heart.

The sentence

The explanation of his behavior is that he is greedy

is equivalent to

That he is greedy is the explanation of his behavior.

More generally, if S is a content-giving sentence for N, then the (equivalent) sentence of form (i) is reversible in the sense that the phrase following the form of "to be" can be transported to the front of the sentence, followed by a form of "to be."

Putting this together, then, we have noted four features of content-giving sentences. In particular, let us say that

A sentence S is a *content-giving sentence for the noun N* if and only if
(a) N is abstract from a physical and agent point of view;
(b) S is, or is equivalent to, a non-distributive sentence of form (i);
(c) The (equivalent) sentence of form (i) has a this-is equivalent of form (ii);
(d) The (equivalent) sentence of form (i) is reversible.

For example,

The function of the heart is pumping the blood

satisfies these conditions. "Function" is abstract in the appropriate way. The sentence in question is a non-distributive one of form (i). It has a this-is equivalent, viz.

This is the function of the heart: pumping the blood.

And it is reversible, since it is equivalent to

Pumping the blood is the function of the heart.

Note that for S to be a content-giving sentence for a noun N that noun can, but need not, appear in S itself. The sentence

The reason that Nero fiddled is that he was happy

is a content-giving sentence for the noun "reason," which appears in the sentence itself. (All four conditions above are satisfied.) Since that sentence is equivalent in meaning to

Nero fiddled because he was happy,

the latter (according to the definition above) is also a content-giving sentence for the noun "reason."

We can now say that N is a content-noun if and only if there is a sentence which is a content-giving sentence for N. The nouns on the list at the beginning of the present section are all content-nouns by this criterion.

Let us consider sentences that do not satisfy the conditions of the definition of a content-giving sentence for a noun N. Here are ones that violate all four conditions with respect to the emphasized nouns:

The *father* of Isaac is Abraham;

The *height* of the Matterhorn is 14,700 feet;

The *river* in North America which is the largest is the Mississippi;

The *color* of the Taj Mahal is white;

The *force* that Newton discovered is gravity.

The emphasized nouns are not abstract from a physical and agent point of view. And these sentences are not (and are not equivalent to ones) of the form (i) in condition (b). Therefore, they also fail to satisfy conditions (c) and (d), each of which requires the satisfaction of (b). (To be sure, these sentences are reversible, but they are not reversible sentences of form (i).) These are not content-giving sentences for the emphasized nouns.

There are also sentences that satisfy condition (b) for content-giving sentences but fail to satisfy one or more of (a) or (c) or (d), e.g.,

The execution now occurring is the killing of the prince;

The ceremony of marriage that took place in the chapel at 2 P.M. was a uniting of feuding families;

An earthquake is a trembling of the earth;

The decline in the G.N.P. is the commencing of a recession.

The first and third violate (a) but satisfy the remaining conditions; the second satisfies (b) but violates the others; the fourth satisfies (a) and (b) but not (c) or (d). Neither the satisfaction of (a) nor of (b) guarantees that of (c) or (d), while (c) and (d), of course, presuppose (b).

Even if the noun in a sentence of form

The + noun N + phrase + to be _____

is a content-noun, that does not suffice to make the sentence a content-giving one for that noun. Thus,

The reason that Nero fiddled is difficult to grasp,

The cause of the devastation at Hiroshima [viz. the detonating of the atomic bomb] was the beginning of a new age,

The cause of the explosion is what I am trying to discover,

The reason that Nero fiddled is that he was either happy or bored, violate the conditions for being content-giving sentences for the nouns "reason" and "cause." The first, third, and fourth are not equivalent to non-distributive sentences of form (i). (In the third sentence what follows "is" is not a nominal but a phrase implicitly containing a relative clause.) The second has no this-is equivalent. In its most natural interpretation it is equivalent to "what caused the devastation at Hiroshima began a new age," but not to "this is what caused the devastation at Hiroshima: the beginning of a new age." The first ascribes something to the reason without giving it. The second and third ascribe something to the cause without giving it. The fourth gives a distributive disjunction of reasons.

Psychological nouns, such as "belief," "fear," and "desire," are content-nouns by the above definition. There are sentences such as

The belief of Jones is that the earth will be destroyed,

which are content-giving sentences for their respective nouns. The only difficulty that may arise here is over whether certain sentences with such nouns are content-giving sentences. For example, it seems plausible to say that

(5) The desire of John is for money

is equivalent to

The desire of John is that John should obtain money.[26]

If so, then (5) is a content-giving sentence for the noun "desire." But consider (the somewhat awkward)

The hatred of John is of Castro,

which seems to have no equivalent of form (i). If not, then this is not a content-giving sentence for the noun "hatred."

6. COMPLETE CONTENT-GIVING PROPOSITIONS

Let us call a proposition a *content-giving proposition* (for the concept expressed by the noun N) if and only if it is expressible by a content-giving sentence for N. Thus, assuming that

(1) The reason that Nero fiddled is that he was happy

and "Nero fiddled because he was happy" express the same proposition, they both express a content-giving proposition (for the concept expressed by "reason").

I shall now say that

---

26. See Peter Geach, "Teleological Explanation," in Stephan Körner, ed., *Explanation* (Oxford, 1975).

p is a *complete content-giving proposition with respect to question Q* if and only if
- (a) p is a content-giving proposition (for a concept expressed by some noun N);
- (b) p is expressible by a sentence obtained from a complete answer form for Q (whose ϕ-term is N) by filling in the blank; and
- (c) p is not a presupposition of Q.

Thus, the proposition expressed by

(1) The reason that Nero fiddled is that he was happy

is a complete content-giving proposition with respect to the question

(2) Why did Nero fiddle?

(a) The proposition expressed by (1) is a content-giving proposition for the concept expressed by "reason." (b) This proposition is expressible by sentence (1) which is obtained from a complete answer form for (2) whose ϕ-term is "reason," viz.

The reason that Nero fiddled is _____,

by filling in the blank. (c) The proposition expressed by (1) is not a presupposition of (2).

By contrast, the proposition expressed by

(3) The reason that Nero fiddled is difficult to grasp

is not a complete content-giving proposition with respect to question (2), since it is not a content-giving proposition for the concept expressed by "reason."

Consider the question

What force caused the acceleration?

which I shall take to be equivalent to

(4) What caused the acceleration that is a force?

The proposition expressed by

(5) The cause of the acceleration that is a force is the pressing of Joe's hand on the body

is a complete content-giving proposition with respect to (4). It is a content-giving proposition for the concept expressed by "cause." It is expressible by sentence (5) which is obtained from a complete answer form for (4) whose ϕ-term is "cause," viz.

The cause of the acceleration that is a force is _____,

by filling in the blank. And the proposition expressed by (5) is not a presupposition of (4).

Similarly, if we take the question

What execution is now occurring?

to be equivalent to

What event is now occurring that is an execution?

then

(6) The event now occurring that is an execution is the killing of the prince

expresses a complete content-giving proposition with respect to that question. (Although (6) is not a content-giving sentence for the noun "execution," it is so for the noun "event.")

By contrast, consider

(7) What person caused the acceleration? (Alternatively, what caused the acceleration that is a person?)

The propositions expressed by

The *person* that caused the acceleration is Joe

and

The *cause* of the acceleration that is a person is Joe

are not complete content-giving propositions with respect to (7). Neither is a content-giving proposition for the concepts expressed by the emphasized nouns.

For another contrast consider

(8) What event is now occurring?

and the following "answer":

(9) The event now occurring is what event is now occurring.

The sentence (9) does not express a complete content-giving proposition with respect to (8). For one thing, (9) is not a content-giving sentence for "event." ((9) is to be construed as "the event which is now occurring is the event which is now occurring"; the which-clauses are relative clauses, not nominals.) For another thing, the proposition (9) expresses is a presupposition of (8).

Finally, consider

(10) What university does Mary intend to aid?

A sentence obtained from a complete answer form for (10) by filling in the blank is

(11) A university that Mary intends to aid is Johns Hopkins University.

And this sentence might be said to express the same proposition as

(12) An intention that Mary has is to aid Johns Hopkins University,

which is a content-giving sentence for "intention." But for (12) to count as expressing a complete content-giving proposition for (10) it is required that the content-noun in (12) be "university"—the ϕ-term in (10)—which is not the case.

In rough terms, we may think of a complete content-giving proposition with respect to a question Q as a content-giving proposition which constitutes an answer to Q that entails all of Q's presuppositions but is not entailed by any of them.

7. UNDERSTANDING AND CONTENT

We are now in a position to return to understanding. It is my contention that A understands q only if A knows a correct answer to Q *which is a complete content-giving proposition with respect to Q*. That is, A understands q only if

(1) ($\exists$p)(A knows of p that it is a correct answer to Q, and p is a complete content-giving proposition with respect to Q).

Thus, one who knows of the proposition

(2) The reason that Nero fiddled is that he was happy

that it is a correct answer to

(3) Why did Nero fiddle?

satisfies this condition for understanding why Nero fiddled, since (2) is a complete content-giving proposition with respect to (3). By contrast, one who knows the truth of

(4) The reason that Nero fiddled is difficult to grasp

does not thereby understand why Nero fiddled. For one thing, (4) is not a correct answer to (3). For another, (4) is not a complete content-giving proposition with respect to (3).

Let me mention four other cases in which (1) is violated. First, consider the question

(5) How high is the Matterhorn?

and the proposition

(6) The height of the Matterhorn is 14,700 feet.

Someone might know of (6) that it is a correct answer to (5). Moreover, proposition (6) is expressible by a sentence obtained from a complete answer form for (5)—"The height of the Matterhorn is―――"—by filling the blank with a number plus a term for a distance. Nevertheless, (6) is not a complete content-giving proposition with respect to (5), since it is not a content-giving proposition (for the concept expressed by "height"). (The sentence "The height of the Matterhorn is 14,700 feet" violates all four conditions for content-giving sentences given in Section 5.) Therefore, someone could not be said to understand how high the Matterhorn is in virtue of the fact that he knows of (6) that it is a correct answer to (5). By contrast, one who knows of (6) that it is a correct answer to (5) can be said to know how high the Matterhorn is. In general, knowing q, by contrast with understanding q, does not require knowledge of content-giving propositions.

Second, consider the question

(7) What execution is now occurring?

and the proposition

(8) The execution now occurring (alternatively: the event now occurring that is an execution) is the one that is being covered by the Associated Press.

Someone might know of (8) that it is a correct answer to (7). Yet it seems wrong to conclude that, in virtue of this, such a person understands what execution is now occurring. And, indeed, (1) thwarts such a conclusion, since (8) is not a complete content-giving proposition with respect to (7).

For a third violation of (1), consider the question

(9) What caused the explosion?

and the true proposition

(10) The cause of the explosion is what caused the explosion.

For the sake of argument, let us suppose that (10) can be counted as a correct answer to (9). (In Chapter 4 a criterion of correctness will be proposed which precludes this answer.) Proposition (10) is expressible by a sentence obtained from a complete answer form for (9)—"The cause of the explosion is ―――."—by filling the blank; and "cause" is a content-noun. But surely we would deny that someone who knows of (10) that it is a correct answer to (9) in virtue of this fact understands (or even knows) what caused the explosion. Such a denial is indeed justified by condition (1) for understanding. Proposition (10) is not a complete content-giving one with respect to (9). For one thing, it is not expressible by a content-giving sentence for "cause." ("What caused the explosion" in (10) is not a nominal.) For another, (10) is presupposed by (9).

Fourth, consider again the question (9) and the true proposition

> The cause of the explosion was either the overheating of the boiler in the library, or the detonating of a bomb in the administration building, or the igniting of leaking gas in the chemistry building.

One who knows the truth of this proposition, but not the truth of any disjunct, does not yet have enough information to be said to understand what caused the explosion. This conclusion accords with condition (1) for understanding. The disjunctive proposition known is not expressible by a *non-distributive* sentence of the required form.

Suppose that knowing of p that it is a correct answer to Q does not suffice for understanding q. Can we conclude that q is not understandable? Of course not. (9) is understandable—one can understand what caused the explosion. The point is just that one cannot understand this in virtue of knowing of (10) that it is a correct answer to (9). Indeed, condition (1) does not even preclude someone from understanding (7), viz. what execution is now occurring. As we saw in Section 6, "The event now occurring that is an execution is the killing of the prince" expresses a complete content-giving proposition with respect to (7). One who knows of this proposition that it is true satisfies condition (1) for understanding. We cannot, of course, conclude that such knowledge entails understanding, since condition (1) is not being claimed to be sufficient. (See the last part of Section 10.)

Finally, it might be objected, it will be possible for p to be a complete content-giving proposition with respect to Q when p and Q are expressed in one way but not another. Consider

(11) What person was elected U.S. President in 1980?

On my account the proposition expressed by the sentence

(12) The person who was elected U.S. President in 1980 is Reagan

is not a complete content-giving proposition with respect to (11). By contrast, with respect to

(13) What outcome of the 1980 U.S. Presidential election obtained?

the sentence

(14) An outcome of the 1980 U.S. Presidential election is that Reagan was elected

expresses a complete content-giving proposition. But, it seems, (11) and (13) express the same question, and (12) and (14) the same proposition.

More generally, there will be questions expressed without content-nouns that are expressible using such nouns. Thus, "What number will win the lottery?" can be rephrased as "What will be the *result* of the lottery?"; "Will it rain today" as "What *possibility* concerning its raining today will actually obtain?"; and so forth.

My response is twofold. First, if a question and answer are expressible in such a way that the latter is shown to be a complete content-giving proposition with respect to the former, then we do have a candidate for understanding. But, second, I am dubious that the cases mentioned above are illustrations of this. In particular, e.g., I am dubious that interrogative sentences (11) and (13) express the same question. (13) is broader than (11); it can be answered in ways not open to the latter, e.g., by

> An outcome of the 1980 U.S. Presidential election is that the Democrats lost control of the Senate.[27]

To be sure, there are cases in which an interrogative sentence beginning with "what" which lacks a content-noun is transformable into an equivalent interrogative with such a noun. ("What execution is now occurring?" we took to be equivalent to "What event is now occurring that is an execution?") But what I am now claiming is that this is not universally the case. When such transformations are made the new interrogative will not always be equivalent to the old. Thus, "The result of the lottery will be that the state will go broke" expresses an answer to "What will be the result of the lottery?" but not to "What number will win the lottery?"; and "A possibility concerning its raining today that will actually obtain is that my airplane flight will be delayed" is an answer to "What possibility concerning its raining today will actually obtain?" but not to "Will it rain today?".

8. ELLIPSES

My contention is that A understands q only if

> (1) ($\exists$p)(A knows of p that it is a correct answer to Q, and p is a complete content-giving proposition with respect to Q).

In accordance with this, one cannot understand how high the Matterhorn is simply in virtue of knowing of "The height of the Matterhorn

---

27. Even if we try to make (13) more specific by writing "What outcome involving the U.S. Presidency (or what Presidential outcome) obtained in the 1980 Presidential election?" we get similar results. "An outcome involving the U.S. Presidency (or a Presidential outcome) that obtained in the 1980 election is that Jimmy Carter lost" is an answer to this question but not to (11).

is 14,700 feet" that it is a correct answer to "How high is the Matterhorn?" Still, there are situations in which we might speak of understanding how high the Matterhorn is. If John knows that it is sufficiently high to require equipment and a guide for climbing we might well say that he understands how high the Matterhorn is. In such a case we are treating "John understands how high the Matterhorn is" as elliptical for something like

 (2) John understands what significance the height of the Matterhorn has for the prospective climber.

The constituent question in (2) is

 (3) What significance does the height of the Matterhorn have for the prospective climber?

John knows of the proposition

 (4) The significance of the height of the Matterhorn for the prospective climber is that this height is sufficient to require equipment and a guide for climbing

that it is a correct answer to (3). But (4) is a complete content-giving proposition with respect to (3). Therefore, by (1), John satisfies a necessary condition for understanding with respect to (3).

Bobby who knows nothing about physics sees a certain instrument for the first time and is informed that it is a bubble chamber. Does he now understand what instrument this is simply in virtue of knowing of the proposition

 This instrument is a bubble chamber

that it is a correct answer to the question

 What instrument is this?

No, but if he is now informed that this instrument is used in physics to record the tracks of subatomic particles, then we might conclude that now he understands what instrument this is. If so, our claim is elliptical for

 Bobby understands what the function of this instrument is.

The constituent question here is one for which there is an answer that is a complete content-giving proposition, viz.

 The function of this instrument is to record the tracks of subatomic particles.

More generally, "A understands q" may be elliptical for "A understands q'" where condition (1) is satisfied with respect to Q' but not Q.

Another case of this sort involves questions of the form "What is (an) X?" (where X is a type of thing or substance). Do I understand what copper is in virtue of knowing of the proposition

(5) Copper is a metal

that it is a correct answer to the question

(6) What is copper?

If so, then we have a violation of condition (1) since proposition (5) is not a complete content-giving proposition with respect to (6). (All the conditions for content-giving sentences are violated.) It might be concluded that this result is welcome, since one cannot be said to understand what copper is simply in virtue of knowing (5). But suppose I know of the proposition

(7) Copper is the metal of atomic number 29 which has the melting point 1083° C, is reddish in color, is malleable and ductile, etc.

that it is a correct answer to (6). It might now be concluded that, in virtue of this wealth of knowledge, I understand what copper is. Yet (7), like (5), is not a complete content-giving proposition with respect to (6).

This case can be accommodated by construing "understanding what (an) X is" as elliptical for (something like) "understanding what fact about X is important (in the context in question)," or perhaps for "understanding what significance or importance X has (in the present context)." On such a construction, one who knows of the proposition

(8) A fact about copper that is important (in the context) is that copper is the metal of atomic number 29 which has the melting point 1083° C, etc.

that it is a correct answer to

(9) What fact about copper is important?

might be said to understand what copper is, in virtue of (8)'s being a complete content-giving proposition with respect to (9). Moreover, there will be contexts in which a fact about copper that is important (or in which what significance copper has) is that it is a metal. In such contexts it will be possible to say that one who knows the relevant information understands what copper is. (The relativity of understanding is an idea that will be explored in Section 10.)

Finally, there are cases in which, although we may be reluctant to speak of understanding q, there is something about the situation— some related q'—that we do understand, even though "understanding q" is not elliptical for "understanding q'." Mary intends to leave her money to a certain university. Do I understand what university Mary intends to aid in virtue of knowing of the proposition "The university Mary intends to aid is Johns Hopkins" that it is a correct answer to "What university does Mary intend to aid?"? Not on the account of understanding presented here. Nevertheless, on this account, there is something about the situation that I do understand, viz. what Mary's intention is.

9. EXPLAINING REVISITED

Returning to the concept of an illocutionary act of explaining, we are now in a position to formulate some further conditions. So far no restrictions have been imposed on the indirect question q being explained, or on u, what is uttered by the explainer. (The conditions speak only of S's intentions and beliefs concerning q and u.) Various positions might be advocated. According to one, no restrictions whatever should be imposed on q or u; i.e., condition (3) of Section 1 should stand as it is. One can explain anything by uttering anything, so long as one has the right intentions and beliefs. A position at the other end of the spectrum would be that q must always be a why-question and that u should cite causes and laws. The problem we face is how to draw the line between something which is an explaining act (however bad the product), and something which is no such act at all. To some extent this will be arbitrary; I doubt that there is a precise dividing line. However, I do think that the concept of an illocutionary act of explaining is somewhat narrower than that suggested so far. In what follows I shall suggest some restrictions to be added to condition (3) of Section 1 that still allow a broad range of explaining acts.

Consider once again the question

(1) How high is the Matterhorn?

I claimed that someone could not be said to understand how high the Matterhorn is simply in virtue of the fact that he knows of the proposition

(2) The height of the Matterhorn is 14,700 feet

that it is a correct answer to (1). Now I suggest that an analogous claim can be made for explaining: By uttering (2) a speaker S is not explaining how high the Matterhorn is. Unless (1) is being construed as elliptical for something else, it is not the right sort of question to fill the q-position in "S explains q by uttering u." Still, intentions being what they are, it seems possible for some misguided speaker S to utter (2) with the intention described in condition (3) for explaining given in Section 1: that of rendering q understandable ($Q=(1)$) by getting others to know of the proposition expressed by (2) that it is a correct answer to Q. In such a case we would not say that S is explaining q by uttering (2), but that he *intends* to be doing this.

Let us call Q a *content-question* if and only if $(\exists p)(p$ is a complete content-giving proposition with respect to Q). A content-question has a complete answer form whose blank can be filled with content-giving expressions that will transform the result into a sentence that expresses a content-giving proposition. A question such as "Why did Nero fiddle?" is a content-question, but one such as "How high is the Matterhorn?" is not. Now I suggest that the q-position in "S explains q by uttering u" can be filled only by interrogatives expressing content-questions, or by interrogatives that in "S explains q by uttering u" are elliptical for ones expressing content-questions. Since (1) is not a content-question, if, by uttering (2), S is said to be explaining how high the Matterhorn is, what is said is either false or else elliptical for something in which the explanatory question is a content-question.

I turn next to a proposal for a restriction on u, what is uttered in an act of explaining q.

u-restriction for Q: what S utters is, or in the context is transformable into, a sentence expressing a complete content-giving proposition with respect to Q.

Suppose that S explains why Nero fiddled, by uttering

u: Nero was happy.

In the context of utterance S's utterance u is transformable into (what S said in that context can also be expressed by) the sentence

The reason that Nero fiddled is that he was happy,

which expresses a complete content-giving proposition with respect to the question "Why did Nero fiddle?"

The u-restriction will allow us to exclude a number of kinds of cases. For example, with it S cannot be explaining how Jones escaped from prison, by uttering

The reason that Nero fiddled is that he was happy

(unless, in the context of utterance, this is to be understood as expressing some different proposition). Nor can S be explaining why Nero fiddled, by uttering "The reason that Nero fiddled is difficult to grasp," or nonsense words like "glip glop."

However, the u-restriction still allows a broad range of cases. Thus it permits S to explain why atoms emit discrete radiation, by uttering

The reason that atoms emit discrete radiation is that God is love, provided, of course, that the other conditions on explaining are satisfied (in particular, e.g., that S believes that this sentence expresses a correct answer to the question). Will the u-restriction countenance explaining acts involving utterances even more bizarre than this? That depends upon which u's we take as expressing propositions.

I construe propositions broadly to be expressible by that-clauses following a range of psychological verbs (such as "believe," "fear," and "hope") and illocutionary verbs (such as "say," "state," "propose," and "suggest"). Such verbs can be followed by that-clauses that are grammatically or semantically deviant, even though the resulting sentence is neither, and indeed is true, e.g.,

Heidegger believed that *the nothing noths*,

John said that *numbers speak silently*.

I shall say that the emphasized words in these sentences express propositions. More generally, u can be said to express a proposition if and only if it can appear in (many)[28] contexts of the form

(3) Subject term + psychological or illocutionary verb + that + u.

(Here the "that" is to be construed as associated with the verb, not as modifying u.) Although by this criterion syntactically and semantically deviant utterances can express propositions, not all utterances can. Thus, since

John said (believes, hopes) that *glip glop*,

John said (believes, hopes) that *go table chair*,

are deviant, the emphasized words do not express a proposition. ("John said 'glip glop' (or 'go table chair')" is not deviant, but this is not the relevant context.)

To adopt this criterion is not to be committed to the view that every true sentence of form (3) requires the subject to have *de re* knowledge (by acquaintance) of the proposition expressed by u. If Sam, who knows no physics, is told by a reputable physicist that strangeness is conserved in strong interactions, then it may be true to say that

(4) Sam believes that strangeness is conserved in strong interactions,

which is a sentence of form (3). But what makes (4) true in this case is not (among other things) Sam's *de re* knowledge of the proposition that strangeness is conserved in strong interactions, since he has none. Rather it is his *de re* knowledge of the sentence "Strangeness is con-

---

28. Obviously, some such contexts will be inappropriate, e.g., "S predicts that yesterday it snowed."

served in strong interactions." Sam believes, of this (or some equivalent) sentence, that it expresses a true proposition. In the present case this is sufficient to make (4) true. If Sam understood physics, what could make (4) true is his belief, of the proposition in question, that it is true—which involves acquaintance with the proposition itself. Accordingly, even though propositions are expressible by that-clauses in sentences reporting beliefs (etc.), this does not necessitate an analysis of belief sentences that requires the believer to be acquainted with some proposition.

Nevertheless, if sentences such as "The nothing noths" and "Numbers speak silently" do express propositions, then one ought to be able to have *de re* knowledge by acquaintance with the propositions they express. But how is this possible if these sentences are meaningless? In Section 3, I suggested that to be acquainted with a proposition p is to be in a certain mental state that can be functionally characterized by typical causes and effects. Typically, such a state is caused by having seen or heard certain sentences that express p whose meaning one knows, and it results in the ability to produce the same or equivalent sentences, of which one knows the meaning and knows that they express p. Now, just as "proposition" is being construed broadly, so is "meaning." Not every syntactically or semantically deviant sentence is utterly without meaning; there are degrees of deviation. A sympathetic reader of Heidegger knows something about the sentence "The nothing noths" that the non-reader does not. Within the Heideggerian corpus it has some meaning (I am inclined to suppose) despite its deviance, and the sympathetic reader knows what this meaning is. Even if for such a person the sentence is sufficiently opaque to lack a truth-value, it is not in the same league, e.g., as "glip glop." On my suggestion, those who suppose that it is should refuse to assent to "Heidegger believed that the nothing noths," but should accept only sentences such as "Heidegger wrote (or said, or believed true, the sentence) 'the nothing noths' " (just as we are willing to assert "John said 'glip glop' " but not "John said that glip glop"). However, it is not my purpose here to argue that this particular Heideggerian sentence (or the sentence "Numbers speak silently") is not complete nonsense; I assume only that a range of syntactically or semantically deviant sentences are not. (For the sake of the argument I will continue to use these sentences.)

With the present understanding of what can express a proposition, the u-restriction for Q allows S to be explaining by making deviant utterances. Heidegger can be explaining something by uttering a sentence of the form

The reason that _____ is that the nothing noths.

And John can be explaining why numbers cannot be heard, by uttering

> The reason that numbers cannot be heard is that numbers speak silently.

However, John cannot be explaining why numbers cannot be heard, by uttering

> The reason that numbers cannot be heard is that glip glop.

Thus, although the u-restriction (in conjunction with a criterion permitting deviant sentences to express propositions) does exclude various utterances from explaining acts, it is sufficiently broad to allow some that violate rules of syntax or semantics. But this is as it should be. At least in certain cases we do, I think, want to describe the philosopher, the scientist, or the nonspecialist as engaging in an act of explaining, even if we criticize the act, the explanation, or both for insufficient intelligibility. Some will insist that we have allowed too broad a class of explaining acts. Perhaps, but while this may be somewhat arbitrary, I am inclined to stop at this point and treat further restrictions as proposals for *evaluating* explaining acts and their products.

We may now formulate the resulting conditions for explaining, as follows:

> "S explains q by uttering u" is true if and only if either
> (a) Q is a content-question, the u-restriction for Q is satisfied, and condition (3) of Section 1 holds (i.e., S utters u with the intention that his utterance of u render q understandable by producing the knowledge, of the proposition expressed by u (in that context), that it is a correct answer to Q); *or*
> (b) "S explains q by uttering u" is elliptical for "S explains q' by uttering u," Q' is a content-question, the u-restriction for Q' is satisfied, and condition (3) obtains with respect to Q'.

The parenthesized words "in that context" in (a) allow S to explain q by uttering words that do not normally express a proposition, but do so in the explaining context. Thus, S might explain why Othello killed Desdemona by uttering simply "uncontrollable jealousy"—which, in the context, can be taken to express the proposition that Othello was uncontrollably jealous.

Just as the u-restriction permits a wide range of explanatory utterances, though not every concatenation of words, so the restriction of the interrogative to one expressing a content-question also permits a wide range of interrogatives, but not just any. It permits interrogatives with false presuppositions. (Mary can explain why helium is the lightest element.) Indeed, it permits ones with presuppositions expressed by sentences that are grammatically or semantically deviant. (Heidegger can explain why the nothing noths.) But not every interrogative is possible. As noted, one cannot explain how high the Matterhorn is (unless this is elliptical for something else). Nor can one explain why go table chair, since nothing of the form "the reason that go table chair is that————" expresses a proposition; which is required for q to express a content-question.[29]

10. INSTRUCTIONS

In Section 7 the following condition was proposed as a necessary one for understanding: A understands q only if

(1) ($\exists$p)(A knows of p that it is a correct answer to Q, and p is a complete content-giving proposition with respect to Q).

Is this also sufficient for understanding?

What complicates the issue is that often a question can be correctly answered in different ways by providing various kinds and amounts of information. A person might be said to understand q in one way but not another. This idea can be explicated by introducing the concept of *instructions* for a question.

Consider the question

(2) What caused Smith's death?

Each of the following, let us assume, is a correct answer to (2):

(a) The cause of Smith's death was his contracting a disease;
(b) The cause of Smith's death was his contracting a disease involving a bacterial infection;
(c) The cause of Smith's death was his contracting legionnaire's disease.

Someone who replies to (2) in one of these ways may be following

---

29. Herbert Walker has noted another bizarre case allowed by (a) and (b) above: one in which the explainer S, believing that u expresses a correct answer to Q, utters u with an intention appropriate for explaining, although S does not know what u means. (Say S reads the answer u to Q in a textbook and then utters u to his students believing that they will come to understand q by this means.) This could be precluded by adding a further condition in (a) and (b) that S has *de re* knowledge of the proposition expressed by u. However, I am inclined to regard this type of case as on or near the borderline, admit it under the liberal concept of explaining given above, and say that it is to be precluded, if necessary, by means of conditions for evaluating explaining acts.

Ia: Say in a very general way what caused Smith's death, e.g., whether it was caused by contracting a disease, or by some accident that befell him, or by an act of suicide;

Ib: Follow Ia, and if a disease is cited indicate something about what it involves, e.g., whether it is bacterial or viral;

Ic: Follow Ia, and if a disease is cited identify it by using a common name for it.

Instructions are rules imposing conditions on answers to a question. They govern the proposition that is the answer, and the *act* of answering only in so far as they do this. ("Answer in a low voice" imposes a condition on the act of answering, but not on the proposition that is the answer.) Various instructions are generally possible for one and the same question. Some instructions will be vague, some precise. Some will be appropriate for science, others not. Some will be quite general, others very specific. Each of the theories of explanation cited in Chapter 1, in effect, proposes a set of instructions for answers to explanatory questions, at least in science. (In Chapter 4 I shall consider whether there can be a universal set of such instructions in science.) Talk of "a way of understanding q" will be construed by reference to a set of instructions for Q. Someone who knows that (a) above is a correct answer to (2), but does not know that (b) or (c) are, can be said to understand q in a way that satisfies instructions Ia but not Ib or Ic.

More generally, utilizing (1) above, we can say that *A understands q in a way that satisfies instructions I*, or, more briefly, A understands q_I, only if

(3) $(\exists p)$(p is an answer to Q that satisfies instructions I, and A knows of p that it is a correct answer to Q, and p is a complete content-giving proposition with respect to Q).

For example, A has fulfilled this condition with respect to question (2) and instructions Ia, if A knows of proposition

(a) The cause of Smith's death was his contracting a disease

that it is a correct answer to

(2) What caused Smith's death?

An answer to question (2) will be said to *satisfy* instructions Ia above if and only if that answer says (and does not merely purport to say) in a very general way what really did cause Smith's death, i.e., if it cites a true cause. In such a case, if the instructions have been satisfied then question (2) has been correctly answered. But consider the following instructions for (2):

Id: Give an answer that George accepts.

An answer to (2) satisfies Id if and only if it is an answer to (2) that George accepts. The satisfaction of these instructions, unlike the satisfaction of Ia, does not guarantee that (2) has been correctly answered. In general, some, but not all, questions and instructions are such that if the instructions are satisfied the questions will have been correctly answered. Of course, because of false assumptions, inconsistencies, excessive vagueness, or just sheer irrelevance, some instructions cannot be satisfied at all with respect to a given question. In such a case, understanding q in a way that satisfies I will be impossible.

In Section 3, I said that A is in a knowledge-state with respect to Q if $(\exists p)$(A knows of p that it is a correct answer to Q). If condition (3) above is satisfied—i.e., if A knows of some complete content-giving proposition with respect to Q which satisfies I that it is a correct answer to Q—I shall say that A is in a *complete knowledge-state with respect to Q_I*. Being in such a state, I am claiming, is at least a necessary condition for understanding q_I. By this criterion, if A knows of (a) above (which is a complete content-giving proposition with respect to (2) that satisfies Ia) that it is a correct answer to (2), then A fulfills a necessary condition for understanding what caused Smith's death, in a way that satisfies instructions Ia.

Can a claim of the form

(4) A understands q,

where there is no explicit appeal to any instructions, be construed in terms of understanding q_I? Three possibilities suggest themselves. First, it might be that (4) is true if and only if A understands q in a way that satisfies some instructions or other, i.e.,

$(\exists I)$(I is a set of instructions for Q, and A understands q_I).

But this would render claims about understanding very weak, since their truth would then require only a complete knowledge-state with respect to an answer satisfying the weakest instructions for Q. My knowing that Smith's death was caused by his contracting a disease would always suffice to say that I understand what caused his death. But it seems doubtful that we would want to say that such knowledge is always sufficient for understanding.

This leads to the second proposal, which is that when a speaker utters a sentence of form (4) his claim is always elliptical for

A understands q_I

where I is some contextually implicit set of instructions intended by the speaker, and I may vary from one context of utterance to another.

Sentences of form (4) are, I think, sometimes used in this way. We may say that someone understands something, meaning that he understands it in a way that we have in mind. Let me call this the "implicit-instructions" use of (4).

There is, however, another more likely possibility. Suppose that I hear speaker S assert (4). I may not know what instructions, if any, the speaker S intended, and this may not be clear from the context of his utterance. Moreover, I may have no idea what particular instructions A's understanding satisfies. Still it seems possible for me to assert (4), because I believe that A understands q in some way that is *appropriate*—even if I do not know what this is.

This leads to the third suggestion. When a speaker utters (4) his claim is elliptical for

(5) $(\exists I)$(A understands q_I, and I is a set of appropriate instructions for Q).

Let me call this the "appropriate-instructions" use of (4). It is, I suggest, the most typical use of (4). Various views are possible about how to decide whether I is a set of appropriate instructions for Q. On one, there are universal standards of appropriateness, at least in science. On another, the standards of appropriateness can vary, depending on contextual features of A's situation. In Chapter 4 these issues will be discussed. I shall provide a general characterization of the concept of appropriate instructions and will ask whether there are, or ought to be, universal instructions applicable to all scientific contexts.[30]

Recall now that A is said to be in a complete knowledge-state with respect to Q_I provided that

(3) $(\exists p)$(p is an answer to Q that satisfies instructions I, and A knows of p that it is a correct answer to Q, and p is a complete content-giving proposition with respect to Q).

If we use (5) above—the appropriate-instructions use of "understand"—then on the basis of (3) we may conclude that

A understands q only if $(\exists I)$(I is a set of appropriate instructions for Q, and A is in a complete knowledge-state with respect to Q_I).

On the appropriate-instructions use, then, knowing of a complete content-giving proposition that it is a correct answer to Q is not sufficient for (non-relativized) understanding; the proposition known must also satisfy appropriate instructions for Q.

---

30. Another possible use of (4) combines the previous two, viz. A understands q_I, where I is a contextually implicit set of appropriate instructions for Q.

11. ARE THERE ADDITIONAL CONDITIONS FOR UNDERSTANDING?

The conditions for (instructions-relative) understanding formulated in the previous section are embodied in the following principle:

A understands q_I only if $(\exists p)$(p is an answer to Q that satisfies I, and A knows of p that it is a correct answer to Q, and p is a complete content-giving proposition with respect to Q).

Or, as I put it, A understands q_I only if A is in a *complete knowledge-state with respect to* Q_I. This condition, I now suggest, is not only necessary but sufficient as well. In order to argue for this claim, several further conditions will be discussed and shown to be unnecessary.

a. Breadth of knowledge. It might be supposed that understanding q_I requires being in a complete knowledge-state not only with respect to Q but with respect to related questions as well. Thus, it has been claimed by Jane Martin that understanding the Congress, or the First World War, or the atomic nucleus, involves "seeing connections" among the parts of these items (what she calls "internal understanding") and also between the item itself and others ("external understanding").[31] Martin places no limits on the kinds of connections that can be appropriate for understanding. Nor does she try to analyze the relevant sense of "seeing" except to say that it is an "intellectual, not visual, confrontation."[32] But whatever such connections and seeing amount to, the point is that understanding involves at least some breadth of knowledge.

No doubt when one is said to understand the Congress, or the First World War, or the atomic nucleus, one has a breadth of knowledge. This is because understanding, in these cases, is understanding q_1, ..., q_n, which involves being in a complete knowledge-state with respect to several different questions about these items. However, it is by no means obvious that understanding q_I—where q is a single question and I is a set of instructions for Q—requires knowing answers to questions related to Q. To be sure, in order for A to know that p is a correct answer to Q, A may need to know that p' is a correct answer to Q'. What I am saying is that, for A to understand q_I, A need not know answers to any questions other than those he needs to know to be in a knowledge-state with respect to Q_I.

b. Coherence. A second proposal is that understanding involves coherent knowledge. It requires knowing a correct answer to Q that

---
31. Jane Martin, *Explaining, Understanding, and Teaching* (New York, 1970), chapter 7.
32. *Ibid.*, p. 165.

coheres with one's other beliefs. This means, at least, that the answer must be logically compatible with one's other beliefs, or perhaps that these other beliefs must not render this answer improbable.[33]

Suppose that I walk into an automobile showroom, and seeing a car that I like, I ask the salesman whether I can test drive it. He cheerfully agrees, and getting underneath the car, he proceeds to lift it up by himself and carry it to the street. Naturally I am surprised and puzzled. The car, I have good reason to believe, weighs over 3000 pounds. The salesman, who is short and skinny, did not use any mechanical lifting device for this purpose, nor did he receive aid from any other source. I do not understand how the salesman got the car to the street. Yet with respect to the question

Q: How did the salesman get the car to the street?

I know of the answer

(1) The method the salesman used to get the car to the street was lifting it and carrying it on his back

that it is correct. I know that it is correct because, let us assume, it is, and I saw him do it. (Unbeknownst to me, the car is made of cardboard and weighs 30 pounds, though it looks 100 times heavier.) Since (1) is a complete content-giving proposition with respect to Q, and I know of (1) that it is a correct answer to Q, I am in a complete knowledge-state with respect to Q. Yet I do not understand q. My lack of understanding, it would seem, derives from the fact that the answer I know to be correct is incompatible with other beliefs I hold, viz. that this car weighs over 3000 pounds and that no person can lift a 3000-pound car all by himself.[34]

In assessing this claim it is important once again to notice that Q can be answered in different ways, e.g., by following one or the other of these instructions:

I_1: Give the method the salesman used;

I_2: Follow I_1 and say how it was possible for the salesman to use the method he did in getting the car to the street.

In the above example it seems plausible to say that since I know that the salesman got the car to the street by lifting it all by himself, there

33. In *Law and Explanation* (Oxford, 1971), chapter 4, I suggest this requirement, which I now believe is unnecessary.
34. It might be objected here that since (1) contradicts other beliefs I hold, I do not really know of (1) that it is a correct answer to Q. To be sure, I might not believe my eyes the first time the salesman performs this feat; but after several repetitions I am finally convinced, and justifiably so. I know that some of my other beliefs therefore must be incorrect (though not necessarily which). But I could, I think, readily get myself in the position of *knowing* that (1) is true. I needn't come to know that (1) is true only after I discover that the car is made of cardboard.

is a way in which I do understand how he got the car to the street; but since I do not know how it was possible for him to use that method, there is a way in which I do not understand how he got the car to the street. I understand it in a way that satisfies one set of instructions (I_1) but not another (I_2). And the conditions for understanding that I am proposing are sufficient to justify this last claim without introducing a further coherence requirement. I am in a complete knowledge-state with respect to Q_{I_1} (in virtue of the fact that (1) is an answer to Q that satisfies I_1, that (1) is a complete content-giving proposition with respect to Q, and that I know of (1) that it is a correct answer to Q). But I am not in a complete knowledge-state with respect to Q_{I_2}. And I fail to be in the latter state not in virtue of the fact that I know a correct answer to Q which is incompatible with my other beliefs, but in virtue of the fact that I do not know a correct answer to Q that satisfies I_2.

To take this stand regarding coherence is not necessarily to reject its relevance to understanding. What is being argued is only that in order for one to understand q_I no more coherence is required among one's beliefs than is required for one to be in a complete knowledge-state with respect to Q_I.

c. Knowledge of knowledge-states. Understanding q_I, it might be suggested, requires not simply being in a complete knowledge-state with respect to Q_I but knowing that one is. The latter is not guaranteed by the former since being in the former state does not require knowing that the answer one knows satisfies instructions I. Suppose that A knows of a certain answer to Q, which is a complete content-giving proposition with respect to Q, that it is correct; and suppose that this answer happens to satisfy instructions I, although A does not know this. A is in a complete knowledge-state with respect to Q_I although he does not know that he is. But is it plausible to demand that A know this in order to understand q_I?

Let I be a set of very complex instructions for answering a question Q in a physics textbook. A has worked out an answer to Q which he knows to be correct, although, due to the complexity of instructions I, A believes that his answer does not satisfy I. But, it turns out, A is mistaken in this latter belief: his answer does in fact satisfy I. The physics teacher who examines A's answer would surely conclude that A does understand q in a way that satisfies I, a conclusion that seems justified. One can be mistaken about one's understanding of q, not only by failing to understand q in a way in which one believes one understands it, but also by understanding q in a way which one believes one does not understand it.

The present suggestion might then be weakened to require only that one know that one knows a correct answer to Q—not that one know that this answer satisfies instructions I. Epistemologists ask whether knowing that p entails knowing that one knows that p (the so-called K-K thesis). If the K-K thesis is to be rejected, then the present suggestion should probably be as well. If one can have knowledge of which one has no knowledge, it seems plausible that one could have understanding of which one has no knowledge. On the other hand, if the K-K thesis is correct, then the present suggestion would be redundant. Being in a complete knowledge-state with respect to Q_I would entail knowing that one knows a correct answer to Q.

d. Puzzlement. It might be supposed that if A understands q then A must not be puzzled over anything. But this will not do as it stands, since A might understand q while being puzzled over q' ($\neq$ q). Moreover, A might be puzzled over the very same q with respect to one set of instructions but not another. (Witness the automobile showroom example.) So the proposal might go: if A understands q_I then he is not puzzled over q with respect to I. An even stronger condition might require not simply the absence of puzzlement over q with respect to I but the presence of some "Aha" or "hat-doffing" feeling or experience which precludes puzzlement and is something over and above being in a knowledge-state with respect to Q_I.[35] Let me concentrate just on the absence of puzzlement, which is entailed by this stronger view. Is it necessary for understanding?

It has already been noted that being in a complete knowledge-state with respect to Q_I does not require knowing that one is, since it does not require knowing that the answer to Q that one knows satisfies instructions I. The case was mentioned in which A has discovered an answer to Q which he knows to be correct, but, because of the complexity of I, he believes that this answer does not satisfy I. In view of this, A might well come to be puzzled over q *with respect to I*—the emphasized words expressing the object of his puzzlement. He might be puzzled over how Q can be correctly answered in a way that satisfies I. Despite his puzzlement, the physics teacher who notes that A's answer to Q does in fact satisfy I—and concludes that A is in a complete knowledge-state with respect to Q_I—seems justified in saying that A understands q_I. Puzzlement over how Q can be correctly answered in a way that satisfies I does not preclude understanding q_I.

Nevertheless, it might be claimed that understanding q_I is not compatible with puzzlement over q itself (where this is not puzzlement

---

35. Herbert Feigl speaks of the "Aha" experience in "Some Remarks on the Meaning of Scientific Explanation," in H. Feigl and W. Sellars, eds., *Readings in Philosophical Analysis* (New York, 1949), p. 512; Stephen Toulmin speaks of the "hat-doffing" experience in *Philosophy of Science* (London, 1953), p. 117.

with respect to any particular instructions). That is, suppose that A is in a complete knowledge-state with respect to Q_I, but that A is puzzled over q itself. It is not at all clear to me that such a situation is possible. If A is in a complete knowledge-state with respect to Q_I and he claims to be puzzled over q itself, we might well reply: "you claim (or seem) to be puzzled, but you are not really, since you know a correct answer to Q."[36] Indeed, even if we accept A's claim to puzzlement, a reasonable response would be: "there is no justification for such puzzlement, since you understand q perfectly well." Even if puzzlement over q were possible in the case of someone who is in a complete knowledge-state with respect to Q, more than this would be needed to show that such a person fails to understand q.

The four criteria discussed in the present section seem to me the most promising candidates to suggest as additional conditions for understanding. None of them turns out to be necessary. This does not *prove* that being in a complete knowledge-state with respect to Q_I is sufficient for understanding q_I. But unless counterexamples can be produced, I am inclined to suggest that it is sufficient. If it is, then on the appropriate-instructions use, we can conclude that A understands q (non-relativized) if and only if $(\exists I)(I$ is a set of appropriate instructions for Q, and A is in a complete knowledge-state with respect to Q_I).

12. NOT UNDERSTANDING: n-STATES

If A is not in a complete knowledge-state with respect to Q_I, then it is not the case that A understands q_I. But we cannot, in general, infer from this that A does not understand q_I (just as we cannot move from "It is not the case that the Washington Monument is intelligent" to "The Washington Monument is not intelligent"). The second negation statement (in both cases) carries certain implications or presuppositions not borne by the first. (Possibly, in ordinary speech "A fails to understand q" would better express the idea of concern to me in the present section. But I will continue to use "A does not understand q.") The additional implications are these:

First, Q is a content-question. (There is a proposition that is a complete content-giving proposition with respect to Q.) Even if I do not know whether the Matterhorn is over 14,000 feet high, it would not be concluded that I do not understand whether it is. (The question "Is the Matterhorn over 14,000 feet high?" violates this condition.)

Second, Q_I is a sound question, i.e., it admits of an answer that is

---

36. This is compatible with the case in which A is in a complete knowledge-state with respect to Q_I but is puzzled over q with respect to I, since in the latter case A does not know that the answer p satisfies I.

correct and that satisfies instructions I. It would be a mistake, I believe, to say that A does not understand why hydrogen has the atomic number 2, even though A is not in a knowledge-state with respect to this question. The question has no correct answer, since it contains a false presupposition.

Third, suppose that although A is not in a complete knowledge-state with respect to Q_I, he has conceived of several answers to Q_I, and he knows that one of these is correct, but he does not know which one is. I shall then say that A is in an *alternation-state* with respect to Q_I. The further condition which must be satisfied if A is to be said not to understand q_I is that A is not in an alternation-state with respect to Q_I. If I know that the burglar entered the house either through the front door or through the bedroom window—but not which—then the claim that I do not understand how he entered the house would be too strong to make, since I am in an alternation-state with regard to that question. Alternation-states are intermediate between understanding and not understanding. They involve too much knowledge to permit the claim that those who are in them do *not* understand q, but not enough to permit the claim that they do. If one is in an alternation-state with respect to Q_I, then one knows the truth of some proposition expressible by a distributive disjunctive sentence, without knowing the truth of any of the disjuncts. But this, we recall from earlier sections, is not sufficient for understanding.

Sylvain Bromberger has proposed a condition even stronger than not being in an alternation-state. He suggests that if A does not understand q, then either (i) none of the answers to Q that A can conceive (conjure up, invent) is one that he can accept; or (ii) none of the answers to Q that A can conceive (conjure up, invent) is correct.[37] This condition, however, is too strong. To revert to a previous example, consider

    Q: What caused Smith's death?

A correct answer is his contracting legionnaire's disease. Suppose, however, that Dr. Robinson believes that Smith's death was caused by pneumonia. He therefore does not satisfy Bromberger's condition (i), since he can conceive an answer to Q that he can accept. Suppose also that Dr. Robinson has heard of legionnaire's disease, although he is unfamiliar with its symptoms. Then Bromberger's condition (ii) is not satisfied, since Dr. Robinson can conceive (conjure up) an answer to Q that is correct. Nevertheless, it seems possible to say that since Dr. Robinson thinks that Smith's death was due to pneumonia, he does

---

37. Bromberger, "An Approach to Explanation," p. 83. Bromberger's more general theory of explaining will be discussed in Section 14.

not understand what caused Smith's death.[38] The reason is that he is not in a complete knowledge-state, or in an alternation-state, with respect to Q, which is a sound content-question.

More generally, I suggest, A does not understand q_I if and only if

(i) A is not in a complete knowledge-state with respect to Q_I;
(ii) Q is a content-question;
(iii) Q_I is sound;
(iv) A is not in an alternation-state with respect to Q_I.

If A satisfies these conditions, I shall say that he is in an *n-state* (for "non-understanding") with respect to Q_I. One can, of course, remove A's n-state with respect to Q_I by getting him into an alternation-state. (We supply various possible answers to Q_I and get him to realize that one of these, but not which, is correct.) But this by itself would not make A understand q_I, since if A is in an alternation-state with respect to Q_I, then he is not in a complete knowledge-state with respect to Q_I. To remove A's n-state in such a way as to get A to understand q_I, it is necessary to get A into a complete knowledge-state with respect to Q_I.[39]

13. EXPLAINING AND UNDERSTANDING

Let me now relate my discussion of understanding to the concept of explaining. When S explains q by uttering u he intends to render q understandable; he intends to enable others to understand q. (Occasionally, we explain things to people who, we believe, already understand them; but then we act as if they did not.) Since several uses of sentences of the form "A understands q" have been distinguished, we might be tempted to distinguish analogous uses of "S explains q by uttering u." In one, S intends to render q understandable in a way that satisfies some particular set of instructions I that is contextually implicit. In another, S intends to render q understandable in a way that satisfies some appropriate instructions. (As noted, S may have both intentions.) Can S explain q without intending to render q understandable in a way that satisfies some appropriate instructions? Per-

---

38. More precisely, he does not understand this in a way that satisfies instructions Ic of Section 10 (though he might understand it in a way that satisfies some different instructions). But Bromberger's account does not encompass understanding that is relativized to instructions.

39. This account of not understanding is relativized to instructions. In Section 10, when considering unrelativized understanding, I distinguished two uses of "A understands q": the "implicit-instructions" and "appropriate-instructions" uses. Analogous procedures can be adopted for unrelativized sentences of the form "A does *not* understand q."

haps, but there is something untoward, unserious, about this. I suggest that the most central and typical cases of explaining involve an intention with respect to the appropriate-instructions use of "understand." These will be the cases of concern in what follows.

If S intends to enable others to understand q, then—employing our account of understanding—S intends to enable others to be in a complete knowledge-state with respect to Q and some appropriate set of instructions or other. (One is in such a state if one knows of some answer to Q, which satisfies appropriate instructions and which is a complete content-giving proposition with respect to Q, that it is a correct answer to Q.) We can think of *understanding q* as a property (or state) of persons. If S intends to enable other persons to understand q, then

(1) $(\exists P)(P =$ the property of understanding q, and S intends of P that others be enabled to have P).

On the present account,

(2) The property of understanding q = the property of being in a complete knowledge-state with respect to Q and some appropriate instructions.

Therefore, if S intends to enable other persons to understand q, then

(3) $(\exists P)(P =$ the property of being in a complete knowledge-state with respect to Q and some appropriate instructions, and S intends of P that others be enabled to have P).

The intentions in (1) and (3) are *de re*. If S intends to enable others to understand q, it is not required that S know that (3) is true, since S may not know that (2) is true, even if he knows that (1) is.

Returning now to the conditions for explaining formulated in Section 1 (as modified in Section 9), and unpacking the concept of understanding, we can say this. Where Q is a content-question and the u-restriction for Q is satisfied,

(4) S explains q by uttering u

is true if and only if S utters u with these intentions: (a) that his utterance will enable others to be in a complete knowledge-state with respect to Q and some appropriate instructions; and (b) that his utterance will do this by producing the knowledge, of the proposition expressed by u, that it is a correct answer to Q.[40]

Condition (b) indicates the way in which S must intend to produce the complete knowledge-state if S is to be explaining q by uttering u.

---

40. Again, the intention is *de re*.

But how, it might be asked, can the explainer of q produce in others a complete knowledge-state with respect to Q simply by uttering u? Granted that by uttering u, where u expresses a proposition that is a correct answer to Q, he can get others to know, i.e., be acquainted with, a correct answer to Q. But how can he get them to know, of this answer, that it is correct, simply by uttering u?

The simplest reply is to say that he is not required to *get others* into a complete knowledge-state with respect to Q; he is only required to *intend* to do this, in order to be explaining. But then a similar question can be raised at the level of intending. How can the explainer intend to get others to know of the proposition expressed by u that it is a correct answer to Q simply by uttering u?

An explainer assumes that when he utters u his audience will be, or become, acquainted with the proposition expressed by u. Thus, when an explainer explains q by uttering u he takes it for granted that his audience knows what u means (and is therefore in a position to have knowledge-by-acquaintance of the proposition expressed by u when u is uttered). Otherwise he would have chosen some other sentence; or before uttering u he would first have gotten the audience in a position to know what u means. Moreover, when he explains q by uttering u the explainer intends others to take his word that the proposition expressed by u is a correct answer to Q. In the act of explaining the explainer assumes the role of the authority and intends the fact that he intends to provide a correct answer to Q, and that he is willing to communicate this answer to others, to be sufficient grounds for others to believe that the answer is correct. Indeed, for many explainers on many occasions this fact provides sufficient grounds for others to be said to know that the answer supplied is correct.

Suppose that the intention of an explainer is satisfied, and that, as a result of an explaining act, the audience comes to know of some proposition p that it is a correct answer to Q. Now it is part of my account that for S to be explaining q by uttering something that expresses proposition p, it need not be the case that S know that p is a correct answer to Q. But if S does not know this, then, it may be objected, we have the following paradoxical situation: S's intention of getting the audience into a certain epistemic state with regard to p (that of knowing that p is a correct answer to Q) can be satisfied by his act of explaining even if S himself is not in that state.

My reply is to deny that such a paradoxical situation can arise. If S does not know that p is a correct answer to Q, and if S's audience comes to believe that p is a correct answer to Q solely on S's authority, then the audience too lacks the same knowledge that S does. As a result of the explaining act this audience may come to be acquainted with proposition p, and it may come to believe that p is a correct answer to Q (and it may even be justified in this belief). But in the situation envisaged it does not come to *know* that p is a correct answer to Q. Hence the explainer's intention is not in fact satisfied.

14. BROMBERGER'S THEORY

Bromberger's remarks about not understanding have been cited. But these are only part of his larger and very suggestive theory of explanation, which, like the one just developed, focuses on explaining acts. I shall briefly describe this theory, and then say why I believe it is not completely successful.

Bromberger is concerned primarily with situations in which someone explains something to someone else. We may take his basic locution to be "S explains q to A," in which q is an indirect question. Six conditions are cited for the truth of sentences of this form.[41] To formulate them Bromberger introduces the notions of a *p-predicament* and a *b-predicament*. A is in a p-predicament with respect to a question Q if and only if, on A's view, Q has a correct answer, but none of the answers to Q that A can conceive is one that he can accept. A is in a b-predicament with respect to Q if and only if Q has a correct answer, but none of the answers to Q that A can conceive is correct.[42] When S explains q to A, S assumes that A is in one of these two predicaments which S then tries to remove.

Bromberger's conditions for "S explains q to A" can be formulated as follows:

1. Q is a sound question, i.e., it admits of a correct answer.
2. S knows a correct answer to Q.
3. S assumes that A is in a p-predicament with respect to Q; or S assumes that A is in a b-predicament; or S assumes that A is in one of these two predicaments.
4. S presents the facts that, in his opinion, A must learn to know a correct answer to Q.
5. S provides A with such instruction as S thinks necessary to remove whatever predicament cited in (3) that S believes A to be in.
6. At the end of the episode all of the facts mentioned in (4) and (5) have been presented to A by S.

It is not unfair, I believe, to classify Bromberger's view as illocutionary, although he himself does not do so, and his conditions do not explicitly mention S's intentions. But the following condition for the

---

41. Bromberger, "An Approach to Explanation," pp. 94–95.
42. Note the relationship between this and Bromberger's two conditions for not understanding given in Section 12. If condition (i) obtains and A believes that Q has a correct answer, then A is in a p-predicament with respect to Q. If (ii) obtains and Q has a correct answer, then A is in a b-predicament.

truth of "S explains q to A" seems roughly to capture the spirit of his account:

> S, who knows a correct answer to Q, cites facts and provides instructions with the intention that these remove A's p-predicament or b-predicament with respect to Q.

In what ways is this account unsuccessful? To begin with, Bromberger's first two conditions are not necessary for explaining. S can explain q (e.g., why the earth is flat) even though Q is not a sound question and S does not know a correct answer to it. However, this criticism may not be completely fair, since Bromberger seems to be concerned not simply with explaining, but with correct explaining; and for this his first two conditions are necessary (provided that the second condition requires only that S know, i.e., be acquainted with, an answer to Q that is correct, and not that he know that it is correct). His third condition, however, is not necessary. Reverting to the example of legionnaire's disease, S may explain to Dr. Robinson what caused Smith's death, even though S does not assume that none of the answers Dr. Robinson can conceive is one he can accept or that is correct. All that S needs to do is assume (or act as if) Dr. Robinson is in an n-state (a state of not-understanding) with respect to Q (see Section 12), which does not require him to be in a p- or a b-predicament. Bromberger's p- and b-predicaments do not adequately capture the idea of non-understanding.

However, even if we broaden Bromberger's third condition to cover n-states, his conditions remain insufficient, for two reasons. First, although they require that S know a correct answer to Q, they—like Matthews's conditions discussed in Section 2—do not require that S actually present that answer when he explains. All they require is that S present facts which S believes that A must learn to know a correct answer and remove his p- or b-predicament (or his n-state, if this modification is accepted). But it is at least possible for S to do this without actually presenting what he believes to *be* a correct answer to Q. Reverting to an earlier example, S might believe that in order to know a correct answer to the question Q, "Why do the tides occur?," A must learn that truth is beauty; moreover, S might believe that learning this fact causes one to know a correct answer to Q, and is both necessary and sufficient for removing n-states with respect to Q. By presenting this fact, however, S has not explained to A why the tides occur, unless he believes that the proposition that truth is beauty is a correct answer to the question "Why do the tides occur?" It was for such a reason that my account requires not simply that S utter u with the intention that his utterance render q understandable, but that S believe that what he utters expresses a correct answer to Q.

Second, as was also shown in Section 1, for S to explain q it is not sufficient for S to present what he believes to be a correct answer to Q with the intention of rendering q understandable; he must present what he believes to be a correct answer with the intention of rendering q understandable *by producing the knowledge of this answer that it is a correct answer to Q*. Since Bromberger's conditions do not require that S actually present what he believes to be a correct answer to Q, a fortiori, they do not require that S do so with this intention.

15. NON-ILLOCUTIONARY VIEWS OF EXPLAINING

Let me now contrast my view of explaining acts with non-illocutionary accounts. In Chapter 1, I pointed out that standard theories of explanation, such as Aristotle's doctrine of the four causes, and the theories of Hempel, Salmon, and Brody, concentrate on explanations (products) and their evaluations rather than explaining acts. Nevertheless, taking a cue from such theories, I did formulate conditions for explaining which proponents of these theories might conceivably wish to suggest. For example, where Q is a question of the form

Q: Why does X have P?

it is conceivable that Aristotle, or a defender of the Aristotelian view, might say that

(1) S explains q by uttering u if and only if S utters u, and u expresses a proposition of the form "X has P because r" in which r purports to give one or more of Aristotle's four causes of X's having P.

Suppose that S utters the sentence

(2) This metal expanded because it was heated.

If account (1) is accepted then S has explained why the metal expanded. He has done so in virtue of the fact that he has uttered a sentence of the form "X has P because r" in which r purports to cite an (efficient) cause of the metal's expanding. Why is to do this to explain? The answer is that to explain is to utter a sentence, such as (2), that expresses a proposition citing a cause.

The illocutionary account I have presented rejects these claims. To utter a sentence expressing a proposition citing a cause of an event is not necessarily to explain that event. It depends on the intentions of the person doing so. In response to the question

What expanded because it was heated?

I might utter (2). Although I have uttered a sentence expressing a proposition that cites a cause of the expansion, I have not performed the illocutionary act of explaining why this metal expanded, since I did not utter (2) with the intention appropriate for such an act. My intention was only to identify what it was that expanded because it was heated. (My intention was to provide a correct answer to "What expanded because it was heated?" not to "Why did this metal expand?") No doubt the audience can use the information I have provided in (2) to explain why this metal expanded. But my own act in this case was not that of explaining why this metal expanded.[43]

Even if uttering sentences expressing causal propositions is not sufficient for explaining, people often do explain by (doing something which involves) uttering sentences that express causal propositions. What makes it possible to explain by doing this? The illocutionary theory has a reply.

A question of the form

Q: What is the cause of X?

is a content-question. Moreover, it is often a sound one; i.e., it admits of a correct answer. Furthermore, people are frequently not in a complete knowledge-state with respect to such a question, without being in an alternation-state with respect to it. The satisfaction of these conditions means that people are frequently in n-states—states of nonunderstanding—with respect to q (see Section 12). Uttering a sentence expressing a proposition that cites the cause of X can remove these n-states without producing alternation-states. Since such a proposition is a complete content-giving one with respect to Q, producing it can get such people into a complete knowledge-state; therefore, it can get them to understand what the cause of X is. Uttering a sentence expressing a causal proposition can accomplish this by getting people to know, of such a proposition, that it is a correct answer to Q. Therefore, an explainer can utter a sentence expressing a causal proposition with this intention: that his utterance enable others to understand q by producing the knowledge, of the proposition expressed by the sentence he utters, that it is a correct answer to Q. In short, he can *explain* q by uttering sentences expressing causal propositions.

It is possible to do so not because causal propositions are "intrinsic explainers" the mere expressing of which is necessarily to explain, or because "to explain" means "to utter a sentence expressing a causal proposition," but because causal questions are content-questions: they

---

43. It must be stressed that the present illocutionary remarks are not intended to demolish Aristotle's theory of explanation, but only (1) above. Aristotle does not explicitly propose a theory of explaining acts, but only of products and their evaluation. (For the latter the considerations in Chapters 3 and 4 will be particularly relevant.) My claim is that non-illocutionary versions of Aristotelianism (or Hempelianism, etc.) with respect to explaining acts are inadequate.

can be associated with states of non-understanding; and because people are frequently in such states with respect to causal questions.

If "to explain" does not mean "to utter sentences expressing causal propositions," why, a defender of (1) may still ask, are causes so frequently sought when explanations are demanded? The answer of the illocutionary view is that events have causes. But nature and our own finite minds conspire to produce in us n-states with respect to the question of what caused many of those events. We so often cite causes when we explain because doing so will alleviate n-states with which we are frequently plagued.

Still, it might be asked, why do we explain by appeal to causes of things rather than, say, by appeal to colors? The answer is that we do explain by appeal to colors, e.g., "I don't like that dress because that dress is red." What we don't do is explain what color that dress is by saying

(3) The color of that dress is red,

whereas we can explain what caused Smith's death by saying

(4) The cause of Smith's death was his contracting legionnaire's disease.

The illocutionary theory can account for this difference. The question

Q_1: What color is that dress?

is not a content-question. Accordingly, n-states with respect to Q_1 are impossible. One can neither understand nor fail to understand what color that dress is (unless "understanding what color that dress is" is being construed as elliptical for something else). Therefore, (3) is not a proposition the citing of which can render q_1 understandable.

By contrast, the question

Q_2: What caused Smith's death?

is a content-question (e.g., (4) is a complete content-giving proposition with respect to Q_2). Assuming that Q_2 is sound, one can understand, or fail to understand, what caused Smith's death. If (4) is a correct answer to Q_2, it is a proposition the expression of which can render q_2 understandable.

Other non-illocutionary accounts of explaining acts outlined in Chapter 1 are subject to the same treatment. Thus, if I utter sentences expressing the D-N argument

This metal was heated
All metals expand when heated
    Therefore,
This metal expanded,

I may not be explaining anything. I may be uttering these sentences solely with the intention of illustrating a point in logic, or with the intention of defending a conclusion, or with any of a number of intentions different from ones associated with explaining. In general, there are no sentences the mere uttering of which constitutes explaining.

Nevertheless, one does utter sentences when one explains; and these may include ones of the sort the theorists in Chapter 1 advocate (ones they call "explanations"—that invoke causes, ends, deductive-nomological arguments, probabilistic laws, essential properties, or whatever). So the question becomes how it is possible to explain by uttering sentences of that particular type. The answer of the illocutionary theory is that such sentences can express complete content-giving propositions which are correct answers to questions with respect to which states of non-understanding are possible. To utter such sentences can alleviate those states; so one can utter them with this intention.

16. A CIRCULARITY OBJECTION

Some critics might seek to reject the illocutionary analysis of explaining on grounds of circularity, or at least on grounds that it invokes concepts in the analysis that are dangerously close to the concept of explaining itself.

Thus consider a question such as

Q: Why did that occur?

Suppose that speaker S utters a sentence of the form

u_1: The reason that occurred is that p,

or

u_2: That occurred because p,

or

u_3: The explanation of why that occurred is that p,

with the intention of rendering q understandable by producing the knowledge, of the proposition expressed by u_1 or u_2 or u_3, that it is a correct answer to Q. Since Q is a content-question, and the u-restriction is satisfied by u_1 through u_3, on the illocutionary analysis, S is explaining q. But sentences u_1, u_2, and u_3 utilize the terms "reason," "because," and "explanation," which are too akin to the concept of explaining itself to say that progress has been achieved.

This charge is unwarranted. Neither the definition of explaining offered by the illocutionary theory, nor that of understanding (nor, indeed, the definition of a correct answer to a content-question, to be given in Chapter 4, Section 1), invokes terms such as "reason," or "because," or "explanation." Rather, what is invoked is the general notion of a complete content-giving proposition with respect to Q, whose characterization, as given in Sections 5 and 6, is not dependent on explanatory notions. True, u_1 through u_3 all express complete content-giving propositions with respect to Q. But so do

> The importance of that occurrence is that p,
>
> The danger of that occurrence is that p,
>
> The penalty for that occurrence is that p,
>
> The function of that occurrence is f,

with respect to the question

> Q': What is the importance of (danger of, penalty for, function of) that occurrence?

One might explain q' by uttering a sentence expressing one of these propositions. Yet it cannot be concluded from this that the illocutionary definition of explaining appeals to the concepts of importance, danger, penalty, or function. To be sure, in explaining q by uttering u one could attempt to satisfy instructions that bar the use of terms such as "reason," "because," and "explanation" in the sentence u giving the explanation. (In Chapter 5 we will consider models of explanation that propose just such instructions.) What am I claiming now is simply that the use of such terms in particular explaining episodes is not sufficient to render circular the general characterization of these episodes offered by the illocutionary theory.

17. CONCLUSION

The present chapter has provided conditions for explaining and for the related idea of understanding. To this end special concepts were introduced, including that of a content-giving sentence (and various derivative notions) and that of instructions for answering a question. Reasons were proposed for saying that, where Q is a content-question and the u-restriction for Q is satisfied, then (barring ellipses) S explains q by uttering u if and only if S utters u with the intention that his utterance render q understandable by producing the knowledge, of the proposition expressed by u, that it is a correct answer to Q. And it was also argued that A understands q if and only if there is a set of appropriate instructions for Q and A is in a complete knowledge-state with respect to Q and those instructions. (In Chapter 4 the concept of "appropriate instructions" will be discussed.) The illocutionary theory of explaining was contrasted with Bromberger's seminal account and also with possible non-illocutionary positions. In the next chapter my concern will be with the products of explaining acts: explanations. I want to consider what kinds of entities, if any, these are, and whether, and if so how, they can be characterized independently of the concept of an explaining act.

CHAPTER 3

What Is an Explanation?

Most theories of explanation focus not on explaining acts but on what I have been calling the products of such acts. They make claims about the ontological status of such products and about their evaluation. What sort of entity, if any, is an explanation? Views frequently expressed are that explanations are sentences, or propositions, or arguments. It is my aim here to examine these and other possibilities. I want to show why the simpler candidates will not do, and that if explanations are entities they are more complex than is generally believed. I also plan to ask whether explanation products can be understood independently of the concept of an explaining act. At the end I shall raise the question of how dependent on its ontology is a theory that provides conditions for evaluating explanations. Would criteria for evaluating explanations supplied by the D-N theory, e.g., be precluded by the adoption of an ontology different from that proposed by the theory itself?

1. SENTENCE AND PROPOSITION VIEWS

Let us suppose that some explaining act has occurred, e.g., Dr. Smith explained why Bill got a stomach ache, by uttering the sentence

(1) Bill ate spoiled meat.[1]

Consider the explanation of why Bill got a stomach ache which is a product of Dr. Smith's explaining act. (In what follows I shall use

---

1. The characterization of explaining acts given in Chapter 2, Section 9, allows what is uttered to be something other than a sentence. But to simplify the discussion in this chapter I shall assume that it is a sentence. However, there is nothing in the conditions for explaining given in Chapter 2, or in the denotation condition to be formulated below, that requires an explainer to utter only sentences that express complete content-

product-expressions of the form "S's explanation of q" or "the explanation of q given by S" in the sense of "the explanation of q which is the product of S's act of explaining q.") To what sort of entity, if any, does the product-expression "Dr. Smith's explanation of why Bill got a stomach ache" refer?

A simple answer is that it refers to a sentence. Which one? Obviously, the sentence that the doctor uttered, viz. (1). In general, if S explains q by uttering a sentence, it might naturally be supposed that the product-expression "the explanation of q given by S" denotes the sentence S utters. Assuming that such a product-expression is uniquely referring, we can formulate the following condition:

Denotation condition (sentence view): "The explanation of q given by S" denotes u if and only if
  (i) u is a sentence;
  (ii) S explained q by uttering u;
  (iii) (v)(S explained q by uttering $v \supset v = u$). (I.e., whatever S explained q by uttering is identical with u.)[2]

In accordance with this view, *explanations are sentences* (including conjunctions) *by the uttering of which explainers explain*. (To infuse this doctrine with more of the spirit of the ontological theories described in Chapter 1, various conditions could be imposed on the sentence u— e.g., that it contain causes, laws, etc.; in Section 8 specific comparisons will be made.)

Two major problems beset the sentence theory, but since these are common to other views I shall deal with them later. However, a point will now be mentioned which suggests the preferability of a proposition view. Suppose that Dr. Smith explained Bill's stomach ache (why Bill got a stomach ache) by uttering

(1) Bill ate spoiled meat,

while Dr. Robinson explained it by uttering

(2) Bill ate meat that was spoiled.

If (1) is the only sentence by the uttering of which Dr. Smith explained Bill's stomach ache, and (2) is the only sentence by the uttering of which Dr. Robinson did so, then, on the present view, "the

giving propositions with respect to Q; indeed, in what follows, it will be useful to consider sentences that are not of this type.

2. The product-expression "the explanation of q given by S" may be used to refer uniquely even if S gave different explanations of q; for example, when we use this expression we might be referring to the explanation of q given by S during time t, in which case the expression "S explained q by uttering u" in (ii) and (iii) becomes "S explained q by uttering u during t." Again, the product-expression may be used to refer uniquely even if S explained q on many occasions provided he gave the same explanation on each.

explanation of Bill's stomach ache given by Dr. Smith" denotes sentence (1), while "the explanation of Bill's stomach ache given by Dr. Robinson" denotes sentence (2). But since sentence (1) ≠ sentence (2),

the explanation of Bill's stomach ache given by Dr. Smith ≠
the explanation of Bill's stomach ache given by Dr. Robinson,

which seems unsatisfactory. Intuitively, both doctors have given the same explanation. Both have attributed the stomach ache to the eating of spoiled meat, despite the fact that the particular sentences used by each are not the same.

This multiplication of explanations can be avoided by identifying explanations with propositions, not sentences.[3] Since sentences (1) and (2) express the same proposition we can conclude that the explanations given by the doctors are the same. Let us then formulate the following

Denotation condition (proposition view): "The explanation of q given by S" denotes x if and only if
 (i) x is a proposition;
 (ii) (∃u)(S explained q by uttering u);
 (iii) (u)(S explained q by uttering u ⊃ u expresses x).

On this view *explanations are propositions expressed by sentences by the uttering of which explainers explain.*

Two considerations, however, make this (as well as the sentence view) untenable.

2. THE ILLOCUTIONARY FORCE PROBLEM

To formulate the first, let me begin with a general observation about illocutionary acts and products, viz.

(1) The product of S's illocutionary act is an (illocutionary product) F only if S F-ed.

For example, the product of S's illocutionary act is a promise, or a warning, or a criticism, or an explanation, only if S promised, or warned, or criticized, or explained. This does not preclude there being several illocutionary products when S utters u, so long as, by uttering u, S is performing several illocutionary acts.

Let us suppose, now, that by uttering the sentence

(2) Bill ate spoiled meat

Jane criticized Bill for eating spoiled meat. Then

---

3. It could also be avoided by identifying explanations with *sets* of sentences that are equivalent in meaning. This is subject to the same difficulties as the proposition view to be considered next.

(3) The criticism of Bill given by Jane is that Bill ate spoiled meat.

By analogy with the proposition view of the product of explanation, the product-expression "the criticism of Bill given by Jane" will be taken to denote the proposition expressed by the sentence Jane uttered in giving her criticism, viz. the proposition expressed by (2).[4] On the proposition view, then, the product-expression in (3) denotes the same proposition as the product-expression "the explanation of Bill's stomach ache given by Dr. Smith" in our earlier example. Therefore, the explanation of Bill's stomach ache given by Dr. Smith = the criticism of Bill given by Jane. But the criticism of Bill given by Jane was a criticism. Therefore,

(4) The explanation of Bill's stomach ache given by Dr. Smith was a criticism.

Now in accordance with (1)—the general claim about illocutionary acts and products—the product of Dr. Smith's illocutionary act is a criticism, i.e., (4) is true, only if Dr. Smith criticized. However, when Dr. Smith performed the illocutionary act of explaining Bill's stomach ache (we may suppose) he was not criticizing anyone. So (4) is false. What is true is only that what Dr. Smith uttered in explaining Bill's stomach ache is what Jane uttered in criticizing Bill. But (4) does not follow from that.

This will be called the illocutionary force problem. The proposition expressed by what is uttered in an act of explaining may be the same as the proposition expressed by what is uttered in other illocutionary acts, such as an act of criticizing. If the product-expression denotes such a proposition then we will have to conclude that the explanation given by S is a criticism, and hence that when S explained he criticized, even when this is not so. The proposition view (as well as the sentence view) is beset by the illocutionary force problem. It cannot distinguish explanations from the products of other illocutionary acts, where these products are not explanations.

To this someone might reply by rejecting assumption (1) regarding products of illocutionary acts. Suppose Jane in criticizing Bill utters "Bill ate spoiled meat." Dr. Smith hearing her utterance replies

(5) That is an explanation of Bill's stomach ache.

It would seem that (1) is violated since the product of Jane's illocutionary act is an explanation even though Jane was only criticizing and was not explaining anything.

Has (1) really been violated in such a case? In response to (5) Jane can reply

---

4. The denotation condition would be this: "the criticism given by S" denotes x if and only if (i) x is a proposition, (ii) ($\exists$u)(S criticized by uttering u), (iii) (u)(S criticized by uttering u $\supset$ u expresses x).

(6) That is not an explanation of anything. (I didn't even know that Bill got a stomach ache.) It is simply a criticism of Bill—people shouldn't eat spoiled meat.

Suppose the tables are turned and Jane, hearing Dr. Smith utter "Bill ate spoiled meat," asserts

(7) That is a criticism of Bill.

To this Dr. Smith replies

(8) That is not a criticism of anyone. It is simply an explanation of Bill's stomach ache.

I believe that both (5) and (6) (and (7) and (8)) are in the imagined contexts reasonable responses. The "dispute" here is more apparent than real since what is being referred to by "that" in (5) and (6) (and in (7) and (8)) is not the same. In (5) the doctor is referring to the sentence Jane uttered (or perhaps to the proposition it expresses). He is telling us, in effect, that this sentence (or proposition) can be used in providing an explanation of Bill's stomach ache. And this is perfectly true. In (6) Jane is referring to the product of her illocutionary act. She is telling us that this product is not an explanation but a criticism, since her act was a criticizing one, not an explaining one. And this is also true. If (5) is not referring to the product of an illocutionary act, whereas (6) is, then the general assumption (1) is not violated.

3. THE EMPHASIS PROBLEM

The second problem with the proposition view is that its denotation condition assumes that the u-position in sentences of the form "S explains q by uttering u" is referentially transparent when this position is filled by expressions for sentences. That is, it assumes that, from "S explains q by uttering u" and "u = v," we may infer "S explains q by uttering v." But this assumption can be seen to be unjustifiable by appeal to the notion of emphasis.[5] (A more general discussion of emphasis that can be used to defend the claims about to be made will have to await Chapter 6.)

There are expressions for sentences which are such that adding or changing emphases in them will not alter the sentence referred to. For example, suppose I refer to

(1) The sentence "Bill ate spoiled meat on Tuesday,"

5. For a seminal discussion of emphasis, see Fred Dretske, "Contrastive Statements," *Philosophical Review* 81 (1972), pp. 411–37.

and you, who are hard of hearing, think I was referring to the sentence "Bill ate spoiled meat on Monday." I might reply that I am referring to

(2) The sentence "Bill ate spoiled meat *on Tuesday.*"

By using emphasis in (2) I am not referring to a different sentence but to the same one as in (1). Only I use emphasis in (2) and not in (1) to correct your mistake about which sentence I am referring to. In Chapter 6 this is what I call a non-semantical use of emphasis. When such a use of emphasis occurs neither the meaning nor the reference of the referring expression is altered with a change in emphasis. Accordingly, the sentence "Bill ate spoiled meat on Tuesday" = the sentence "Bill ate spoiled meat *on Tuesday.*" Similarly, the sentence "Bill ate spoiled meat on Tuesday" = the sentence "Bill ate *spoiled meat* on Tuesday." Hence

(3) The sentence "Bill ate *spoiled meat* on Tuesday" = the sentence "Bill ate spoiled meat *on Tuesday.*"

Suppose now that

(4) Dr. Smith explained Bill's stomach ache by uttering the sentence "Bill ate *spoiled meat* on Tuesday."

Dr. Smith emphasizes "spoiled meat" to indicate that he believes that this aspect of the situation is relevant for explaining Bill's stomach ache. He believes that it was the spoiled meat Bill ate that caused his stomach ache. Now if the u-position in sentences of the form "S explained q by uttering u" is referentially transparent, then from (4), in virtue of (3), we may infer

(5) Dr. Smith explained Bill's stomach ache by uttering the sentence "Bill ate spoiled meat *on Tuesday.*"

But (5) is false since it entails that Dr. Smith believes that the day on which Bill ate the spoiled meat was relevant for his getting a stomach ache. And I am supposing that Dr. Smith had no such belief. His claim was that Bill got a stomach ache because of what he ate and not because of when he ate it. This distinction is expressed by the difference in emphasis between "The explanation of Bill's stomach ache given by Dr. Smith is that Bill ate *spoiled meat* on Tuesday," which is true if (4) obtains, and "The explanation of Bill's stomach ache given by Dr. Smith is that Bill ate spoiled meat *on Tuesday,*" which is true if (5) obtains.

The emphases in (2) and

(6) The sentence "Bill ate *spoiled meat* on Tuesday"

are non-semantical. A shift of emphasis in these expressions will not change their meanings or references. But when these expressions are embedded in explanation-sentences such as (4) the emphases can assume a semantical role. The emphasized words can become "captured" by the term "explained," indicating that a particular aspect of Bill's eating spoiled meat on Tuesday is claimed to be explanatorily relevant for his getting a stomach ache. A shift in emphasis in (6) as this appears in (4) can transform (4), which is true, into (5), which is false.

This is not to deny that the emphases in both (4) and (5) could be used to play non-semantical roles. (One might assert (5) in response to someone who thinks that Dr. Smith explained Bill's stomach ache by uttering the sentence "Bill ate spoiled meat on Monday.") But if so (4) and (5) would have different readings from the ones being given here. We are using emphasis in (4) in the same way that Dr. Smith did, viz. to indicate a particular aspect of the situation Dr. Smith takes to be explanatorily relevant, not to correct a mistake about some of the words he uttered. When emphasis is understood as playing such a semantical role—which is possible in (4) and (5)—we get readings for these sentences under which (4) is true and (5) is false.

Assuming, then, that (2) and (6) denote the same sentence, and that the substitution of (2) for (6) in (4) turns a true sentence into a false one, we must conclude that the u-position in sentences of the form "S explains q by uttering u" is referentially opaque if that position is filled by sentence-expressions. There are readings for sentences of this form such that the substitution of co-referring u-terms will lead from truths to falsehoods. But the denotation condition of the proposition (as well as the sentence) view assumes that the u-position is transparent.

This problem can be avoided by introducing the notion of an e-sentence ("e" for emphasis), which is a sentence together with its emphasis, if any. To give an e-sentence that S uttered in explaining q one must supply S's words, in the order he gave them, and with any emphasis he used. And we will now suppose that what is uttered by speakers in acts of explaining are e-sentences. The problem above is thus avoided since

(7) The e-sentence "Bill ate *spoiled meat* on Tuesday"

is not identical with

(8) The e-sentence "Bill ate spoiled meat *on Tuesday.*"

Therefore from the fact that Dr. Smith explained Bill's stomach ache by uttering the e-sentence denoted by (7) it does not follow that he explained Bill's stomach ache by uttering the e-sentence denoted by (8).

Nevertheless there is a residual emphasis problem for the proposition view, which must now assume that e-sentences express propositions. Although the e-sentences denoted by (7) and (8) are not identical, the propositions these e-sentences express are. (7) and (8), respectively, denote the e-sentences

(9) Bill ate *spoiled meat* on Tuesday,
(10) Bill ate spoiled meat *on Tuesday*.

A shift in emphasis in (9) and (10) changes the e-sentence but not the proposition expressed. If I utter the e-sentence (9)—using emphasis where I do to indicate my surprise—and you think I am expressing the proposition that Bill ate spoiled meat on Monday, I may correct your mistake by uttering (10). By doing so I am not expressing a different proposition but the original one in a manner that will correct your misconception (i.e., by using a different e-sentence).

Suppose then that Dr. Smith explained Bill's stomach ache by uttering the e-sentence (9), while Dr. Jones explained it by uttering the e-sentence (10). And in both cases assume that requirement (iii)—the uniqueness requirement—of the proposition theory's denotation condition is satisfied. It follows from this denotation condition that "the explanation of Bill's stomach ache given by Dr. Smith" denotes the same proposition as "the explanation of Bill's stomach ache given by Dr. Jones," viz. the proposition expressed by the e-sentences (9) and (10). Therefore

(11) The explanation of Bill's stomach ache given by Dr. Smith = the explanation of Bill's stomach ache given by Dr. Jones,

which we ought to reject since according to Dr. Jones's explanation, but not Dr. Smith's, the date on which Bill ate the spoiled meat is relevant for explaining Bill's stomach ache. If the date was irrelevant—if Bill would have gotten a stomach ache from eating spoiled meat on any day of the week—then, we may suppose, Dr. Smith's explanation is correct while Dr. Jones's is incorrect. We must conclude that these explanations are different and hence that (11) is false.

4. THE ARGUMENT VIEW

Before pursuing a remedy for these problems let me turn briefly to a different product view, suggested by the D-N model, according to which explanations are arguments.

Denotation condition (Argument view): "The explanation of q given by S" denotes an argument one of whose premises is (the proposition expressed by) u if and only if
(i) S explained q by uttering u;

(ii) (v)(S explained q by uttering v⊃v = u (or v and u express the same proposition)).

On one version of this view arguments are composed of propositions, and on the other of sentences. Since the proposition version will allow us to count explanations as identical even if they use different sentences, it will be considered. To ensure referential transparency in the u-position in "S explained q by uttering u" let us assume that the u's are e-sentences. Suppose, then, that Dr. Smith explained Bill's stomach ache by uttering the e-sentence

(1) Bill ate *spoiled meat* on Tuesday.

Assuming that condition (ii) above is satisfied, the product-expression

(2) "the explanation of Bill's stomach ache given by Dr. Smith"

denotes some argument one of whose premises is the proposition expressed by (1). What argument?

Following the D-N model we may assume that it contains a law among its premises and that its conclusion, which is entailed by the premises, describes the event referred to in the object-expression q. We might suppose that the product-expression (2) denotes an argument composed of propositions expressed by sentences such as these:

(3) Bill ate *spoiled meat* on Tuesday. Anyone who eats spoiled meat gets a stomach ache. Hence, Bill got a stomach ache.

On the D-N view an argument such as (3) might be construed as an ordered set of propositions (or sentences), or perhaps better, as an ordered pair whose first member is the conjunction of premises and whose second is the conclusion.

The argument view fares no better than the proposition view of Section 1. First, there is an illocutionary force problem. Poor Sam dies one hour after being operated on by Dr. Smith, who offers an excuse by uttering

(4) Sam had disease d at the time of his operation. Anyone with disease d at the time of an operation dies one hour after the operation. Hence, Sam died one hour after his operation.

Suppose further that Dr. Jones explains Sam's death by uttering (4). I will assume that, on the argument product-view, both "the excuse given by Dr. Smith" and "the explanation of Sam's death given by Dr. Jones" denote the argument given by (4), i.e., the ordered pair whose first proposition is expressed by the conjunction of the first two sentences in (4) and whose second proposition is expressed by the last sentence. If so then the excuse given by Dr. Smith = the explanation of Sam's death given by Dr. Jones. But the excuse given by Dr. Smith was an excuse. Therefore,

(5) The explanation given by Dr. Jones was an excuse,

which is false if Dr. Jones has no intention of excusing anything. Indeed Dr. Jones may believe the operation and resulting death to be inexcusable. What is true is only that what Dr. Smith uttered in giving his excuse was what Dr. Jones uttered in explaining Sam's death. But (5) does not follow from that.

The emphasis problem can also be shown to be present, but I will omit the details.

5. PROPOSITION CONSTRUED AS EXPLANATORY

A view will now be developed which avoids the illocutionary force problem. Later it will be shown how it can be modified to avoid the emphasis problem as well.

The first difficulty with the proposition (and the argument) view is that S may explain q by uttering what is uttered by someone performing a different type of illocutionary act, such as criticizing or excusing. If the product-expression denotes the proposition expressed by the sentence S utters in his act of explaining, then S's explanation can turn out to be a criticism or an excuse even when it is not. The proposition view fails to take proper account of the fact that to explain is to utter something with a certain illocutionary force. Perhaps then the product of an explanation should be taken to be what is expressed by that utterance—a proposition—construed with the illocutionary force of explaining.

Searle in a discussion of illocutionary acts criticizes Austin for failing to distinguish two senses of "statement": the act of stating and what is stated (which Searle calls the statement-object and I call the product of the act of stating).[6] He goes on to identify the latter as a "proposition construed as stated."[7] His discussion here is very brief and we are not told what sort of thing this is. But if we knew we could better understand products of explanation. That is, following Searle's discussion of statements, having distinguished the act of explaining from the product of the act, we now identify the latter as a proposition construed as explaining. But what sort of entity is that?

The proposal I shall now consider is that it is an ordered pair consisting of a proposition and a type of illocutionary act, in the present case an explaining type of act. So, e.g.,

---

6. John Searle, "Austin on Locutionary and Illocutionary Acts," *Philosophical Review* 77 (1968), pp. 405–24.
7. *Ibid.*, p. 423.

(1) (The proposition that Bill ate spoiled meat; explaining Bill's stomach ache)

is a product of explanation, since it is an ordered pair whose first member is a proposition and whose second is the act-type explaining Bill's stomach ache. The product expression

(2) "The explanation of Bill's stomach ache given by Dr. Smith"

could be said to denote the ordered pair (1) provided that Dr. Smith explained Bill's stomach ache by uttering something, and whatever the doctor explained Bill's stomach ache by uttering expresses the proposition that Bill ate spoiled meat. More generally, we have the following

Denotation condition (preliminary ordered pair view): "The explanation of q given by S" denotes (x;y) if and only if
(i) x is a proposition;
(ii) y = (the act-type) explaining q;
(iii) ($\exists$u)(S explained q by uttering u);
(iv) (u)(S explained q by uttering u $\supset$ u expresses x).[8]

This view avoids the illocutionary force problem. The product-expression (2) denotes the ordered pair

(1) (The proposition that Bill ate spoiled meat; explaining Bill's stomach ache),

while

(3) "The criticism of Bill given by Jane"

denotes

(4) (The proposition that Bill ate spoiled meat; criticizing Bill).

But (1) $\neq$ (4). Therefore, neither "The explanation of Bill's stomach ache given by Dr. Smith = the criticism of Bill given by Jane" nor "The explanation of Bill's stomach ache given by Dr. Smith was a criticism" can be derived. This view avoids the illocutionary force problem by taking into account not simply the proposition expressed in the act but also the type of illocutionary act itself. The types of illocutionary acts associated with the products denoted by (2) and (3) are different even though the propositions expressed are the same.

On the other hand, the emphasis problem remains, as can be seen if we suppose that Dr. Smith explained Bill's stomach ache by uttering "Bill ate *spoiled meat* on Tuesday," while Dr. Jones explained it by ut-

---

8. Analogous conditions are possible for other illocutionary product-expressions such as "the criticism of q given by S."

tering "Bill ate spoiled meat *on Tuesday*." On the present view, the explanation given by each doctor is the ordered pair (The proposition that Bill ate spoiled meat on Tuesday; explaining Bill's stomach ache), since the e-sentences uttered by the doctors express the same proposition. Hence, the explanation of Bill's stomach ache given by Dr. Smith = the explanation of Bill's stomach ache given by Dr. Jones, which is false.

Can the ordered pair view be modified to avoid this problem? In the following section a proposal will be suggested for doing so.

6. A NEW ORDERED PAIR VIEW

In Chapter 2, the conditions for S performing an act of explaining q by uttering u require that S intend that u express a correct answer to question Q. A view of explanation-products as answers to questions will now be explored.

To do so let us recall some concepts from Chapter 2. A sentence such as

(1) The reason that Bill got a stomach ache is that he ate spoiled meat

is a content-giving sentence for the noun "reason." It expresses a content-giving proposition (for the concept expressed by "reason"). Assuming that (1) expresses the same proposition as

(2) Bill got a stomach ache because he ate spoiled meat,

the proposition expressed by (2) is also a content-giving proposition. Furthermore, the propositions expressed by (1) and (2) are complete content-giving propositions with respect to the question

(3) Why did Bill get a stomach ache?

The proposition expressed by "Bill ate spoiled meat" is not a complete content-giving proposition with respect to (3), since a complete content-giving proposition with respect to Q entails all of Q's presuppositions.

Finally, question Q was called a content-question if and only if there is a complete content-giving proposition with respect to Q. Since (1) and (2) express complete content-giving propositions with respect to (3), the latter is a content-question. In Chapter 2, it was claimed that the object-expression q in "S explains q by uttering u" always expresses a content-question (or else in that sentence q is elliptical for an interrogative that expresses a content-question).

Returning now to the ordered pair view, my proposal is that the constituent proposition in the ordered pair is a complete content-giving proposition with respect to a certain question. This question will be given by the object-expression q. As in the case of "S explains q by uttering u" it will be supposed that any product-expression of the form "the explanation of q given by S" is, or is transformable into, one in which q is an indirect interrogative whose direct form expresses a content-question Q. Thus the product-expression "the explanation of Bill's stomach ache given by Dr. Smith" could be reformulated as "the explanation given by Dr. Smith of why Bill got a stomach ache." And the latter will be taken to denote an ordered pair one of whose constituents is a complete content-giving proposition with respect to question (3). But which proposition will this be and how is it related to Dr. Smith's act of explaining?

Recall from Chapter 2 the u-restriction imposed on explaining acts: what S utters is, or in the context is transformable into, a sentence expressing a complete content-giving proposition with respect to Q. If p is such a proposition, let us say that p is *associated with* S's explaining act. Equivalently, p is associated with S's act of explaining q by uttering u if

(a) p is a complete content-giving proposition with respect to Q;

(b) The act was one in which p was claimed to be true;

(c) p entails the proposition expressed by u.[9]

According to these conditions, if Dr. Smith explained why Bill got a stomach ache, by uttering

(4) The reason that Bill got a stomach ache is that he ate *spoiled meat* on Tuesday,

then the proposition (4) is associated with the act of explaining performed by Dr. Smith. But equally if Dr. Smith explained why Bill got a stomach ache, by uttering

(5) Bill ate *spoiled meat* on Tuesday,

then proposition (4) is also associated with Dr. Smith's act of explaining, even though proposition (4) is not expressed by the e-sentence which Dr. Smith uttered. If by uttering (5) Dr. Smith explained why Bill got a stomach ache, his act of explaining was one in which the truth of proposition (4) was still being claimed. Since proposition (4) is a complete content-giving one with respect to question (3), and it entails the proposition expressed by sentence (5) which Dr. Smith uttered, proposition (4) is associated with his act of explaining.

9. If u is not a sentence, then, by the u-restriction, it is transformable into one that is; and p will entail the proposition expressed by the latter. In what follows only u's that are sentences will be considered.

Now the constituent proposition in the explanation of q given by S will be one *associated with* S's act(s) of explaining q. Accordingly, we may formulate the following

Denotation condition (new ordered pair view): "The explanation of q given by S" denotes (x;y) if and only if
  (i) Q is a content-question;
  (ii) x is a complete content-giving proposition with respect to Q;
  (iii) y = explaining q;
  (iv) (∃a)(∃u)(a is an act in which S explained q by uttering u);
  (v) (a)(u)[a is an act in which S explained q by uttering u ⊃ (r)(r is associated with a ≡ r = x)]. (I.e., x is the one and only proposition associated with every act in which S explained q by uttering something.)[10]

Since (4) is a complete content-giving proposition with respect to the content-question (3), "the explanation given by Dr. Smith of why Bill got a stomach ache" denotes (proposition (4); explaining why Bill got a stomach ache), provided that there was an act in which Dr. Smith, by uttering something, explained why Bill got a stomach ache, and for any such act proposition (4) was the one and only associated proposition.

Returning now to the definition of association, why are such propositions and only these taken to be constituent propositions of explanations? That is, why are conditions (a) through (c) for association imposed? The first, by requiring that p be a complete content-giving proposition with respect to the content-question Q, guarantees that p is an answer to Q; and this relates the product of explanation to the act in which the explainer intends to provide an answer to Q. Moreover, by requiring that answer to Q to be a complete content-giving proposition the emphasis problem is avoided (as demonstrated in the next section). The reason for condition (b) can be shown by an example in which Dr. Smith by uttering the e-sentence

  (5) Bill ate *spoiled meat* on Tuesday

explains why Bill got a stomach ache. The proposition

  (6) The reason Bill got a stomach ache is that Bill ate spoiled meat *on Tuesday*

is a complete content-giving one with respect to question (3), and it entails the proposition expressed by the e-sentence Dr. Smith uttered, viz. (5). But (6) could not be the constituent proposition in Dr. Smith's explanation. (Otherwise his explanation would be identical with Dr.

---

10. Given the definition of "association," (i) and (ii) are redundant since they are entailed by (iv) and (v).

Jones's, which it isn't.) Condition (b) precludes this on the grounds that (6) is not a proposition claimed to be true in Dr. Smith's act of explaining. By uttering (5) in explaining why Bill got a stomach ache Dr. Smith does not claim the truth of (6).

The reason for condition (c) can also be given by means of the above example in which Dr. Smith by uttering (5) explained why Bill got a stomach ache. The proposition

(7) The reason Bill got a stomach ache is that he ate spoiled meat

is a complete content-giving one with respect to question (3), and it is also a proposition claimed to be true in Dr. Smith's act of explaining. Both propositions (4) and (7) satisfy conditions (a) and (b), and without condition (c) there would be no unique proposition to associate with Dr. Smith's explaining act. Condition (c) precludes (7) as a proposition associated with this act, since (7) does not entail the proposition expressed by the e-sentence Dr. Smith uttered, viz. (5).

In choosing the constituent proposition to be one satisfying the conditions of *association,* the product of explanation is seen to be intimately related to the act of explaining. The constituent proposition is an answer to a question (a question which the explainer in his act intends to answer); it is one claimed to be true in the explaining act; and either it is one expressed by what the explainer utters in that act or it entails the proposition expressed by what the explainer utters.

The new ordered pair view can also supply conditions for being *an* explanation of q given by S. Here since uniqueness is not required, we need not suppose that the same proposition is associated with every act in which S explained q. Accordingly, we can write

(x;y) is an explanation of q given by S if and only if
  (i) Q is a content-question;
  (ii) x is a complete content-giving proposition with respect to Q;
  (iii) y = explaining q;
  (iv) ($\exists$a)($\exists$u)(a is an act in which S explained q by uttering u, and x is associated with a).

As will be shown in Section 8, the ordered pair view can be generalized to explanations that are not given by anyone (i.e., that are not products of any explaining acts). What first needs to be asked, however, is whether it is subject to the difficulties of the previous views.

7. ARE THE ILLOCUTIONARY FORCE AND EMPHASIS PROBLEMS AVOIDED?

Since the second constituent of the product of explanation is an explaining type of act we avoid the illocutionary force problem. Even if the constituent proposition of Dr. Smith's explanation is the same as that of Jane's criticism it will not follow that "Dr. Smith's explanation of why Bill got a stomach ache" and "Jane's criticism of Bill" denote the same entity.

Second, if the constituent proposition of an explanation is a complete content-giving one with respect to Q then the emphasis objection is avoided. Suppose that the explanation given by Dr. Smith of why Bill got a stomach ache is that Bill ate *spoiled meat* on Tuesday, and that the explanation given by Dr. Jones is that Bill ate spoiled meat *on Tuesday*. The emphasis problem arises if we take the constituent propositions of these explanations to be ones expressed by the e-sentences

(1) Bill ate *spoiled meat* on Tuesday,

(2) Bill ate spoiled meat *on Tuesday*.

The propositions expressed by (1) and (2) are identical, from which we would have to conclude that the explanations given by the two doctors are identical. However, the propositions expressed by (1) and (2) are not complete content-giving ones with respect to "Why did Bill get a stomach ache?" Therefore, on the ordered pair view of Section 6 these propositions are not the constituent propositions of the explanations in question. But we can suppose that the constituent proposition of the explanation given by Dr. Smith is

(3) The reason Bill got a stomach ache is that he ate *spoiled meat* on Tuesday,

while that of the explanation given by Dr. Jones is

(4) The reason Bill got a stomach ache is that he ate spoiled meat *on Tuesday*.

(3) and (4) are complete content-giving propositions with respect to "Why did Bill get a stomach ache?"; and these propositions are not identical. Shifting emphasis in (3) and (4), unlike (1) and (2), means changing the propositions expressed. If Dr. Smith by uttering (1) explained why Bill got a stomach ache, he put the emphasis where he did to indicate an explanatorily relevant aspect of an event. But the emphasis in (1) does not indicate this. By contrast, the emphasis in (3) does, since it is captured by "reason." In general, in a complete content-giving proposition any explanatory emphasis used by the explainer will be so captured.

Why is this so? Consider a content-giving sentence of form (i) of Chapter 2, Section 5 (any content-giving sentence is equivalent to one of this form):

(i) $\begin{Bmatrix} \text{The} \\ \text{A} \end{Bmatrix}$ + content-noun N + phrase + to be + (preposition) + nominal.

Let us also consider a question of the form (or equivalent to one of the form)

What + is + the + content-noun N + phrase?

which is constructed from (i). Now instead of answering a question of this form by uttering a sentence of form (i), one might answer simply by uttering the nominal in (i). (If the nominal is a that-p clause, one might simply utter p.) For example, suppose that the question is

Q: What is the reason that Bill got a stomach ache?

or, its equivalent, "Why did Bill get a stomach ache?" In explaining q, instead of uttering (something equivalent to)

(5) The reason that Bill got a stomach ache is that Bill ate *spoiled meat* on Tuesday,

which has form (i), someone might utter

(1) Bill ate *spoiled meat* on Tuesday,

which is just the p-sentence in the nominal in (i). But when a speaker by uttering (1) explains why Bill got a stomach ache, he is implicitly claiming that (5) is true. (The proposition expressed by (5) is associated with his act of explaining.)

Now in (5) the emphasized words are captured by the content-noun "reason." In general, to anticipate the discussion in Chapter 6, many (though not all) content-nouns in content-giving sentences of form (i)—as well as in content-giving sentences of other forms—are emphasis-selective: in sentences of form (i), emphasized words in the nominal become selected or captured by the content-noun, indicating that a particular aspect of the situation described by the nominal is operative (is the reason, the cause, what is important, etc.). A shift of emphasis, as from (3) to (4) above, results in a shift in the operative aspect; and this can change the truth-value of the resulting sentence. If the content-noun is emphasis-selective (as is "reason"), then a content-giving sentence of form (i), or any equivalent content-giving sentence, will be able to incorporate operative aspects emphasized by explainers who may utter abbreviated versions of such sentences.

Thus, when Dr. Smith utters the abbreviated (1) in explaining why Bill got a stomach ache, he is selecting the item that Bill ate as an explanatorily operative feature. This feature is explicitly selected by the content-noun "reason" in the sentence (3) which expresses a proposition associated with Dr. Smith's act of explaining. The proposition expressed by (3) is a complete content-giving one with respect to the question "Why did Bill get a stomach ache?" More generally, in an explaining act a speaker may employ explanatory emphasis in such an abbreviated sentence. This emphasis can be explicitly captured in a sentence that expresses a complete content-giving proposition that is associated with this explaining act.

8. A GENERALIZATION OF THE ORDERED PAIR VIEW AND COMPARISONS WITH OTHER THEORIES

Expressions of the form "the explanation of q given by S" have been said to denote something only if S has explained q. But we can also refer to explanations that are not products of any acts of explaining— explanations no one has given—using expressions of the forms "the explanation of q" and "the explanation of q that is F." The use of the term "product-expression" will be extended to cover expressions of these forms which may denote entities even if no explaining acts have been performed. We might say that they denote entities that if not actually products of explaining acts are so potentially. Can the ordered pair view provide denotation conditions for this more general class of product-expressions?

The only difference between "the explanation of q" and "the explanation of q given by S" is that the use of the latter but not the former entails that there was some act in which S explained q. This entailment is reflected in (iv) of the ordered pair denotation condition for "the explanation of q given by S." If (iv) is dropped and (v) is suitably generalized so as not to refer to a particular explainer, we obtain

Denotation condition: "The explanation of q" denotes (x;y) if and only if
 (i) Q is a content-question;
 (ii) x is a complete content-giving proposition with respect to Q;
 (iii) y = explaining q;
 (iv) (a)(S)(u) [a is an act in which S explains q by uttering $u \supset (r)(r$ is associated with $a \equiv r = x)$].[11]

(iv), which expresses a uniqueness condition, does not require that any particular act of explaining q has occurred but only that if any does then x is the only proposition associated with it. To obtain a denotation condition for "the explanation of q that is F" (e.g., "the explanation of q that is the least plausible") we retain (i), (ii), and (iii) above, replace (iv) with

(a)(S)(u) (a is an act in which S explains q by uttering u, and x is the one and only proposition associated with $a. \supset a$ is an act in which S explains q in the F-manner),

---

11. "The explanation of q" in which no further clause is added is often used to mean "the correct explanation of q." In such a case "explains" in condition (iv) can be understood as "correctly explains."

and add

(a)(S)(a is an act in which S explains q in the F-manner ⊃ a is an act for which x is the one and only associated proposition).

The ordered pair theory can provide conditions for sentences of the form

(1) An explanation of q is that c.

Let us call such a sentence *restructured* if q is an indirect interrogative expressing a content-question Q and c is a sentence expressing a complete content-giving proposition with respect to Q. I shall assume that any unrestructured sentence of form (1) is paraphrasable into a restructured sentence of this form. For example,

An explanation of why Bill got a stomach ache is that Bill got a stomach ache because he ate spoiled meat

is a restructured paraphrase of

An explanation of Bill's stomach ache is that Bill ate spoiled meat.

On the ordered pair view, a restructured sentence of form (1) is true if and only if

(2) $(\exists x)(x = $ (the proposition expressed by c; explaining q)).

We can also say that

(3) E is *an* explanation of q if and only if
  (i) Q is a content-question;
  (ii) E is an ordered pair whose first member is a complete content-giving proposition with respect to Q and whose second member is the act-type *explaining q*.

For (x;y) to be *an* explanation of q we need not suppose that any particular explaining acts have occurred. Nor must we suppose, as was done earlier to ensure uniqueness, that in any act in which q is explained the constituent proposition x is associated with that act. On the present view, an explanation of q is an ordered pair whose second member is the act type "explaining q" and whose first member is an answer to Q that is a complete content-giving proposition with respect to Q.

It is at this point that the present account of explanation-products is most usefully compared with several in Chapter 1. Aristotle and Hempel, e.g., can be construed as providing conditions not for E's being *someone's* explanation of q, but for E's being *an* explanation of q, as follows:

Aristotle: E is an explanation of q if and only if
(i) Q is a question of the form "Why does X have P?"
(ii) E is a proposition of the form "X has P because r" in which r purports to give one or more of Aristotle's four causes of X's having P.

Hempel: E is an explanation of q if and only if
(i) Q is a question of the form "Why is it the case that p?"
(ii) E is a D-N (or inductive) argument, whose conclusion is p, that contains lawlike sentences (and that satisfies other conditions Hempel imposes).

It can readily be seen that (3) above provides much broader conditions than either of these. (3) allows questions associated with explanations to be content-questions of any sort, and not simply content-questions of the form "Why does X have P?" (Aristotle), or "Why is it the case that p?" (Hempel), or "Why is that X, which is a member of A, a member of B?" (Salmon). Explanations answer a variety of questions of forms other than these.

Second, allowing the constituent proposition of an explanation of q to be any complete content-giving proposition with respect to Q permits us to identify as explanations a variety not recognized by the above accounts. There is no need to restrict constituent propositions of explanations to ones of the form "X has P because r" in which r gives a cause, or "p, therefore r," where Hempel's conditions are satisfied. For example, the following count as explanations, according to (3), even if their constituent propositions are not expressible by sentences of these forms:

(The event now occurring in the bubble chamber is that alpha particles are passing through; explaining what event is now occurring in the bubble chamber);

(The significance of that document is that it is the first to proclaim the rights of animals; explaining what the significance of that document is);

(The purpose of the flag is to warn drivers of danger; explaining what the purpose of that flag is).

Finally, and very importantly, without associating illocutionary act-types with explanations, as is done in the second condition of (3)—or at least without some such proviso—we have been unable to distinguish explanations from products of other (possible) illocutionary acts. For example, what makes the proposition "Bill got sick because he ate spoiled meat" an explanation rather than an excuse, or a complaint, or a simple identification of who got sick because he ate spoiled meat?

Similar questions are possible with respect to arguments of the form "p, therefore q." This is the illocutionary force problem discussed in Section 2. What the ordered pair theory is proposing is that to distinguish an explanation from any one of a number of other illocutionary products we invoke illocutionary acts (i.e., types). The propositions expressed in such acts will not suffice to distinguish the products, since the propositions may be identical even though the products are not. The conditions in (3) make essential reference to a type of explaining act, viz. explaining q, where q is whatever it is that is being explained. These conditions appeal to a concept for which we proposed an analysis in Chapter 2. By contrast, the theories of explanation outlined in Chapter 1—theories such as Aristotle's and Hempel's—provide conditions for being an explanation that do not invoke the concept of an explaining act. Such theories do not adequately distinguish explanations from other illocutionary products.

9. A NEW PROPOSITION VIEW

Is the ordered pair theory, which invokes both propositions and types of explaining acts, the only product view that will avoid the illocutionary force and emphasis problems? In light of the ordered pair theory perhaps it is possible to suitably modify the proposition view of Section 1, as follows. An explanation will be construed as a proposition, but not just any proposition expressed by an explainer in an explaining act.

Suppose that by uttering

(1) Bill ate spoiled meat

Dr. Smith explained why Bill got a stomach ache. Instead of taking Dr. Smith's explanation to be the proposition expressed by (1), as in the original proposition view, let us take it to be the proposition expressed by

(2) An explanation of why Bill got a stomach ache is that Bill got a stomach ache because he ate spoiled meat.

More generally, let us say that

E is an explanation of q if and only if
(i) Q is a content-question;
(ii) E is a proposition expressible by a sentence of the form "an explanation of q is that c," in which c expresses a complete content-giving proposition with respect to Q.

We can also supply the following

Denotation condition (new proposition view): "The explanation of q given by S" denotes x if and only if
(i) Q is a content-question;
(ii) x is a proposition expressible by a sentence of the form "an explanation of q is that c," in which c expresses a complete content-giving proposition with respect to Q;
(iii) (∃a)(∃u)(a is an act in which S explained q by uttering u);
(iv) (a)(u)(a is an act in which S explained q by uttering u ⊃ (r)(r is associated with a ≡ r = the proposition expressed by c)).

Let us apply this condition to the example above. Proposition (2), on this view, is supposed to be what is denoted by "the explanation given by Dr. Smith of why Bill got a stomach ache." Is it? (i) The explanatory question "Why did Bill get a stomach ache?" is a content-question. (ii) Proposition (2) is expressible by a sentence of the form "An explanation of q is that c" in which c expresses a complete content-giving proposition with respect to Q. (iii) There was an act in which by uttering something (viz. (1)) Dr. Smith explained why Bill got a stomach ache. (iv) Any act in which Dr. Smith explained this by uttering something had associated with it the proposition expressed by "Bill got a stomach ache because he ate spoiled meat." So the conditions are satisfied.

Is the illocutionary force problem avoided? As in the case of explanations, other illocutionary products will not be construed as just any proposition expressed by a speaker performing an illocutionary act. For example, a criticism will now be construed as a proposition expressible by a sentence of the form

A criticism of x is that————.

Therefore, if Jane criticized Bill by uttering

(1) Bill ate spoiled meat,

and if Dr. Smith by uttering (1) explained why Bill got a stomach ache, it will not follow that the explanation of Bill's stomach ache given by Dr. Smith = the criticism of Bill given by Jane. This is so since, on the new proposition view, the explanation of Bill's stomach ache given by Dr. Smith is the proposition

(2) An explanation of why Bill got a stomach ache is that he got a stomach ache because he ate spoiled meat.

The criticism of Bill given by Jane is the proposition

(3) A criticism of Bill is that he ate spoiled meat (or that he got a stomach ache because he ate spoiled meat).

And proposition (2) ≠ proposition (3). So the illocutionary force problem of Section 2 is avoided. It is avoided because the proposition with which an illocutionary product is identified itself expresses the kind of product it is.

The emphasis problem of Section 3 is also avoided. Suppose Dr. Smith by uttering

> Bill ate *spoiled meat* on Tuesday

explains why Bill got a stomach ache; and Dr. Jones explains it by uttering

> Bill ate spoiled meat *on Tuesday*.

Their respective explanations are not the propositions expressed by these sentences (which express the same proposition), but the propositions

> An explanation of why Bill got a stomach ache is that he got a stomach ache because he ate *spoiled meat* on Tuesday,
>
> An explanation of why Bill got a stomach ache is that he got a stomach ache because he ate spoiled meat *on Tuesday*.

And these propositions are not identical. Shifts in emphasis here can alter meanings.

Is the new proposition view, then, as successful as the ordered pair view? Are types of explaining acts really avoided in favor of propositions alone? According to the new proposition view, an explanation of q is a certain proposition expressible by a sentence of the form

> (4) An explanation of q is that c.

What proposition is this? What are the truth-conditions for sentences of this form? Taking part of our cue from the ordered pair theory, we might say that a sentence of form (4) is true if and only if it is, or is transformable into, a sentence satisfying these conditions:

> (5) (i) q expresses a content-question;
> (ii) c expresses a complete content-giving proposition with respect to Q.[12]

But these conditions are not sufficient to distinguish propositions expressible by (4) from ones expressible by sentences such as

> (6) A *claim* about q is that c;
> A *theory* about q is that c;
> A *belief* about q is that c;

---

12. Keep in mind that we are not proposing truth-conditions for "a correct (or good) explanation of q is that c."

A *hope* about q is that c;
A *fear* about q is that c.

To be sure, the q's in (6) need not be content-questions. But then we can restrict our attention to those sentences of the above forms in which q does express a content-question.

Consider, e.g.,

(7) A claim about q is that c.

Shall we say that a sentence of this form is true if and only if the conditions in (5) are satisfied? That is, shall we say that the same conditions that make a sentence of form (4) true make one of form (7) true? If so, then if

(2) An explanation of why Bill got a stomach ache is that Bill got a stomach ache because he ate spoiled meat

is true, then so is

A claim (theory, belief, hope, fear, etc.) about why Bill got a stomach ache is that he got a stomach ache because he ate spoiled meat.

And such an inference seems unjustified.

The ordered pair view avoids this problem by the introduction of explaining act-types as parts of explanations. On this view, a sentence of form (4) is true if and only if it is, or is transformable into, one for which the following condition is satisfied:

(8) $(\exists x)(x = $(the proposition expressed by c; explaining q), and Q is a content-question, and the proposition expressed by c is a complete content-giving proposition with respect to Q).

By analogy, if we restrict q and c as we have done, then, on the ordered pair view, a sentence of one of the forms in (6) is true only if

$(\exists x)(x = $[the proposition expressed by c; making a claim about q (or, having a belief, theory, hope, fear, about q)]).

And there is no reason to suppose that whatever suffices to guarantee the existence of one of these ordered pairs suffices to guarantee the existence of the others.

To be sure, one might hold, with the new proposition view, that an explanation is a certain proposition expressible by a sentence of form (4); and instead of proposing (5) as a sufficient set of truth-conditions for such sentences one might propose the stronger (8). But if so, then types of explaining acts are no longer avoided. The truth-conditions for sentences which express propositions that are supposed to be explanations will make essential reference to such types. A proposition theorist may be able to supply conditions for sentences of the form (4) that make no appeal to types of explaining acts; but if so, I am not prepared to say how.

10. A NO-PRODUCT VIEW

A basic assumption of our discussion has been that explanation products are entities. And the most successful theory so far is the ordered pair view of Section 6. But an ontological purist may wonder whether he needs to postulate ordered pairs, or indeed anything else, as referents of explanation product-expressions. He may ask whether a sentence containing "the explanation of q given by S" can be understood without presupposing such entities but only particular acts of explaining. In this section a "reductionist" view will be considered according to which

> Any sentence containing an explanation product-expression is paraphrasable into a sentence that contains no such expression, but does contain one or more expressions of the form "a is an act in which S explained q by uttering u."

When this thesis is combined with the view that such paraphrases are more fundamental than sentences with product-expressions we get the no-product view.

To see how the paraphrases are supposed to work let me begin with sentences of the form

(1) The explanation of q given by S is that c.

Recalling a term introduced in Section 8, such a sentence is *restructured* if q is an indirect interrogative expressing a content-question Q and c is a sentence expressing a complete content-giving proposition with respect to Q. It is assumed that any unrestructured sentence of form (1) is paraphrasable into a restructured sentence of this form. A restructured sentence of form (1) can then be understood as

(2) $(\exists a)(\exists u)$(a is an act in which S explained q by uttering u) and (a)(u)(a is an act in which S explained q by uttering u $\supset$ (r)(r is associated with a $\equiv$ r = the proposition expressed by c)),

i.e., there is an act in which S explained q by uttering something, and for any such act c expresses the one and only associated proposition. To understand (2) we need to suppose that S performed one or more explaining acts, but no products of those acts need to be postulated.

A restructured sentence of the form "*An* explanation of q given by S is that c" can be paraphrased simply as "$(\exists a)(\exists u)$(a is an act in which S explained q by uttering u, and c expresses a proposition associated with a)," in which no product of explanation is invoked.

On the ordered pair view a sentence of the form

(3) The explanation of q given by S_1 = the explanation of q given by S_2

is to be understood as a genuine identity in which the expression on the left is being said to denote the same product as the one on the right. On the no-product view (3) is not a genuine identity but is to be understood as

$(\exists a)(\exists b)(\exists u_1)(\exists u_2)$(a is an act in which S_1 explained q by uttering u_1, and b is an act in which S_2 explained q by uttering u_2), and $(\exists p)(a)(b)(u_1)(u_2)$(a is an act in which S_1 explained q by uttering $u_1 \supset p$ is the one and only proposition associated with a; and b is an act in which S_2 explained q by uttering $u_2 \supset p$ is the one and only proposition associated with b).

A sentence such as "The explanation of q given by S is correct" can be interpreted as

$(\exists a)(\exists u)$(a is an act in which S explained q by uttering u) and $(\exists p)(a)(u)$[a is an act in which S explained q by uttering $u \supset$(p is the one and only proposition associated with a, and p is a correct answer to Q)].

That is, there is an act in which S explained q by uttering something, and there is a proposition uniquely associated with all such acts, and that proposition is a correct answer to Q.

Finally, how can we understand a restructured sentence of the form

(4) An explanation of q is that c

in which no uniqueness is assumed and no reference to a particular explainer is made? One way is to construe (4) as asserting the possibility of an act in which someone explains q by uttering something that expresses the proposition expressed by c, i.e.,

(5) Possible $(\exists a)(\exists S)(\exists u)$(a is an act in which S explains q by uttering u, and u and c express the same proposition).

(Compare this with (2) of Section 8, which is proposed by the ordered pair view.) A stronger way of interpreting (4) above is as saying that a *correct* explanation of q is that p, which can be construed as a conjunction of (5) with "c is true" (provided that (4) is restructured—see Chapter 4, Section 1).

The present no-product view and the ordered pair view of Section 6 differ over whether explanations are entities. According to the no-product view, a restructured sentence of the form (1) is to be understood as a conjunction of an existentially general and a universally general sentence describing acts of explaining, viz. (2). According to the ordered pair view, (1) is a singular sentence in which the product-expression denotes an ordered pair. Nevertheless, there is an interesting similarity between these views. On the ordered pair view a restructured sentence of form (1) will be true if and only if the product-expression in (1) denotes an ordered pair consisting of the proposition expressed by c and the act-type explaining q, and (2) above obtains. (See the denotation condition of Section 6.) On the no-product view a restructured sentence of form (1) is paraphrased into (2). So what is a paraphrase of (1) on the no-product view is one of the truth-conditions for (1) on the ordered pair view.

Is one of these views superior? One consideration is how well each can justify inferences that are drawn from sentences containing explanation product-expressions. For example, from a sentence of form (1) we may quite naturally infer "S explained q by uttering something." Such an inference is sanctioned by the no-product view which paraphrases (1) as (2), which entails

(6) $(\exists a)(\exists u)$(a is an act in which S explained q by uttering u).

But the ordered pair view is equally successful, since (6) is derivable from (1) via the denotation condition for "the explanation of q given by S." More generally, it can be shown that for every inference sanctioned by one view there is an identical or closely corresponding one sanctioned by the other.

Is one of these views preferable on ontological grounds? Both the ordered pair and no-product views invoke acts of explaining and propositions as entities. But the former in addition requires types of acts and ordered pairs consisting of these and propositions. Ontologists who crave simplicity may give an edge to the no-product view, provided that paraphrases such as those above are adequate. On the other hand, the no-product paraphrase of (4), viz. (5), uses a modal operator, which some may wish to avoid. In any case, a more liberal ontological position is possible. Two views may be equally satisfactory, even if one invokes more entities than the other, provided there are no special problems with these entities. Liberal ontologists who find types of acts no more problematic than propositions, and who have no reasons for rejecting ordered pairs or modal operators, may well conclude that there is little to choose between the ordered pair and no-product views. Finally, both views require the concept of an explaining act. Neither characterizes explanations independently of such a concept.

11. IMPLICATIONS FOR STANDARD THEORIES OF EXPLANATION

If one holds either the ordered pair view of Section 6 or the no-product view outlined in the preceding section, then at least parts of the theories of explanation of those philosophers discussed in Chapter 1 must be rejected. These theories—as I have been interpreting them—are product theories: they assert that explanations are entities of certain sorts—propositions or arguments. The no-product view rejects this idea. And the ordered pair view, while it accepts the idea that explanations are entities, refuses to identify them as the entities that Aristotle, Hempel, and others suggest. However, these philosophers make not only ontological claims about explanations but evaluative ones as well. Do the evaluative parts of their theories depend on the ontological ones? I suggest that they do not, for it is possible to retain their evaluative claims within a more successful ontological view.

To see this, it will suffice to consider Hempel's D-N theory. It is possible to reformulate this theory so that it will make claims not about the ontological character of explanations but only about their goodness. On this construction, the D-N theorist could accept the ordered pair view of Section 6. He could say that an explanation is an ordered pair (p; explaining q) whose first member is a complete content-giving proposition with respect to Q and whose second member is the act-type explaining q, where Q is a content-question. His theory of evaluation will now be restricted to explanations of certain types only, as follows.

D-N condition of evaluation (using ordered pair view): If Q is a content-question expressible by a sentence of the form "Why is it the case that x?" and (p; explaining q) is an explanation of q in which proposition p is expressible by a sentence of the form "The reason that x is that y," then (p; explaining q) is a *good explanation of q* if and only if
 (i) y is a true sentence;
 (ii) y is a conjunction at least one of whose conjuncts is a law;
 (iii) y entails x.

For example, by this condition the ordered pair

(The reason that this metal expanded is that this metal was heated and all metals expand when heated; explaining why this metal expanded)

is a good explanation of why this metal expanded.

Once the D-N theory is reformulated as a theory without particular ontological commitments but only as a theory for evaluating explanations, it is also possible to use the D-N theory in connection with the no-product view:

D-N condition of evaluation (using no-product view): If Q is a content-question expressible by a sentence of the form "Why is it the case that x?" then the explanation of q given by S is good if and only if

(∃a)(∃u)(a is an act in which S explained q by uttering u), and (∃p)(a)(u)(∃x)(∃y)(a is an act in which S explained q by uttering u ⊃ p is the one and only proposition associated with a, and x and y are sentences, and the sentence "the reason that x is that y" expresses p, and y is a conjunction at least one of whose conjuncts is a law, and y entails x).

The D-N theory, as formulated by Hempel, has quite definite ontological commitments. These are incompatible with the ordered pair and no-product views. However, it is possible to reformulate the D-N theory so that it provides conditions only for the evaluation of explanations. So reformulated, it is neutral between product and no-product views. In general, the same is true of all of the theories described in Chapter 1. This is not to say, of course, that the resulting evaluative theories are acceptable. The question of their adequacy as evaluative theories will be addressed in the next two chapters.

12. CONCLUSIONS

On the usual views, explanations are entities of certain sorts—sentences, propositions, or arguments—that are to be characterized independently of the concept of an explaining act. These views are subject to the illocutionary force problem: they are unable to distinguish explanations from other illocutionary products. They are also subject to the emphasis problem: they are unable to capture the idea that shifts of emphasis in what is asserted in an explanation can change the meaning of what is said, and hence the identity of the explanation. These problems are avoided by the ordered pair and no-product views, both of which appeal to the concept of an explaining act. One holding either of these views must reverse the usual order of procedure. Such a person must begin with the concept of an illocutionary act of explaining and characterize explanations, by reference to this, rather than conversely. Finally, we saw how the evaluative aspect of a typical theory of explanation—the D-N model—can be retained while abandoning its ontological commitments in favor of those of the ordered pair and no-product views.

CHAPTER 4

The Evaluation of Explanations

PART I   GENERAL CONDITIONS FOR EVALUATIONS

At the beginning of Chapter 1, three questions were raised: What is an explaining act? What is an explanation? How should explanations be evaluated? So far the theory I have developed has addressed the first two questions. The third is the subject of the present chapter.

1. CORRECT EXPLANATIONS

Let Q be a content-question and u a sentence satisfying the u-restriction. The act of explaining q by uttering u is one in which u is uttered with the intention of rendering q understandable by producing the knowledge, of the proposition expressed by u, that it is a correct answer to Q. The explanation that results—the product—is correct if and only if the proposition expressed by u is a correct answer to Q. The notion of being a correct answer to a question has so far been left unanalyzed. Restricting questions to wh-questions, and using the concept of a complete answer form introduced in Chapter 2, the following condition might be suggested:

> (1) p is a correct answer to Q if and only if p is true and p is expressible by a sentence obtained from a complete answer form for Q by filling in the blank.

Let Q be

> (2) Why did that metal expand?

and let p be

> (3) The reason that metal expanded is that it was heated.

The latter is obtained from a complete answer form for Q,

(4) The reason that metal expanded is ———,

by filling the blank with "that it was heated." If (3) is true, then by (1), it is a correct answer to (2).

Unfortunately, (1) is inadequate for two reasons. First, it is not sufficient. A proposition such as

(5) The reason that metal expanded is easy to grasp

is not a correct answer to (2). Yet it satisfies (1): it is true, and it is expressible by a sentence obtained from a complete answer form for (2), viz. (4), by filling the blank with "easy to grasp."

Second, (1) is not a necessary condition. The proposition

(6) That metal was heated

should, it seems reasonable to say, be counted as a correct answer to (2). Yet it does not satisfy (1) since it is not expressible by a sentence obtained from a complete answer form for (2) by filling in the blank.

I do not know how to supply general conditions for being a correct answer to a (wh-)question. However, a more limited problem does have a solution. As emphasized in Chapter 2, the kinds of questions that can appear in contexts of the form "A understands q" and "S explains q by uttering u" are content-questions. If we restrict questions to these then a simple condition can be formulated:

(7) p is a correct answer to a content-question Q if p is true and and p is a complete content-giving proposition with respect to Q.

This condition allows us to count

(3) The reason that metal expanded is that it was heated

but not

(5) The reason that metal expanded is easy to grasp

as a correct answer to

(2) Why did that metal expand?

(2) is a content-question, and although both (3) and (5) are true, (3), but not (5), is a complete content-giving proposition with respect to (2). (3) gives the reason that metal expanded, (5) ascribes something to that reason without giving it.

Although (7) does not permit counting (5) as a correct answer to (2), which is desirable, it also does not permit counting

(6) That metal was heated,

which is undesirable. Can (7) be broadened to allow (6) but exclude (5)?
Notice that (3) above has the following equivalent:

(8) This is the reason that metal expanded: that metal was heated,

in which the sentence following the colon is (6). (Since (3) is a content-giving sentence so is (8).) More generally, where c is a sentence expressing a proposition, if a sentence of the form

(9) This + is + the + content-noun N + phrase: c

is a content-giving sentence for N, I shall say that c is a *truncated version* of that content-giving sentence. Thus, (6) is a truncated version of (8). And if a proposition is expressible by means of a content-giving sentence of form (9), I shall say that c expresses a truncated version of that proposition. Thus sentence (6) expresses a truncated version of proposition (3). Similarly,

(10) The heart pumps the blood

is a truncated version of

This is the function of the heart: the heart pumps the blood.

We can now say that

(11) p is a correct answer to a content-question Q if and only if *either*
  (i) p is a complete content-giving proposition with respect to Q, and p is true; *or*
  (ii) ($\exists p^*$)(p^* is a complete content-giving proposition with respect to Q, p^* is true, and p^* is expressible by a content-giving sentence whose truncated version expresses p).

According to this definition, if (3) is true, then (in virtue of (i)) it is a correct answer to (2). And if (3) is true, then (in virtue of (ii)), (6)—a truncated version of (3)—is also a correct answer to (2). But (5), even if true, is not a correct answer to (2). Again, if

The function of the heart is pumping the blood

is true, then (in virtue of (ii)), (10)—a truncated version—is a correct answer to

(12) What is the function of the heart?

But

The function of the heart was discovered by Harvey,

even if true, is not a correct answer to (12).

Note that in accordance with (11) p can be a correct answer to Q even if p is false. This can happen, however, only when p is expressible by a truncated version of a complete content-giving proposition with respect to Q, where the latter proposition is true. Thus, the proposition "The radius of the orbit of any planet, in astronomical units, is $R = .3 \times 2^{(n-2)} + .4$," although false, is a correct answer to "What is Bode's law?" What makes this a correct answer, according to (11), is the truth of the following complete content-giving proposition with respect to that question: "Bode's law is that the radius of the orbit, etc."

Throughout the book I have been construing answers to questions as propositions. There is, however, a broader concept of answer which includes non-propositions as well. Thus, we might say that

(13) Pumping the blood

expresses a correct answer to question (12), even though (13) does not express a proposition. For this broader concept of answer we can extend the notion of a truncated version of a content-giving sentence by also counting the nominal in a content-giving sentence of the form

This + is + the + content-noun N + phrase: nominal

as a truncated version of that sentence. On this construction—again using definition (11)—(13) expresses a correct answer to the content-question (12), since the complete content-giving proposition with respect to Q,

This is the function of the heart: pumping the blood,

is true and is expressible by a content-giving sentence whose truncated version is (13).

In what follows, however, I will continue to construe answers in the narrower propositional sense. For the theory of explanation being developed, and specifically the ordered pair view of Chapter 3, this will suffice. Returning to the latter we can say that

If (p; explaining q) is an explanation, then it is *correct* if and only if p is a correct answer to Q.

But from Chapter 3 we know that (p; explaining q) is an explanation only if Q is a content-question and p is a complete content-giving proposition with respect to Q. Therefore, in virtue of definition (11), it follows that

If (p; explaining q) is an explanation, it is correct if and only if p is true.

2. ILLOCUTIONARY EVALUATIONS

Using the ordered pair view, a very simple criterion has just been formulated for the correctness of explanations:

(1) If (p; explaining q) is an explanation, it is correct if and only if p is true.

When we evaluate an explanation, therefore, why not simply ask whether it is correct in this sense? There are two reasons this will not suffice.

First, an explanation may be correct without being particularly good (enlightening, informative, "scientific"). Consider

(2) (The reason the patient died is that he had a bacterial infection; explaining why the patient died).

Using criterion (1), this explanation is correct if the proposition that is the first member of the ordered pair is true. But in many circumstances explanation (2) would be judged inadequate, since it is too uninformative. This explanation, it might be urged, is correct, or correct as far as it goes, but it does not go far enough.

A criterion such as (1) will not suffice for theorists of the sort discussed in Chapter 1. Their aim might be said to be to provide necessary and sufficient conditions for being a correct explanation *of a certain kind* (e.g., of a kind appropriate in science). And explanation (2), though correct, is not the kind of correct explanation they have in mind (since, e.g., it lacks laws, a deductive structure, etc.). In addition, some of these theorists will say, the criterion (1) on which it is based, although it is not circular, is not sufficiently illuminating. It allows us to utilize terms such as "reason," "because," and "cause" in explanations, and leaves us with the question of when propositions expressed by sentences with forms such as these are true:

The reason that p is that r,

p because r,

The cause of its being the case that p is that r.

What a number of these theorists want (in particular, Hempel and Salmon) is a definition of "correct explanation" which disallows such propositions in explanations. The prospects for producing a definition of this sort will be examined in Chapter 5. In the present chapter I will continue to consider explanations whose constituent propositions can be ones expressible by sentences of the forms above.

The second reason (1) will not suffice for evaluating explanations is that an explanation may be a good one without being correct. Although Ptolemy's explanation of the motions of the planets is incorrect, it has considerable merit. The goodness or worth of an explanation is multidimensional; correctness is only one dimension in an evaluation.

An explanation is evaluated by considering whether, or to what extent, certain ends are served. The ends may be quite varied. They may concern what are regarded as universal ideals to be achieved, particularly in science, e.g., truth, simplicity, unification, precision. Other ends are more "pragmatic." Thus an explanation might be judged to be a good one for stimulating scientists to think about a new problem, or for introducing a subject to students. In what follows the end I propose to consider is one which, by definition, an explainer has when he performs an illocutionary act of explaining q, viz.

(3) Rendering q understandable by producing the knowledge, of the answer one gives, that it is a correct answer to Q.

My concern is with evaluating an explanation from the viewpoint of its capacity to serve this end. I shall put this by asking: is the explanation a good one for *an explainer to give in explaining q*? And I shall speak of this as an illocutionary evaluation. Such an evaluation, it will be shown, can take into account both "universal" and "pragmatic" ends.

There is a related type of evaluation which I shall also classify as illocutionary. An explanation of q might be a good one for an explainer to give in explaining q not only if it is capable of accomplishing (3), but also if it is reasonable for the explainer to believe that it is capable of doing this, even if it is not. A scientist's micro-explanation might be a good one, despite its failure to cite a correct microstructure, but because, given the information then available, it was reasonable for the scientist to believe that it did.[1]

In Chapter 2 the idea of ways of understanding was introduced by reference to instructions for answering a question. I claimed that S explains q only if S intends to render q understandable in a way that satisfies appropriate instructions (see Chapter 2, Section 13). Accordingly, when we offer an illocutionary evaluation of an explanation we consider whether it is (reasonable to believe it) capable of accomplishing (3) with respect to such instructions. In what follows the idea of appropriate instructions will need to be discussed.

3. REASONS FOR FOLLOWING INSTRUCTIONS

To formulate necessary and sufficient conditions for appropriate instructions, and then for good explanations, I shall proceed indirectly.

---

1. We might compare the present distinction to one drawn by Hempel between true and well-confirmed explanations, and to one employed by Salmon between homogeneity and epistemic homogeneity. With the introduction of criteria of confirmation and epistemic homogeneity these writers are acknowledging the possibility of evaluations of the second type.

Let us suppose that in explaining q an explainer has followed instructions I. (Following instructions is, in part, an intentional idea; an explainer has followed I in explaining q only if he intends that his answer satisfy I, and it does in fact do so.) I shall also assume that an explainer is explaining q to, or for the benefit of, some audience. (When the physicist explains something in an article in the *Physical Review* he is explaining it to, or for the benefit of, professional physicists.) What are the explainer's reasons for following instructions I? They may be varied, but they will include at least these:

(A) The explainer believes (or at least assumes) that his audience does not understand q_I (that it is in an n-state with respect to Q_I).[2]

(B) The explainer believes that there is a correct answer to Q, that satisfies I, the citing of which will enable the audience to understand q_I by producing the knowledge, of that answer, that it is correct.

In order to explain why the tides occur Newton followed instructions calling (among other things) for the citing of a cause. He did so, in part at least, because he believed that his audience did not understand why the tides occur; and because he believed that there is a correct causal answer to this question, the citing of which will remove the n-state by producing the knowledge that it is correct.

In explaining the series of lines in the hydrogen spectrum Bohr followed instructions calling for a subatomic theory. He did so, in part at least, because he assumed that his audience of physicists was in an n-state with respect to questions such as

What causes the lines?

Why are they discrete?

Why do they have the wavelengths they do?

and with respect to instructions calling for a subatomic theory. He also did so because he believed that there are correct answers to these questions, satisfying such instructions, the citing of which will remove the n-state by producing the knowledge that these are correct answers.

The clause in (B) beginning "the citing of which" is not redundant. The explainer may believe that his audience does not understand q_I, and that there is a correct answer to Q that satisfies I, without believing that such an answer will enable the audience to understand q_I. (He may believe that the audience is not in a position to know of any such answer that it is correct, since it is not in a position to know the meaning of any sentence expressing that answer.)

2. See Chapter 2, Section 12.

110   The Nature of Explanation

Additional reasons an explainer might have for following instructions I pertain to the interests of the audience, and to the value of the instructions for that audience:

(C) The explainer believes that the audience is interested in understanding q in a way that satisfies I.

(D) The explainer believes that understanding q in a way that satisfies I, if it could be achieved, would be valuable for the audience (whether or not its members are interested in understanding q in a way that satisfies I).

The explainer may recognize that his audience is in an n-state with respect to Q and *many different instructions*. He may choose to explain q in a way that satisfies some particular set of instructions I because he believes that the audience is interested in understanding q in a way that satisfies I, and because he believes that it is valuable for that audience to do so.[3]

Bohr followed instructions calling for a subatomic theory to explain the spectral lines, in part at least, because he believed that his audience of physicists was interested in understanding the spectral lines at the subatomic level, and because he believed that such an understanding, if it could be achieved, would be valuable. In the *Principia* Newton followed instructions calling for an axiomatic presentation of his results because he believed that understanding mechanics in this way is something that would be of interest to, and valuable for, his audience.

Condition (C) can be broadened. A poet in a physics class who is taking the course only to satisfy distribution requirements may express no interest in understanding the hydrogen spectral lines in subatomic terms. Still, one assumes, he wants to pass the course, which, let us suppose, requires that he understand this. Although he has no "intellectual" interest in such an understanding, he does have a practical one. For the physicist who is doing the explaining this practical interest on the part of his audience may (alas) have to suffice. The "interest" in (C) can be an intellectual or a practical one.

Even if the audience has no interest in understanding q in a way that satisfies I, an explainer may still follow I in explaining q to this audience. He may believe that doing so will *create* such an interest. Teachers frequently assume that their own interest in the subject and their explanatory talents will create similar interests on the part of students. (C), then, should be amended to say that the explainer be-

---

3. Condition (D) invokes the idea of a way of understanding that is valuable for an audience without indicating how this is to be determined, even by the explainer. Are there general criteria for deciding this, or must this always vary from one audience to another? This question will be addressed in Part III of the present chapter.

lieves that the audience is interested in understanding q_I or would become so in the course of an explaining act.

An audience may have no interest in understanding q in a way that satisfies I because it is unaware of Q and I, or because it rejects assumptions in I. This need not deter an explainer. He should consider whether the audience would be interested in understanding q_I if it were made aware of Q and I, or if it were to believe that understanding q in a way satisfying I could be achieved. Some members of Bohr's audience in 1913 may still have believed that the spectral lines were due not to subatomic events but to the vibration of the atom (see Section 10 below). But Bohr supposed that any persons of this sort *would* be interested in understanding the spectral lines at the subatomic level if they believed that such an understanding could be achieved. He then proceeded to show how such an understanding is possible. By contrast, suppose Bohr had confronted an audience whose members would have no interest in understanding the spectral lines at the subatomic level, even if they were to believe such an understanding possible. For example, the only type of understanding these persons regard as valuable is one that invokes final causes. While they agree that a subatomic understanding of the spectral lines is possible, they regard this as being of little or no value, and so have no interest in it. With such an audience, following instructions that call simply for a subatomic hypothesis is pointless. What Bohr must do, if he can, is try to change the attitude of the audience regarding the value of microexplanations and final causes.

Accordingly, (C) can be broadened by allowing the explainer to believe that the audience *would be* interested in understanding q_I *if* it were made aware of Q and I or if it believed such an understanding could be achieved. The last clause allows an explainer to follow instructions he believes his audience will reject. But it does not trivialize the condition, since the audience might have no interest in understanding q in a way satisfying those instructions even if it were to come to believe such an understanding attainable.

With these several changes we might reformulate (C) as

(C)' The explainer believes that the audience is or (under certain conditions) would be interested in understanding q in a way that satisfies I.

The conditions in the parentheses will be those noted in previous paragraphs.

(A), (B), (C)', and (D) are not only among the reasons an explainer can have for following I, they are reasons an explainer is committed to if he is seriously engaged in an illocutionary act of explaining q. Keep in mind that the defining intention of such an act is

(1) Rendering q understandable by producing the knowledge, of the answer one gives, that it is a correct answer to Q.

If, for a certain audience, an explainer intends to achieve (1) with respect to instructions I, then he must believe or assume that the audience does not already understand q in a way that satisfies I. Otherwise his aim of *rendering* q_I understandable to that audience (of bringing it about that the audience understands q in a way that satisfies I) is bound to be frustrated. So there is a commitment to (A). Second, if, for a certain audience, an explainer intends to achieve (1) with respect to instructions I, then he is committed to there being a correct answer to Q that satisfies I the citing of which will accomplish (1); hence (B).

Finally, as noted earlier, in explaining q there is an intention to render q understandable in a way that satisfies appropriate instructions. Accordingly, if, for a given audience, an explainer intends to achieve this end with respect to instructions I, then he is committed to the belief that, for this audience, instructions I are appropriate. Now, I am suggesting, the latter belief is true only if that audience has, or would have, an interest in understanding q in a way that satisfies I, and such an understanding would be valuable for it. So there is also a commitment to (C)' and (D). To say that the patient died because he led a dissipated life, even if true, may not be to provide understanding of a sort that would be of interest to, or valuable for, an audience of doctors. On the other hand, an audience may have no interest in understanding q in a way that satisfies I, even if the explainer believes it should have, and even if the audience believes such an understanding possible. The frustrated motorist does not want to understand his tire blow-out by reference to the laws of statistical mechanics. For such a person micro-physical instructions are pointless.

4. APPROPRIATE INSTRUCTIONS

Even if (A) through (D) are, or are among, the reasons an explainer follows the instructions he does, we have not yet formulated conditions for evaluating those instructions or the resulting explanation. Nevertheless, the previous discussion provides a basis for doing both of these things.

I suggest that instructions are appropriate ones for an explainer to follow in explaining q to an audience if and only if either the four types of beliefs to which the explainer is committed when he follows those instructions are true, or it is reasonable for him to believe that they are. That is,

(1) I is a set of appropriate instructions for an explainer to follow in explaining q to an audience if and only if *either*
  (a) The audience does not understand q_I; and
  (b) There is a correct answer to Q, that satisfies I, the citing of which will enable the audience to understand q_I by producing the knowledge, of that answer, that it is correct; and
  (c) The audience is interested in understanding q in a way that satisfies I (or would be under the conditions noted in Section 3); and
  (d) Understanding q in a way that satisfies I, if it could be achieved,[4] would be valuable for the audience;
or
  It is reasonable for the explainer to believe that (a) through (d) above are satisfied.

The second part of (1) is an epistemic criterion. Whether the instructions are epistemically appropriate ones depends on what it is reasonable for the explainer to believe about them and about the audience. The first part of (1) is not epistemic in this way (though, to be sure, it contains epistemic elements in virtue of referring to the understanding and interests of the audience).

(1) is meant to apply to any explainer,[5] audience, and set of instructions, provided that these satisfy the conditions specified. If we are giving a non-epistemic evaluation (one in accordance with the first part of (1)), then our evaluation holds for any explainer whatever, since (a) through (d) impose no conditions on the epistemic state of the explainer. By contrast, in the case of an epistemic evaluation (one in accordance with the second part of (1)), what we say may hold for one explainer but not another: it may be reasonable for one explainer, but not another, to believe that conditions (a) through (d) are satisfied.

When an evaluation of instructions is made without explicit reference to an audience, such a reference is implicit in the context of evaluation. The audience may be broadly or narrowly characterized ("everyone that is literate," "professional physicists," or "the persons now in my office"). To judge instructions non-epistemically is to judge

---

4. Condition (b), if satisfied, entails that such an understanding can be achieved. Since later I want to consider cases in which (d) is satisfied but not (b), (d) is formulated in a manner that does not require that it is possible for the audience to understand q_I.

5. Even if an explainer does not understand the instructions they may still be appropriate ones for him to follow. (He should try to get himself in the position of understanding them.) By analogy, the instructions printed in English on a bottle of medicine may be appropriate for a patient to follow, even if that patient cannot read English or otherwise cannot understand them.

them as appropriate for some audience. And to judge them epistemically one determines whether it is reasonable for the explainer to believe certain things with respect to some audience. The explainer in question may or may not be the one who originally formulated the explanation, and the audience may or may not be the one for which the explanation was originally intended. The instructions Bohr followed in giving his explanation of the spectral lines may be evaluated epistemically as appropriate ones for him to follow with respect to the audience he intended, but inappropriate ones for contemporary physicists to follow with respect to an audience of their peers. No doubt we evaluate instructions without explicit reference to an explainer or audience. We may say simply that these are good instructions to follow for explaining q. But if so, then, I am claiming, reference to an audience is implicit; and reference to an explainer is so too, if the evaluation is epistemic.

The first part of (1), I suggest, provides a set of sufficient conditions. If the audience does not understand q in a way that satisfies instructions I, if there is a correct answer to Q that satisfies I the citing of which will enable the audience to understand q_I by producing the knowledge that it is a correct answer, if the audience is interested in understanding q_I, and if understanding q_I would be valuable for the audience, what more could one ask of instructions?

Does the second part of (1) similarly provide sufficient conditions? Suppose that there is no correct answer to Q that satisfies I, so that (b) is violated. Then although it might be reasonable to believe that I is appropriate for an explainer to follow in explaining q, in fact it isn't (or so it might be said). We do make such (non-epistemic) evaluations. But we also evaluate in accordance with the second part of (1). We say that those were appropriate instructions for the explainer to follow in explaining q because of what it was reasonable for that explainer to believe about them. So I am inclined to say that both parts of (1) provide sufficient conditions, and to keep in mind that the second part, but not the first, furnishes the basis for an epistemic evaluation in the sense indicated.

I am also inclined to suggest that (1) constitutes a set of necessary conditions. Our concern is with evaluating instructions from the viewpoint of the end of an illocutionary act of explaining, viz.

(2) Rendering q understandable by producing the knowledge, of the answer one gives, that it is a correct answer to Q.

What I am now claiming is that, given this end, instructions I are appropriate only if (1) is satisfied. If the audience already understands q_I (or if it is reasonable to believe or assume that it does), then the end described in (2) is defeated by following I (or it is reasonable to believe it is). In such a case it would not be appropriate for an explainer to follow I in explaining q. If (it is reasonable for the explainer to believe that) there is no correct answer to Q that satisfies I, the citing of which will enable the audience to understand q_I by producing the knowledge that it is a correct answer, then once again (it is reasonable to believe that) the end (2) is defeated by following I. Finally, "rendering q understandable" means "rendering q understandable in a way that is appropriate." And, as urged in the previous section, a way of rendering q understandable is appropriate for an audience only if the audience is or would be interested in understanding q in that way, and such an understanding would be valuable for it. Therefore, I is a set of appropriate instructions for an explainer to follow in explaining q to an audience, and hence for achieving (2) with respect to that audience, only if (it is reasonable for the explainer to suppose that) (c) and (d) are satisfied.

In Chapter 2 the appropriate-instructions use of "understand" was introduced, according to which A understands q if and only if

(3) $(\exists I)$(A understands q_I, and I is a set of appropriate instructions for Q).

Can the second conjunct in (3) be explicated by reference to (1)?

We cannot construe "I is a set of appropriate instructions for Q" in (3) simply as "I is a set of appropriate instructions for an explainer to follow in explaining q to A," since, according to the first conjunct in (3), A (already) understands q_I. But we can construe it as "Before A understood q_I I was a set of appropriate instructions for an explainer to follow in explaining q to A." It might be objected that this would introduce circularity into our theory of explanation, since (1) characterizes appropriate instructions by reference to understanding. However, there is no circularity here since what (1) invokes is the concept of *understanding q in a way that satisfies I*. And the latter is defined in Chapter 2 independently of the concept of appropriate instructions. (A understands q_I if and only if $(\exists p)$(p is an answer to Q that satisfies I, and A knows of p that it is a correct answer to Q, and p is a complete content-giving proposition with respect to Q).)

It might also be objected that (3), construed in the present way, affords no motivation for further inquiry. If A comes to understand q_I—where before he did so I was a set of appropriate instructions for an explainer to follow in explaining q to A—why should A seek to improve his knowledge-state with respect to Q? The objection should lose its force if we keep in mind that the truth of (3) is compatible with that of

$(\exists I')$(A does not understand $q_{I'}$, and I' is a set of appropriate instructions for an explainer to follow in explaining q to A).

116   The Nature of Explanation

It is possible for someone who understands q to improve his understanding.

5. GOOD EXPLANATIONS

Can (1) of the previous section be used to characterize good explanations? Let us suppose once again that an explanation is an ordered pair (p; explaining q) satisfying the ordered pair view of Chapter 3. (There is an analogous procedure for the no-product view.) We seek to provide an illocutionary evaluation of (p; explaining q), i.e., to determine whether it is a good one for an explainer to give in explaining q. As in the case of (1), I assume that when such an evaluation is offered there is at least implicit reference to an audience. An explanation might be a good one for an explainer to give in explaining q to one audience but not to another. Since we are interested in epistemic as well as non-epistemic evaluations, we are concerned also with whether it is reasonable for the explainer to believe that (p; explaining q) is capable of accomplishing (2) of the previous section with respect to a given audience. Employing an epistemic evaluation, we may conclude that an explanation is a good one for a certain explainer to give in explaining q to one audience, but not for another explainer to give for the benefit of another audience. (Bohr's explanation was a good one for him to give in 1913 for physicists of the day, although, we might say, it is not a good one for contemporary physicists to give to their professional colleagues.) Using (a) through (d) in the conditions for appropriate instructions (see (1) of Section 4), these ideas can be expressed as follows:

(1) (p; explaining q) is a good explanation for an explainer to give in explaining q to an audience if and only if *either*
   (i) ($\exists$I)(I satisfies (a) through (d) in the conditions for appropriate instructions, and (p; explaining q) is capable of rendering q_I understandable to that audience by producing the knowledge of p that it is a correct answer to Q); *or*
   (ii) It is reasonable for the explainer to believe that (i) obtains.

Accordingly, a necessary, but not a sufficient, condition for (p; explaining q) to be a good explanation for an explainer to give in explaining q is that p is either a correct answer to Q or it is reasonable for the explainer to believe it is. The notion of a correct answer to Q can be understood by reference to (11) in Section 1. Recall that the latter, together with the conditions for the ordered pair theory, allows us to conclude that

If (p; explaining q) is an explanation, then it is correct if and only if p is true.

From (1), then, it follows that (p; explaining q) is a good explanation for an explainer to give in explaining q only if p is true or it is reasonable for the explainer to believe it is. However, correctness is not sufficient for being a good explanation. What is required is appropriate correctness—an idea captured in (1).

Although (1) requires that the constituent proposition p in an explanation (p; explaining q) be a correct answer to Q, or the reasonableness of believing this, we are not thereby precluded from making non-illocutionary evaluations. (p; explaining q) might be a good explanation for causing scientists to think about a new problem, or for achieving some desired unification, even though p is an incorrect answer to Q and it is not reasonable for the explainer to believe it to be correct; or even though, while p is correct, (p; explaining q) will fail to render q understandable in an appropriate way.

As in the case of illocutionary evaluations of instructions, if we give a non-epistemic evaluation of an explanation (i.e., one in accordance with (i) in (1)), then our evaluation holds for any explainer whatever. ((i) imposes no special conditions on the explainer himself.) If we are giving an epistemic evaluation (one in accordance with (ii)), then (p; explaining q) might be a good explanation for one explainer to give, but not another. Both epistemic and non-epistemic evaluations, however, require reference to some audience.

The definitions offered in this and the previous section are insufficient for determining that instructions I are appropriate, or that an explanation is a good one, until it is decided whether understanding q in a way that satisfies I would be valuable for the audience. Criteria for determining this have not yet been mentioned. (In Part III this will be taken up.) Still, even before doing so, it is possible to utilize these definitions to confront an important question.

6. IS THERE A WAY OF EXPLAINING PARTICULARLY APPROPRIATE FOR SCIENCE?

It might be contended that there is a set of universal instructions that scientists follow when they explain things, at least to a scientific audience, and that it is appropriate and even necessary for them to do so. Such instructions determine what constitutes scientific understanding. That is, there is a set of universal scientific instructions I_u which is such that

(1) (p; explaining q) is a good explanation for a scientific explainer to give in explaining q to a scientific audience if and only if the explanation satisfies I_u.

(2) If (p; explaining q) is an explanation, it is capable of rendering q *scientifically* understandable if and only if it satisfies I_u.

What kinds of instructions these are supposed to be, and whether there can be any of this sort, are questions that will be taken up in Parts II and III.

Will this discussion be relevant for assessing standard theories of explanation of the sort outlined in Chapter 1? At the end of that chapter, I noted that one of the most important aims of the theorists in question is to provide conditions for being a good explanation. They seek accounts of this form:

(3) E is a good explanation of q if and only if it satisfies conditions C.

The conditions C do not mention explainers, audiences, or explaining acts. They abstract from "pragmatic" considerations. What sort of evaluation does (3) provide? More particularly, assuming, as we have been doing, that explanations are constructed principally for the purpose of allowing explainers to explain, what, if anything, is supposed to follow from (3) about whether E is a good explanation for that purpose?

The theorists of Chapter 1 are focusing on explanations in science. On one interpretation of their views, if something is a good scientific explanation of q (if it satisfies C), then it is a good one for a scientist to give in explaining q to a scientific audience. A defender of this position would be seeking to provide a basis for (illocutionary) evaluations of form (1), with respect to some proposed universal scientific instructions I_u. These instructions which an explanation is supposed to satisfy (e.g., Hempel's D–N conditions) do not mention explainers, audiences, or explaining acts. Nevertheless, they provide necessary and sufficient conditions for an illocutionary evaluation of an explanation in science. In Part II of this chapter this is the position I want to consider.

Hempel at one point draws an analogy between defining the concept of a (good) scientific explanation and that of a (good) proof in mathematics. The rules for the latter, formulated by the logician or mathematician, do not mention "provers," audiences, or acts of proving. Nevertheless, proofs in mathematics are constructed principally for the purpose of allowing mathematicians to prove theorems for other mathematicians. And (it might be claimed) something is a good proof for a mathematician to give in proving a theorem to an audience of his peers if and only if the rules governing mathematical proofs are satisfied. I propose to examine an analogous claim for scientific explanations.

There are weaker interpretations of the aims of the theorists of Chapter 1. On one, universal scientific instructions are to be thought of as providing necessary but not sufficient conditions for illocutionary evaluations. (The "if and only if" in (1) should be changed to "only if.") This view will be taken up in Part III. On an even weaker interpretation, the theorists of Chapter 1 do not intend to provide either necessary or sufficient conditions for illocutionary evaluations of explanations in science. Rather they are concerned only with non-illocutionary evaluations of certain types. This viewpoint will be discussed in parts of the present chapter, but principally in Chapter 5.

PART II THE EVALUATION OF SCIENTIFIC EXPLANATIONS

7. UNIVERSAL SCIENTIFIC INSTRUCTIONS

To aid the discussion I shall expand upon an example to which reference has already been made—Bohr's 1913 explanation of the visible lines in the hydrogen spectrum. When hydrogen is excited by heat or electricity it emits light, which, when analyzed spectroscopically, is seen to consist of a series of discrete lines at various wavelengths. Bohr explained this phenomenon by indicating that it is produced at the subatomic level of the hydrogen atom. Following Rutherford's model of the atom, Bohr postulated that the hydrogen atom consists of a nucleus around which a single electron revolves in various possible energy states. However, going beyond Rutherford, Bohr assumed that when the electron jumps from one stable orbit to another of lower energy, light is radiated whose wavelengths are those of the lines in the visible spectrum of hydrogen. Bohr introduced two special hypotheses governing the hydrogen atom. One is that the only permissible electron orbits are those for which the angular momentum of the electron is a whole multiple of $\frac{h}{2\pi}$ (h = Planck's constant), and no energy is radiated while the electron is in one of these orbits. The second is that energy is emitted or absorbed by the hydrogen atom in whole quanta of amount $h\nu$ (ν = frequency of the radiation). Bohr also utilized several general laws which he believed are applicable to this system when it is in a stationary state, e.g., Newton's second law of motion and Coulomb's law of electric charges. From these laws, together with his specific postulates governing the hydrogen atom, he showed how to derive Balmer's formula which quantitatively relates the visible lines in the hydrogen spectrum.

Bohr begins his classic paper "On the Constitution of Atoms and Molecules" by summarizing Rutherford's model of the atom and indicating its principal difficulty (viz. that using assumptions of ordinary electrodynamics the electrons in the atom should be unstable, which they clearly are not). He then proposes that if Planck's theory of quantization of energy is introduced the stability can be restored in the Rutherford model. After Bohr puts his basic ideas in quantitative form, he writes:

> We shall . . . [now] show how by the help of the above principal assumptions, and of the expressions (3) for the stationary states, we can account for the line spectrum of hydrogen.[6]

The "principal assumptions" he refers to are that when the atom is in a stationary state ordinary mechanics applies, but not when there is a transition from one energy state to another; and that in the latter case there is emission of radiation for which the relation between the frequency and the energy of the radiation is given by Planck's theory. The expressions to which he refers relate the ionization energy, frequency, and major axis of the orbit, to the mass and charge of the electron and to Planck's constant. Bohr then characterizes his task as one of using these qualitative and quantitative ideas to explain the hydrogen spectrum.

It may seem plausible, then, to say that in explaining the spectral lines Bohr was following instructions such as these:

Instructions B (for Bohr):
(a) Using basic ideas of Rutherford's model describe the hydrogen atom in such a way that events occurring within it produce the spectral lines.
(b) Assume that the hydrogen atom, as described by the Rutherford model, obeys classical mechanics when it is in a stationary state.
(c) Set particular conditions on the hydrogen atom when it is in a non-stationary state which incorporate Planck's idea that energy is quantized.
(d) From (b) and (c) derive a formula that relates the wavelengths of the lines.

Obviously such instructions—involving as they do particular empirical assumptions—are not universal in science. They are not instructions that all scientists have followed, or that it would be appropriate for them to follow, in explaining all phenomena (or even in explaining the spectral lines—contemporary quantum theory would require quite different instructions).

With this defenders of universal scientific instructions will no doubt agree. Of course, they will grant, if you produce very specific instructions like B, you are bound to say that scientific instructions vary considerably. Even if it is right to assume that Bohr was following instruc-

---

6. Niels Bohr, "On the Constitution of Atoms and Molecules," *Philosophical Magazine and Journal of Science* 6 (1913), p. 8.

tions such as B, this is not incompatible with supposing that he was also following ones of a much more general kind as well. Hempel, e.g., would urge that Bohr was also following

> *Instructions H (for Hempel):* Present an argument whose premises contain laws, whose conclusion contains a description of the phenomenon in question, and whose premises entail (or inductively support) its conclusion.[7]

Bohr's explanation can and should be evaluated, from a scientific viewpoint, by reference to instructions H—or some other equally general ones.

Theorists like Hempel, Salmon, Brody, and Aristotle want to supply universal scientific instructions. I shall suppose that the following features are among those desired of such instructions:

U_1: Universal scientific instructions are to be ones the satisfaction of which guarantees explanations that are "scientific" in some broad but recognizable sense of that term. Such instructions should enable us to distinguish scientific from non-scientific explanations.

U_2: Universal scientific instructions are to be truth-guaranteeing. Instructions for answering a question Q will be said to be truth-guaranteeing if an answer that satisfies them is necessarily a correct answer to Q. Hence, the satisfaction of such instructions guarantees that Q has been correctly answered.[8]

U_3: Universal scientific instructions are to provide an adequate basis for evaluating an explanation in science. In particular, they should constitute necessary and sufficient conditions for an explanation to be a good one for a scientist to give in explaining q to a scientific audience. (See (1) of Section 6.)[9] And they should assist in the comparison of explanations. Invoking them will enable us to see why one explanation is better than another. We wish to evaluate an explanation (p; explaining q) from the viewpoint of whether it is capable of rendering q understandable by producing the knowledge of p that it is a correct answer to Q. Now we are concerned with rendering q understandable in a way appropriate to science. The present claim is that an explanation which satisfies universal scientific instructions will accomplish this end. (See (2) of Section 6.)

---

7. This is a simplified version of Hempel's instructions. However, what will be said about H in what follows is not affected by complications Hempel introduces.

8. This is what Hempel seeks in his D-N model, construed as providing a set of sufficient conditions for correct explanation. It is not so for his I-S (inductive-statistical) model. On the other hand, it is supposed to hold for the models of Salmon and Brody. It seems reasonable to associate it also with Aristotle's doctrine of the four causes and with his doctrine in the *Posterior Analytics*.

9. As noted earlier, there are other (weaker) interpretations of such instructions, which will be discussed subsequently.

U_4: Universal scientific instructions are not to incorporate specific empirical assumptions in the instructions themselves. They may require the assumption that there are laws, causes, or ends in nature, but not the existence of specific ones. Accordingly, they are not to mention particular scientific theories, hypotheses, or laws to be used in the explanation.

U_5: Universal scientific instructions are not to vary from one scientific period to another. No matter what scientific theories may be accepted at one time and rejected at another, following such instructions will produce explanations that satisfy U_1 through U_3, i.e., that are genuinely scientific, truth-guaranteeing, and meritorious.

U_6: Universal scientific instructions are ones the following of which is to be justified on a priori grounds alone, not empirical ones. If one were following (Bohr's) instructions B in explaining the hydrogen spectrum, the justification for doing so would need to be, at least in part, empirical: one believes in the basic ideas of Rutherford's model and in certain laws of classical mechanics. By contrast, if one follows (Hempel's) instructions H—instructions which involve no commitment to any particular scientific theory or law—the justification for doing so would presumably be a priori. (Following such instructions is just what it *means* to explain a phenomenon scientifically.) Universal scientific instructions are definitive of the criteria to be used in constructing and evaluating scientific explanations. Understanding in a way that satisfies these instructions is, by definition, what constitutes scientific understanding.

U_7: Finally, universal scientific instructions are not to involve any reference to what Hempel calls "pragmatic" or "contextual" considerations pertaining to the intentions of the explainer or what the audience does or does not understand. Although Hempel recognizes a pragmatic dimension to explanation, as noted earlier, he identifies his task as one of "constructing a non-pragmatic concept of scientific explanation." On the interpretation I am presently giving this, such a concept is, and ought to be, *employed* by scientists when they explain. But the conditions of the concept itself—the universal instructions (e.g., Hempel's H)—are not to mention particular or types of explainers or audiences.

8. ON THE POSSIBILITY AND VALUE OF UNIVERSAL SCIENTIFIC INSTRUCTIONS

Those who wish to formulate instructions satisfying these conditions will be called *universalists*. In this category, until further notice, I shall place all of the theorists mentioned in Chapter 1. Do the instructions such theorists provide satisfy these conditions? If not, is it possible to provide instructions that do so? Is it important to do so?

Conditions U_4 through U_7 are satisfied by the instructions universalists have proposed. Instructions such as Hempel's H, or Aristotle's four causes, or Salmon's statistical model, do not incorporate any fairly specific empirical assumptions. These instructions are not to vary from one scientific period to another. They are to be justified on non-empirical grounds (presumably on the grounds that the instructions define the standards to be used in science, or that they provide what Carnap has called an "explication" of the concept of scientific explanation). And they are free of contextual references.

However, whether such instructions satisfy the first three conditions is another matter. Only brief mention will be given of the first two, since my main concern is with the third.

U_1: *Distinguishing scientific and non-scientific explanations.* Instructions of the sort universalists have in fact provided do not succeed well on this score. Citing one or indeed all of Aristotle's four causes is not by itself sufficient to guarantee an explanation that would generally be regarded as scientific (at least not by today's standards). And there are numerous examples that conform to the instructions of Hempel, Salmon, and Brody which are questionably scientific. Thus it seems farfetched to claim that the following provides a scientific explanation of why crows are black:

Crows are the color of coal
Coal is black
    Therefore,
Crows are black,

even though Hempel's instructions H have been satisfied. If, on the other hand, instructions such as H—or Hempel's more complex version of H—are regarded as supplying only necessary but not sufficient conditions for scientific explanations, there are still two formidable problems. First, as Hempel notes, there will be explanations in science—which may be valuable in various ways—that fail to satisfy these instructions. (Hempel speaks of "partial explanations" and "explanations sketches" and claims that these can be "fruitful and suggestive" in science,[10] and that they can even be "scientifically acceptable.")[11] Second, since numerous *non*-scientific explanations will also satisfy such instructions, the distinction between scientific and non-scientific explanations remains to be drawn.

10. Hempel, *Aspects of Scientific Explanation*, p. 416.
11. *Ibid.*, p. 238.

U_2: *Truth-guaranteeing.* The satisfaction of this criterion will be the principal issue in Chapter 5.

U_3: *Providing an adequate basis for evaluating explanations in science.* I suggest as a general thesis—and the main one of the present section—that if instructions are to satisfy U_3, then they will not satisfy U_4 through U_6. They will not be universal in crucial respects.

Suppose we wish to evaluate Bohr's explanation of the line spectrum of hydrogen. Hempel's instructions H, or comparable ones of other universalists that are supposed to satisfy U_4 through U_6, fail to be adequate for determining whether Bohr's explanation was a good one. To see this, let us suppose that Bohr had proceeded in an entirely different manner. In explaining why the lines in the hydrogen spectrum are discrete and why they are associated with the particular wavelengths they are, suppose that Bohr had simply constructed the argument

(α) (i) Whenever hydrogen is excited thermally or electrically it emits radiation whose spectrum contains lines satisfying the (Balmer) formula

$$\frac{1}{\lambda} = R\left(\frac{1}{2^2} - \frac{1}{n^2}\right)$$

where n = 3 for the first line, 4 for the second, etc.; and λ is the wavelength.
  (ii) R = 109,677.581 cm^{-1}.

Therefore, the hydrogen spectral lines are discrete and have the wavelengths $\lambda_1 = 6562.08$ (x 10^{-7}) cm, $\lambda_2 = 4860.8$ cm, $\lambda_3 = 4340$ cm, $\lambda_4 = 4101.3$ cm.

Hempel's instructions H are satisfied. However, for reasons to be discussed presently, this explanation would not have been judged a good one for Bohr to have given—certainly much inferior to the one he actually gave. Yet instructions H seem to offer no basis for such a judgment.

To this it might be replied that Bohr's actual explanation was a good one because it explains not only why the lines are discrete and why each has the wavelength it does, but also why the lines obey Balmer's formula. It provides laws from which the Balmer formula itself can be derived. By contrast, the above explanation uses Balmer's formula in one of its premises; it does not explain why the lines satisfy this formula.

This reply should be irrelevant if the questions for which we are trying to provide answers in our explanation are simply

Why are the observed lines discrete?

Why do they have the wavelengths they do?

To these questions Bohr's actual explanation and (α) both provide answers that satisfy Hempel's instructions H. It is no part of instructions H that the answers to the explanatory question(s) provide answers to other questions as well. (Hempel in his models does not demand that an explanation provide answers to questions in addition to the ones that prompted the explanation.)

Suppose, however, that instructions H were modified to include such a requirement. The defect in (α), which Bohr's own explanation does not have, is alleged to be that the latter but not the former provides laws from which the Balmer formula itself can be derived. But this defect is easily remedied by generalizing the Balmer formula in the manner accomplished by J. R. Rydberg. Our law, which replaces premise (i) in (α) to form an argument (α)', will then be formulated as follows:

(Part of (α)'): Whenever hydrogen is excited thermally or electrically it emits radiation whose spectrum contains lines satisfying the (Rydberg) formula

$$\frac{1}{\lambda} = R\left(\frac{1}{n_2^2} - \frac{1}{n_1^2}\right),$$

where $n_2 = 2$, and $n_1 = 3, 4, 5$, etc. for the Balmer series; and $n_2 = 1$ and $n_1 = 2, 3, 4$, etc. for the Lyman series (in the ultraviolet region); $n_2 = 3$, and $n_1 = 4, 5, 6$, etc. for the Paschen series (in the infrared region).

From this law Balmer's formula is derivable, as are the formulas for the two series of lines in the non-visible part of the hydrogen spectrum. Again, instructions H are satisfied, this time yielding a description of the lines not only as discrete, and as having the particular wavelengths they do, but as satisfying Balmer's formula.

However, physicists would judge Bohr's explanation of the spectral lines to be a far superior one for him to have given than those offered by (α) or (α)', if indeed they would regard the latter as having any merit at all. Yet in all three cases instructions H have been followed. Why would physicists make such a judgment, and would it be a reasonable one? Possibly a different set of universal instructions will provide an answer.

9. ALTERNATIVE UNIVERSAL INSTRUCTIONS

Let me focus briefly on two, those of Salmon's Statistical-Relevance (S-R) model and Brody's causal model.

Salmon's basic explanatory question, we recall, has the form

Why is X, which is a member of class A, a member of class B?

Salmon's instructions can be formulated as follows:

Instructions S-R: Cite a set of probability laws of the form
$$p(B, A \& C_1) = p_1$$
$$\vdots$$
$$p(B, A \& C_n) = p_n$$

that relate classes A and B, and that are such that the homogeneity condition is satisfied and that the values of $p_1, \ldots, p_n$ are all different; and indicate which one of the mutually exclusive classes $A \& C_1, \ldots, A \& C_n$ contains X as a member.

Reverting to the Bohr example, suppose that the explanatory question (as formulated for the case of a particular sample of hydrogen) is this:

Q: Why did this sample, which is hydrogen, emit radiation whose line spectrum satisfies Balmer's formula?

Let

X = this sample

A = the class of samples of hydrogen

B = the class of things that emit radiation whose line spectrum satisfies Balmer's formula

C_1 = the class of samples excited thermally or electrically

C_2 = the class of samples not excited thermally or electrically.

We can now construct a (Salmonian) explanation for Q, as follows:

(β) $p(B, A \& C_1) = 1$
$p(B, A \& C_2) = 0$
$X \in A \& C_1$.

Since $A \& C_1$ and $A \& C_2$ comprise a partition of A, and both $A \& C_1$ and $A \& C_2$ are homogeneous with respect to B, instructions S-R are satisfied. But this explanation contains laws that are the Salmonian analogues of those in (α) and (α)'. It explains why this sample emits radiation which satisfies Balmer's formula by saying simply that this sample is hydrogen which has been excited thermally or electrically; and that under such conditions the probability of emitting radiation that satisfies Balmer's formula is 1, whereas the probability that a sample of hydrogen which has not been excited thermally or electrically will emit such radiation is 0. Although this explanation provides more information than (α), it too would not have been regarded as a good one for Bohr to have given—far inferior to the one he in fact gave. Yet Salmon's instructions S-R offer no basis for such an evaluation.

The instructions associated with Brody's causal model might seem to be what is required here. Brody might say that the reason that physicists would evaluate Bohr's explanation of the spectral lines as a better one for him to have given than (α) or (β) is that Bohr's explanation cites something which is a cause of the lines, whereas (α) and (β) do not. But this is not really so, since both of these explanations cite the thermal or electrical excitation, which does indeed produce the lines. The problem, physicists would say, is that (α) and (β) do not give a cause *at an appropriate level,* in this case, at the subatomic level of hydrogen. To say this, however, is to evaluate Bohr's explanation from the viewpoint of instructions that are not universal—in particular, ones that do not satisfy conditions U_4 through U_6 for universal scientific instructions. What sorts of instructions will these be?

10. EVALUATING BOHR'S EXPLANATION

Given the information available to Bohr in 1913 it was reasonable for him to believe

(i) that the spectral lines of hydrogen are produced by events occurring within the hydrogen atom, and, in particular, by motions of excited electrons;

(ii) that his audience of physicists did not yet understand the spectral lines in a way that satisfies instructions calling for an answer at the subatomic level of the hydrogen atom, but would do so if a suitable answer satisfying such instructions were produced;

(iii) that this audience was interested in understanding the spectral lines in a way that satisfies such instructions;

(iv) that such an understanding of the spectral lines would be valuable for this audience.

At the time of Bohr's explanation atomic theory was well accepted, though different models of the internal structure of the atom had been proposed. It was considered reasonable to suppose that the spectral lines were produced at the subatomic level by the motions of excited electrons within the atom rather than by the vibration of the atom itself (a hypothesis that had been refuted).[12] This assumption was common to various atomic models in addition to Rutherford's, e.g., to those of J. J. Thomson and of Nagaoka. Physicists were inter-

---

12. See Max Jammer, *The Conceptual Development of Quantum Mechanics* (New York, 1966), p. 68.

ested in using these various models in order to account for the spectral lines. Accordingly, it was reasonable for Bohr to follow instructions calling for an appeal to events within the hydrogen atom.[13]

We need not suppose that he had to accept the particular ideas of Rutherford's model, or the quantization of energy introduced by Planck. But when Bohr's explanation is evaluated as a good one for him to have given in explaining the spectral lines—by contrast with explanations (α) or (β) of Sections 8 and 9—the instructions that provide the basis for such an evaluation will be justified by appeal to assumptions that are, or include, (i) through (iv) above. Such instructions might be the very specific instructions B of Section 7, viz.

(a) Using basic ideas of Rutherford's model describe the hydrogen atom in such a way that events occurring within it produce the spectral lines.

(b) Assume that the hydrogen atom, as described by the Rutherford model obeys classical mechanics when it is in a stationary state.

(c) Set particular conditions on the hydrogen atom when it is in a non-stationary state which incorporate Planck's idea that energy is quantized.

(d) From (b) and (c) derive a formula that relates the wavelengths of the lines.

Or they might be much less specific instructions calling simply for an explanation that appeals to the motions of electrons (or even just to "events") within the hydrogen atom. In either case, conditions U_4 through U_6 for universal scientific instructions will not be satisfied.

U_4 will not be satisfied, since the instructions will incorporate specific empirical assumptions, viz. that there are hydrogen atoms and (electron) events that occur within them which produce the spectral lines. U_5 will not be satisfied, since such instructions are not invariant from one scientific period to another. And U_6 will not be satisfied since following such instructions cannot be justified on a priori grounds alone; it must be justified by appeal to empirical beliefs or assumptions about the world and about the audience of physicists—beliefs or assumptions which are, or include, (i) through (iv) above.

After Bohr's theory was published a number of physicists gave it a very favorable assessment. Einstein described it as an "enormous achievement" and as "one of the greatest discoveries." Jeans called it

---

13. Cf. John Heilbron and Thomas Kuhn, "The Genesis of the Bohr Atom," *Historical Studies in the Physical Sciences* 1, pp. 211–90. See p. 275: "Most physicists in 1912 would have agreed that they [the line spectra] must directly relate to the most basic principles of atomic structure."

"a most ingenious and suggestive, and I think we must add, convincing explanation of the laws of the spectral series."[14] And it is generally recognized today as a good one for Bohr to have given in explaining the spectral lines. Such assessments would not have been made if Bohr had simply offered the "Hempelian" (α) of Section 8 or the "Salmonian" (β) of Section 9. It seems quite inappropriate to claim that Bohr's explanation was a good one for him to have given solely on the grounds that Hempel's instructions H or Salmon's instructions S-R are satisfied. Bohr's explanation was and ought to be rated highly, at least in part, on the grounds that it invokes events at the subatomic level and that Bohr's audience had raised questions about the spectral lines at this level.

These observations can be related to the epistemic condition for being a good explanation given in Section 5, according to which

(1) (p; explaining q) is a good explanation for an explainer to give in explaining q to an audience if it is reasonable for the explainer to believe that $(\exists I)(I$ satisfies (a) through (d) in the conditions for appropriate instructions, and (p; explaining q) is capable of rendering q_I understandable to that audience by producing the knowledge of p that it is a correct answer to Q).

Now I am saying that because it was reasonable for Bohr to believe (i) through (iv) above, as well as various assumptions made by instructions B, it was reasonable for him to believe that instructions B satisfy the conditions for appropriate instructions referred to in (1) above. Let us also assume that it was reasonable of Bohr to believe that the explanation of q that he gave is capable of rendering q understandable to his audience, in a way that satisfies instructions B, by producing the knowledge that his explanation gives a correct answer to Q. If so, then, by (1), this was a good explanation for Bohr to give in explaining to physicists in 1913 how the spectral lines are produced. But the instructions that provide a basis for this evaluation, viz. B, are not universal; they violate U_4 through U_6.

I have been speaking of evaluating Bohr's explanation from his point of view—as a good one for him to have given in the circumstances he did. But illocutionary evaluations from other viewpoints are possible. A contemporary physicist might offer a quite different assessment, saying that

Bohr's explanation of the spectral lines is not a particularly good one because, as quantum mechanics shows, it is only a crude approximation to what actually occurs in the hydrogen atom and it cannot be extended to the case of more complex atoms.

14. See Jammer, *The Conceptual Development of Quantum Mechanics*, p. 86.

Such an evaluation would be made from the viewpoint of an explainer and an audience with information and interests that do not coincide with Bohr's. It would involve the use of standards associated with instructions incorporating modern quantum theory. These instructions, although quite different from Bohr's B, are like the latter in this important respect: they fail to satisfy conditions U_4 through U_6 for universal scientific instructions. They incorporate specific empirical assumptions; they are not invariant from one scientific period to another; and they are not ones the following of which could be justified on a priori grounds alone. The justification of their employment would rest in part on assumptions about the physical world and the current audience of physicists.

11. POSSIBLE REMEDIES

A universalist may respond in various ways. First, he might propose weakening condition U_3 for universal scientific instructions (which requires that such instructions provide an adequate basis for evaluating an explanation as a good one for a scientist to give in explaining q). He might suggest evaluating an explanation only with respect to *correctness*. Indeed, with this modification U_3 could be eliminated altogether in favor of U_2, since the latter requires truth-guaranteeing instructions—ones the satisfaction of which guarantees a correct answer to the explanatory question.

It is difficult to imagine universalists adopting this proposal. Recalling our discussion in Section 1, the truth-guaranteeing requirement is satisfied by instructions such as these:

> Where Q is a content-question, formulate a true proposition which is a complete content-giving proposition with respect to Q.

Thus, if Q is

What causes the spectral lines of hydrogen?

then the proposition

> The cause of the spectral lines of hydrogen is the occurrence of events at the subatomic level

satisfies these instructions. Why should we try to invent more complex instructions, e.g., ones requiring laws, deductive or inductive arguments, or homogeneity?

Universalists may reply that they are concerned with scientific explanations, and therefore that these other factors are demanded by condition U_1 for universal scientific instructions (which requires that the satisfaction of such instructions guarantees explanations that are scientific). But to this we should respond: why should one be particularly interested in *scientific* explanations? (After all, non-scientific explanations can also be correct.) The answer that I take to be implicit in the universalist position is that scientific explanations—at least those satisfying the universalists' instructions—are superior to non-scientific ones. The explanation

> (The cause of the spectral lines of hydrogen is the occurrence of events at the subatomic level; explaining what causes the spectral lines of hydrogen)

which cites no laws, may be correct; but it is not as good (illuminating, enlightening) as one that appeals to laws. If this is the likely reply of the universalist, and I suggest that it is, then the present remedy is not open to him. The universalist requires not simply that an explanation be correct, but that it be correct in the right sort of way.

A second response is to add to instructions such as Hempel's H or Salmon's S-R the following:

> Mi (for "micro"): In the explanation invoke entities, phenomena, or systems at the micro-level.

This rule is not tied to any particular empirical theory, and it would be justified on a priori grounds. (Perhaps this is what is required of scientific explanations.) Furthermore, since Bohr's explanation of the spectral lines follows this rule, but the "Hempelian" explanation (α) and the "Salmonian" explanation (β) do not, we see why the former explanation is superior to the latter.

Some theory-neutral characterization will need to be given of when something can be said to be invoked at the micro-level. Perhaps the following will suffice for our purposes. Suppose that a scientist is trying to explain why X has P (e.g., why hydrogen emits light whose spectrum contains discrete lines at certain wavelengths), and in his explanation he appeals, among other things, to the fact that Y's have Q (e.g., to the fact that hydrogen atoms have a certain structure). If the scientist is assuming that X is *composed of* Y's and that Y's are *unobservable*, then we might say that Y's are entities that the scientist has invoked at the micro-level (and Y's having Q are states of affairs, phenomena, or whatever, invoked at the micro-level).[15] The present characterization of micro-levels commits us to no specific empirical theories about the world; e.g., it does not invoke any notion of atoms or subatomic entities. The proposal we are considering, then, is adding to instructions such as Hempel's H or Salmon's S-R the requirement that an explanation should invoke entities at the micro-level.

---

15. For a fuller account see my "Macrotheories and Microtheories," in P. Suppes et al., eds., *Logic, Methodology, and Philosophy of Science* 4 (Amsterdam, 1973), pp. 533–66.

This idea has two serious flaws. First, it is much too demanding since it would preclude explanations, such as those in classical thermodynamics, which are not given at the micro-level. (A simpler example will be discussed in Section 15.) Second, even if we were to restrict our attention to those scientific explanations which invoke entities at the micro-level, the present proposal will not improve the situation substantially. Thus, suppose that the "Hempelian" explanation (α) of Section 8 is modified by substituting the following for its first premise:

Whenever hydrogen is excited thermally or electrically the hydrogen atoms produce radiation whose spectrum contains lines satisfying the formula $\frac{1}{\lambda} = R\left(\frac{1}{2^2} - \frac{1}{n^2}\right)$.

Entities and phenomena at the micro-level have been invoked since it is being assumed that hydrogen is composed of atoms, which are unobservable. Moreover, the resulting explanation satisfies Hempel's instructions. This explanation, though possibly a slight improvement over (α), would not have been judged a particularly good one for Bohr to have given—still vastly inferior to the one he gave—despite the fact that it yields the lines in the hydrogen spectrum and invokes entities at a micro-level. If, on the other hand, we add to the instructions of Hempel or Salmon a condition calling not simply for entities at a micro-level but for particular kinds (e.g., a condition calling for the particular micro-system postulated by Rutherford's model of the hydrogen atom), then we begin to understand why Bohr's explanation is superior to (α) and to its modification above. But as a consequence we have instructions which are not universal, since they will not satisfy conditions U_4 through U_6 for universal scientific instructions.

What has been said about instructions Mi calling for microexplanations is applicable as well to ones that invoke other general criteria valued in science, e.g.,

Un: In the explanation provide a *unification* of known phenomena.

Even with Un we are not in a good enough position to understand why Bohr's explanation is superior to others that we have been mentioning. The unification that Bohr achieved is limited. For one thing, the principles he used were not themselves particularly unified. He simply utilized principles from classical mechanics and electrodynamics for stationary atomic states and assumed that quite different quantum principles govern transitions between stationary states, without attempting to say why this disparity holds. For another thing, he was not able to use these disparate principles to explain spectral lines of elements more complex than hydrogen. (Presumably one intuitive idea behind Un is that an explanation in science should invoke principles that can be used to explain phenomena of various sorts, not just the one in question.) To be sure, he did something he believed would lead to a significant kind of unification: he related the visible lines in the hydrogen spectrum to phenomena occurring in the hydrogen atom. (Bohr, like others, believed that eventually many macro-events could be explained by appeal to subatomic theory.) But the generalized Rydberg formula which provides the basis for the "Hempelian" $(\alpha)'$ also achieves a significant kind of unification: it enables us to explain several different series of spectral lines: those in the visible spectrum, the ultra-violet region, and the infra-red region. If we add to instructions of Hempel or Salmon a condition not calling simply for unification, but for a particular kind of unification (e.g., one achieved by utilizing Rutherford's model of the hydrogen atom), then we can see why Bohr's explanation is superior to the others given above. But, once again, we will have instructions which are not universal.

This is not to downplay the importance of micro-descriptions and unification in science. (The positive role of such criteria in evaluating explanations will be taken up in Part III of the present chapter.) My claim is only that Mi and Un (as well as other "universal" criteria) do not provide a sufficient basis for judging that an explanation is a good one for a scientist to offer. For this purpose one also needs to make empirical assumptions about the world and the audience that will determine what kind of micro-description and unification is appropriate.

A third possible universalist response is to add to instructions such as those of Hempel or Salmon the following:

Cn (for "contextual"): Determine the particular instructions I which the audience has (or ought to have) imposed on answers to Q, and answer Q in a way that satisfies I.

This rule when combined with instructions such as those of Hempel or Salmon will yield instructions that do not require the making of any specific empirical assumptions; that are not to vary from one scientific period to another; and that would be justified on a priori grounds alone—in short, ones that satisfy conditions U_4 through U_6 for universal scientific instructions. Furthermore, it might be urged, the conjunction of Cn with Hempel's instructions enables us to see why Bohr's explanation of the spectral lines is superior to the "Hempelian" (α). Although both of these explanations satisfy Hempel's instructions, only Bohr's explanation satisfies Cn as well; only Bohr's explanation answers questions about the spectral lines in a way required by physicists of Bohr's day.

Even though instructions Cn satisfy conditions U_4 through U_6 for universal scientific instructions, they do not by themselves satisfy U_1, nor do they enable instructions with which they are conjoined to do so. They are not ones that help to distinguish scientific from non-scientific explanations. On the contrary, they are appropriate for constructing and evaluating any explanation, scientific or otherwise. Since neither Hempel's nor Salmon's instructions by themselves generate explanations that are particularly scientific, neither will the combination of either of these with Cn. The problem is that Cn is too unspecific. It says, in effect, "Follow whatever instructions are (or ought to be) called for by the audience." But the program of the universalist, as I am now envisaging it, is to give a reasonably specific set of instructions for a *scientific* audience. In effect, he wants to *define* a standardized concept of a scientific audience—viz. as one for which such and such instructions are appropriate. For this purpose instructions Cn are of no help whatever.

Most importantly, Cn is not in accord with condition U_7 for universal scientific instructions, since it invokes "pragmatic" considerations. Instructions such as Cn are just the sort that Hempel and others eschew when they construct a non-pragmatic concept of scientific explanation. However, without an appeal to "pragmatic" considerations, I am claiming, we are prevented from evaluating an explanation as a good one for explaining q. To be sure, we can judge whether an explanation is correct. But this by itself does not take us far enough. Both (α) and (β) provide correct answers to the question "Why do the observed spectral lines have the wavelengths they do?" However, they do not correctly answer the question in an appropriate way. And the appropriateness of the way can be determined only by appeal to "pragmatic" considerations.

When we offer an (epistemic) evaluation of an explanation we do so based on information available to some explainer (whether Bohr or someone else) about the world and the audience. This information which the explainer has, or is assumed to have, may be explicitly formulated in a set of instructions—as we did with (Bohr's) instructions B of Section 7. In these instructions there is no reference to "pragmatic" concepts—to an explainer or an audience. These instructions do not violate condition U_7 for universal scientific instructions, which precludes pragmatic considerations. Nevertheless, to justify using these instructions appeal has to be made to an explainer and to an audience. Moreover, condition U_6 is violated, as are U_4 and U_5, since we end up with instructions that incorporate specific empirical assumptions, that are not invariant from one scientific age to another, and that are not ones the following of which can be justified on a priori grounds alone.

In general, if we want a set of appropriate scientific instructions for evaluating explanations a choice is open to us. To avoid "pragmatic" references we can formulate instructions like Bohr's B that violate conditions U_4 through U_6. If we want our instructions to satisfy these conditions, then we can formulate instructions like Cn with its "pragmatic" references, and thus violate condition U_7. But if we want helpful instructions—ones which will enable us to see why Bohr's explanation was better than (α) and (β) for explaining the spectral lines—we cannot have it both ways.

Universalists may now say: "You can have your illocutionary evaluations of explanations and everything you claim about them. Our interest is not in these, but only in non-illocutionary evaluations based solely on criteria that are universal in science (e.g., use of laws, deductive structure, etc.). Since in Sections 2 and 5 you agreed that non-illocutionary evaluations are possible, we have no further quarrel with you."

Universalists ought to be dissatisfied with this response. To be sure, non-illocutionary evaluations are possible. But an illocutionary evaluation is important because it evaluates an explanation using as an end the primary one for which explanations exist: explaining. Explanations are human inventions, serving human purposes. Their most important (though not their only) use is in acts of explaining to achieve a state of understanding in an audience in a way outlined in Chapter 2. On the basis of non-illocutionary evaluations—e.g., ones based solely on the kinds of instructions universalists have so far supplied—nothing follows about what explanation to choose to achieve this end. This will become even more evident in Part III when the role of general methodological values in science is discussed. But we have already seen that if we evaluate an explanation solely according to whether it contains laws, a deductive structure, causal information, unifying principles, and reference to microentities, we are hard pressed to say why Bohr's explanation of the spectral lines is better than the "Hempelian" (α)—modified to contain the generalized Rydberg formula and references to hydrogen atoms—which also satisfies these criteria.

On the other hand, if we offer an illocutionary evaluation, we can readily understand why Bohr's explanation is a good one, by contrast to (α). It is good not simply because it answers causal questions about the spectral lines at a unifying micro-level by deriving a quantitative formula relating the lines from general laws, but because it does these things *at the subatomic level of the hydrogen atom*—a level at which Bohr's audience was interested in understanding the spectral lines. But this consideration is too specific—too "theory dependent" and too contextual—for the kinds of instructions universalists have in mind. As we shall see in Part III, an illocutionary evaluation need not neglect the criteria universalists espouse. It simply does not allow them to provide sufficient conditions for such evaluations.

In short, illocutionary evaluations are essential from the viewpoint of the principal function of an explanation. They are comprehensive, since they take into account not only the situation of the explainer and the audience but also, as will become apparent, general methodological ideals in science. And without them we will be unable to understand why, in many cases, one explanation does, and should, receive higher marks than another.

To this some universalists may respond by accepting the importance of illocutionary evaluations. However, they may state that their concern with such evaluations is not with respect to *scientific* explainers and audiences but *philosophical* ones. They would replace (1) of Section 6, viz.

> (1) (p; explaining q) is a good explanation for a scientific explainer to give in explaining q to a scientific audience if and only if the explanation satisfies I_u

with

> (2) (p; explaining q) is a good explanation for a philosophical explainer (one who proposes to satisfy philosophical standards) to give in explaining q to a (philosophical) audience if and only if it satisfies philosophical instructions I_p.

Or perhaps better:

> (3) (p; explaining q) is a good explanation for any explainer to give in explaining q to any audience if and only if it satisfies I_p—assuming that philosophical standards are to be satisfied.

The philosophical standards will be expressed by instructions I_p which, like I_u, do not mention explainers, audiences, or explaining acts. Presumably these instructions will be ones of the sort suggested by the philosophers we have been considering. They will call for deductive arguments, laws, causes, homogeneity, and so forth. But then I_u and I_p are identical, and (2) and (3) will yield the same problems as (1). Such instructions will not provide a sufficient basis for distinguishing Bohr's explanation from pale substitutes which also satisfy these instructions. Unless universalists are prepared to conclude that, from a philosophical viewpoint, the "Hempelian" $(\alpha)'$, the "Salmonian" (β), and Bohr's explanation are equally meritorious, the present response is not successful.

Furthermore, if I_u and I_p are identical, then it is unclear what the distinction between "philosophical" and "scientific" explainers is supposed to amount to for the universalist. On the original universalist view, according to U_1 (see Section 7), universal scientific instructions I_u are to be ones the satisfaction of which guarantees an explanation that is scientific. On the present amended version, universal philosophical instructions I_p are to be ones the satisfaction of which guarantees an explanation that meets philosophical standards. But if $I_p = I_u$, then any "philosophical" explanation will also be "scientific," and vice versa. If so there is no real difference between the "scientific" universalist, who espouses a form of (1), and the "philosophical" universalist, who espouses a form of (2) or (3).

This leads to a final universalist response, which accepts the importance of illocutionary evaluations in science but says that the instructions universalists provide constitute necessary although not sufficient conditions for such evaluations. (In (1) "only if" should replace "if and only if.") This claim will be examined in Part III. But it is important here to resist the temptation to say that additional universal instructions, if added to the ones universalists have already proposed, will provide a set of sufficient conditions. That is, it may be tempting to think that an appeal, say, to Hempel's instructions, together with certain universal principles requiring micro-theories, or simplicity, or unification, or informational content, or whatever, will provide an adequate basis for evaluating scientific explanations, even if Hempel's instructions alone do not. It is my claim that no set of universal principles—*where these eschew both "pragmatic" features and specific empirical assumptions*—can provide an adequate basis for determining whether an explanation is a good one for a scientist to give in explaining q. The conditions for being a good explanation for explaining q preclude any such set (see (1) of Section 5). Focusing just on non-epistemic evaluations, such conditions require that the explanation be capable of rendering q understandable in a way that satisfies appropriate instructions. But instructions are (non-epistemically) appropriate, we recall, only if conditions (a) through (d) of (1) in Section 4 are satisfied. And one who claims that these conditions are satisfied must make "pragmatic" assumptions about the audience as well as specific empirical assumptions about the world.

PART III THE ROLE OF GENERAL METHODOLOGICAL VALUES

12. GENERAL METHODOLOGICAL VALUES

A scientist's reasons for following the instructions he does may include not only specific beliefs about the world and his audience, but also very general methodological beliefs concerning values reflected in science. Bohr may have followed the instructions he did not only because of specific empirical beliefs about the origin of the spectral lines, but also because of the value he and other scientists accord micro-descriptions. Galileo found fault with explanations of the tides that appeal to the moon's gravitational force on two methodological grounds: an appeal was being made to an "occult" property—one that was unverifiable, or at least for which no observational evidence existed; and the explanation that he was criticizing (which simply invoked gravitational force and nothing else) failed to contain any laws or quantitative descriptions.

In his explanations, it will be said, a scientist seeks, and ought to seek, generality (and thus laws), precision (and thus quantitative hypotheses), unification, simplicity, micro-descriptions, high informational content, and confirmation. A belief in the importance of such values, among others, may be among the reasons that a scientific explainer or evaluator chooses the instructions he does. An explainer may seek laws, quantitative hypotheses, simplicity, or whatever, because he believes that a scientific audience is interested in obtaining such understanding, which, if it could be achieved, would be valuable. He may view these as universal values in science—ones that define the activity of science. We can understand a universalist, then, as someone who focuses on general methodological values, seeking to elevate them to the status of necessary and sufficient conditions for good explanations. And we can put what has been argued so far by saying that general methodological values—in the absence of specific empirical beliefs about the world and the audience—cannot provide a *sufficient* basis for determining whether an explanation is a good one for explaining q. The question I now want to raise is whether they can provide a *necessary* basis.

I will take it for granted here that scientists do value certain general criteria, and that we are concerned with explainers and audiences who share these values. The importance of some of the criteria has changed during the history of science. (Aristotle placed little value on quantitative hypotheses; nineteenth-century positivistic physicists eschewed micro-theories.) For the sake of argument, however, I will simply assume the existence of a set of scientific criteria generally agreed to at a given time, at least within some field. Nor will it be my task here to try to define concepts like "law," "quantitative hypothesis," or "simplicity." I will assume that such concepts can be applied at least in clear cases. What concerns me is the role these and other methodological criteria play in an illocutionary evaluation of explanations. When can we say that because an explanation of q lacks laws (or micro-descriptions, etc.) it is inadequate for explaining q? Must instructions that scientists ought to follow incorporate such values? Are instructions that do so more appropriate in science than ones that do not? In the following sections I consider these questions by means of several examples.

13. THE QUARK MODEL

In Section 4 we said that I is an appropriate set of instructions for an explainer to follow in explaining q to an audience if and only if it is true, or reasonable for the explainer to believe, that

(a) The audience does not understand q_I;
(b) There is a correct answer to Q that satisfies I, the citing of which will enable the audience to understand q_I by producing the knowledge, of that answer, that is is correct;
(c) The audience is, or would be, interested in understanding q in a way that satisfies I;
(d) Understanding q in a way that satisfies I, if it could be achieved, would be valuable for the audience.

Consider simplicity as a universal value in science. Let us suppose that instructions I express demands for some level of simplicity sought by the audience in raising Q, and that understanding q in a way that satisfies I, if it could be achieved, would be valuable for that audience. Does this make I a set of necessary or even appropriate instructions for an explainer to follow in explaining q to that audience?

A striking case in which simplicity instructions are invoked arises in contemporary particle physics in connection with the quark model of hadrons. Hadrons are particles which interact with each other through what physicists call the strong force—that which keeps particles together in the atomic nucleus. They include the proton and neutron as well as over one hundred particles, most short-lived and created in accelerator experiments. Because of the large number of these "fundamental" particles with different properties (e.g., different charges, rest energies, spins, and baryon numbers), physicists began to look for an explanation of hadrons that would simplify physical ontology by reducing the number of different fundamental particles postulated.

Thus the physicist Feinberg writes:

> The system of subatomic particles has become so complex in terms of both the number of objects and their properties, that physicists, who are always searching for simplicity, have become convinced that the observed particles are not the ultimate level of reality, and that something simpler must exist in terms of which the observed particles and their behavior can more easily be understood.[16]

In 1963 Gell-Mann and Zweig, independently, introduced the quark explanation, according to which all hadrons are composed of three types of quarks—the new fundamental particle—which have various

---

16. Gerald Feinberg, *What Is the World Made Of?* (New York, 1978), p. 205.

physical properties including spin, electric charge, baryon number, and strangeness. On the quark theory, the quantum numbers of the hadrons (the numbers representing spin, charge, etc.) can be determined by adding together the quantum numbers of the quarks making up the hadron. Thus, the proton consists of three quarks with electric charges $+2/3$, $+2/3$, and $-1/3$ respectively, which sum to $+1$, the known charge of the proton.

Consider now questions in the set Q:

Why are there so many different types of hadrons?

Why do the hadrons have the properties that they do? (E.g., why does the proton carry a charge of $+1$, and the neutron a charge of 0, though they have the same spin, baryon number, and approximately the same rest energy?)

With respect to these questions, consider instructions calling for answers satisfying some principle appealing to simplicity, e.g.,

I: Provide an answer to the above questions which introduces a simple set of fundamental particles that constitute the hadrons. (The simplicity of the set is to be judged, let us assume, by some criterion taking into account the number of different fundamental particles postulated and the kinds of laws these are supposed to satisfy.)

Is I an appropriate set of instructions for physicists to follow in explaining q? Is it necessary? The answer to these questions is not automatically Yes in virtue of the fact that I calls for simplicity. It depends also on whether it is reasonable to believe that conditions (a) and (b) above for appropriate instructions are satisfied. Even if the audience is interested in understanding q_I, and even if understanding q in a way that satisfies I, if it could be achieved, would be valuable for the audience, it might still not be reasonable to believe that there is a correct answer to Q that satisfies I. It might not be reasonable to believe that there exists a set of fundamental particles constituting hadrons that satisfies some desired level of simplicity. (If such a set does not exist, then not only is condition (b) for appropriate instructions violated but so is (a). Recall from Chapter 2, Section 12, that A does not understand q_I only if Q_I is sound, i.e., only if there exists a correct answer to Q that satisfies I.)

How does one decide whether such a belief is reasonable? By trying to find a correct answer to Q that satisfies the desired simplicity. If physicists attempt to discover correct answers to Q that postulate three such fundamental particles, but are unable to do so, then it may be reasonable to conclude that such standards of simplicity are unattainable—that nature is stubbornly more complex than physicists would like. Indeed, this is just what has happened to the original quark model. A fourth type of quark was introduced with the new physical property called charm. Still later two more types of quarks were postulated, making six so-called "flavors" in all; and each flavor comes in three "colors," for a total of 18 quarks. The original quark model has been complicated considerably and may not accord with the simplicity that physicists would like nature to have. But this fact may not be sufficient to fault the present quark explanation. In view of what physicists now know it may not be reasonable for them to try to follow instructions calling for some level of simplicity which they might otherwise value.

Another possibility is that physicists will come to regard the quark model as unacceptable. It is conceivable that at some point they might abandon attempts to find a correct answer to Q that satisfies I and come to believe that there are no more fundamental particles out of which hadrons are constituted. If so, then explanations of various phenomena which appeal to the existence of a complex realm of hadrons should not be faulted for their complexity. The fault, if any, lies in nature, not in the physicists' explanations.

Defenders of simplicity may now reply by saying that universal instructions should not prescribe some particular level of simplicity. Rather they should say

> Provide an answer to Q which introduces as much simplicity *as possible*. (Don't multiply entities *beyond necessity*.)

What is wanted is maximum simplicity consistent with the evidence. The 3-quark explanation may be simpler than the 18-quark one; but the latter may be the simplest one consonant with what is known.

This is weak and rather unexciting fare. But even more weakening is required. The "possibility" in question is not to be understood as relative only to what is known about the physical world. In determining whether I is an appropriate set of instructions the level of simplicity (or unification, or quantification, etc.) appropriate will depend not only on what information about the world is available to the explainer (or evaluator) but also on information concerning the audience. Would a simpler (or more unifying, or more quantitative) explanation make the audience understand q in the manner required by the second part of (b)? Is the audience interested in receiving such an explanation, and would it be of value for it? The answer to each of these questions is not automatically Yes in virtue of the fact that simplicity is valued in science. Various subquark theories have been proposed which are simpler in important respects than the quark theory. For example, the rishon model of Harari postulates two fundamental particles to account for both quarks and leptons: the T rishon (which has an electric

142   The Nature of Explanation

charge of $+\frac{1}{3}$), and the V rishon (which is electrically neutral). Despite the fact that the rishon model introduces more simplicity (and unification) than the 18-quark theory, it does not follow that it is a better one than the latter for explaining hadrons to a given scientific audience. For one thing, the audience may seek to understand hadrons at the quark level before turning to subquark theories. (An analogous example will be discussed in Section 15.) For another, even if we suppose that both theories are consistent with the evidence, the quark theory is much better supported. The rishon theory is highly speculative.

14. GALILEO'S "SIDEREAL MESSENGER"

Scientists tend to seek quantitative hypotheses and general laws. (Indeed, according to universalists such as Hempel and Salmon, the latter are a *sine qua non* for good scientific explanations.) Must instructions incorporate these values to be appropriate ones for scientists to follow in explaining q? Let us turn to our second example.

Galileo's purpose in the "Sidereal Messenger" was to write a tract "which contains and explains recent observations made with the aid of a new spyglass [the telescope] concerning the surface of the moon" as well as other astronomical phenomena.[17] With his telescope trained on the moon Galileo reported observing several things never before seen. These included numerous small spots all over the lunar surface and the uneven shape of the boundary between the illuminated and unilluminated parts. He also observed that the small spots have "blackened parts directed toward the sun, while on the side opposite the sun they are crowned with bright contours," and that these spots "lose their blackness as the illuminated region grows larger and larger." His explanation for these and other observed phenomena was that

> The surface of the moon is not smooth, uniform, and precisely spherical as a great number of philosophers believe it (and the other heavenly bodies) to be, but is uneven, rough, and full of cavities and prominences, being not unlike the face of the earth, relieved by chains of mountains and deep valleys.[18]

The small spots lose their blackness because the sun rising higher has its rays no longer blocked by the mountains and ridges, just as happens in the valleys on the earth. Galileo's explanations of the various observed phenomena are qualitative rather than quantitative. He may implicitly depend upon certain general laws (e.g., that light rays travel

---

17. Galileo, "Sidereal Messenger," in S. Drake, ed., *Discoveries and Opinions of Galileo* (New York, 1957), p. 27.
18. *Ibid.*, p. 31.

in straight lines), but laws are not explicitly invoked in his explanation. (Generalizations are invoked; but these mention specific objects such as the moon, the sun, and the earth, and would not, I assume, be classified as laws by universalists.) Galileo's procedure, rather, is to explain by assigning specific causes to the observed phenomena and drawing analogies between these causes and similar phenomena on the earth. His explanations might be formulated like this:

> We see small spots throughout the moon's surface because the moon's surface, like the earth's, is not smooth, but contains mountains and craters.
>
> We see dark zones and brightened areas associated with these spots because the brightened areas are prominences that are reflecting light from the sun and the dark areas are craters or valleys which are darkened because the sun's rays are blocked by the prominences.

Galileo, in short, explained the phenomena he observed by attributing their causes to irregularities on the surface of the moon (rather than, say, to irregularities in the operation of his telescope). His explanations were important because they challenged the received cosmological view that all celestial bodies are perfectly smooth spheres, and hence are unlike the earth.

If explicit formulation of quantitative hypotheses and general laws is always required by instructions in science, we shall have to say that Galileo's explanations, despite other possible virtues, suffer because they fail to satisfy such instructions. But to say this is to give insufficient weight to the level of understanding and the interests of Galileo's audience. The members of that audience accepted the traditional doctrine of the perfect smoothness of celestial bodies, including the moon. In the *Dialogue Concerning the Two Chief World Systems,* Galileo has Simplicio formulate the traditional viewpoint and explain the appearances as follows:

> The appearances you speak of, the mountains, rocks, ridges, valleys, etc., are all illusions. I have heard it strongly maintained in public debates against these innovators that such appearances belong merely to the unevenly dark and light parts of which the moon is composed inside and out. We see the same thing occur in crystal, amber, and many perfectly polished precious stones, where from the opacity of some parts and the transparency of others, various concavities and prominences appear to be present.[19]

Galileo replies to this explanation by indicating the differences in reflected light from a perfectly smooth body such as a mirror and that

19. Galileo, *Dialogue Concerning the Two Chief World Systems,* Stillman Drake, ed. (Berkeley, 1967), p. 70.

from a body with a rough surface such as a wall; and he argues that the reflection coming from the moon much more resembles that from the wall than from the mirror.

Galileo, then, in giving his explanations might be said to have followed instructions such as these:

I: Identify what lunar objects, if any, produce the observed appearances;
Indicate, in non-quantitative terms and without invoking general laws, how these objects produce these appearances, by appeal, if possible, to analogous situations on the earth;
Give reasons for rejecting alternative explanations which presuppose the smoothness of the moon's surface.

For each q being explained he followed these instructions because he believed that his audience did not understand q in a way that satisfies I; that there is a correct answer to Q satisfying I that could render q understandable in the manner required for explaining; that the audience would be interested in understanding q in way that satisfies I; and that such an understanding would be valuable for it. Furthermore, I shall assume that it was reasonable for him to believe these things. If so, then it was reasonable for Galileo to follow instructions I. (Recall the conditions (a) through (d) for appropriate instructions given in Section 4.)

Let us also suppose that understanding q in a way that satisfies (different) instructions I* calling for *quantitative hypotheses* and *general laws* would have been valuable for Galileo's audience. This does not mean that understanding q in a way that satisfies I above was *not* valuable for it. Indeed, before an audience can understand q in the more quantitative and general way dictated by I* it may be important for it first to understand q in a way satisfying I. No doubt if the audience had been a very different one—if it already understood q_I and sought to understand q by reference to quantitative hypotheses and general laws—instructions I would have been inappropriate ones for Galileo to follow. But this is not what is at issue.

Since instructions I satisfy conditions (a), (b), and (c) for appropriate instructions with respect to Galileo's audience, what the critic must show is that (d) is violated: he must show that understanding q in a way that satisfies I would be of no value for Galileo's audience. But this seems patently absurd. If one has the misconceptions of that audience—if one believes that the moon and other celestial bodies are perfectly smooth—then coming to understand Galileo's telescopic observations in a way satisfying I is valuable, since these misconceptions will then be removed, even if understanding at a quantitative level involving laws is not achieved.

15. BOYLE'S LAW

Scientists often seek explanations at a micro-level that permit the phenomena to be explained to be related to others with similar micro-causes. Let us suppose that there is a set of instructions calling for q to be explained at that level, and that explanation E fails to satisfy those instructions in virtue of not invoking any micro-events. As an example consider the well-known experiment leading to Boyle's law.

Boyle took a glass J-tube with the short leg sealed off and poured mercury into the long leg until the mercury level was the same in both legs. He then poured mercury into the long leg and recorded the various levels in both legs in the form of a table, part of which is as follows:

A	B	C	D	E
48	00	$29^{1}/_{8}$	$29^{2}/_{16}$	$29^{2}/_{16}$
24	$29^{11}/_{16}$	$29^{1}/_{8}$	$58^{13}/_{16}$	$58^{2}/_{16}$
12	$88^{7}/_{16}$	$29^{1}/_{8}$	$117^{9}/_{16}$	$116^{8}/_{16}$

Column A indicates the distance between the top of the short leg and the mercury level in that leg; column B, the amount of mercury added to the long leg; column C, the number $29^{1}/_{8}$; column D, the addition of numbers in columns B and C. (Column E will be explained presently.) Boyle noted the following regularity. When the number in column A decreases to half its original value, the number in column D increases to approximately double its value; when the number in A decreases to a quarter of its original value the number in D is approximately quadrupled. In short, there is an inverse relationship between the values in A and D. How is this regularity to be explained?

When Boyle added mercury to the long leg he was increasing the pressure on the air above the mercury in the short leg. The pressure on this air is the result of the standard atmospheric pressure and the pressure of the additional mercury added to the long leg. The former pressure, measured in inches of mercury, is $29^{1}/_{8}$, which is the number in column C; the latter pressure, as measured in inches of mercury, is given in column B. The sum of these pressures appears in column D. Column A gives a measurement of the volume of the air in the short leg. The regularity between changes in columns A and D is thus explained by saying that as the pressure on the air, and hence its pressure, increases, its volume decreases proportionately. The more general formulation of this regularity is that the pressure and the volume of a gas are inversely proportional—which is Boyle's law. Column E indicates what, according to this law, the pressure should be if the volume is given in column A. Boyle's law thus explains the observed regularity between the height of the mercury in the long leg and the distance between the mercury and the top of the short leg by indicating what underlying physical quantities these distances reflect, viz. volume and pressure of a gas, and how the latter are related.

In answering the question

> Q: Why is there a regularity between the measurements recorded in columns A and D?

Boyle was following some such (macro-)instructions as these:

> I: Answer Q by reference to macro-properties of the compressed air above the mercury in the short leg and cite a law relating these properties.

Now consider these (micro-)instructions:

> I': Answer Q by reference to *micro*-properties of the compressed air (by reference to particles of which air is composed).

Boyle himself did speculate about micro-mechanisms to account for the compressibility of air. He envisaged two possibilities. According to one, air is composed of micro-particles "lying one upon another, as may be resembled to a fleece of wool."[20] The other supposed that it was composed of particles moving rapidly in a subtle fluid. But Boyle refrained from supporting one or the other of these, and he wrote that he shall

> decline meddling in a subject, which is much more hard to be explicated than necessary to be so by him, whose business it is not, in this letter, to assign the adequate cause of the spring of the air, but only to manifest that the air hath a spring, and to relate some of its effects.[21]

Let us suppose that it was reasonable for Boyle to believe the following: that the audience did not understand $q_{I'}$ (i.e., q at a micro-level); that there is a correct answer to Q that satisfies I', the citing of which will enable the audience to understand $q_{I'}$ by producing the knowledge that it is a correct answer to Q; that the audience would be interested in understanding $q_{I'}$; and that understanding $q_{I'}$ would be valuable for it. We are assuming, then, that micro-instructions I' constitute an appropriate set for explaining q to the given audience. Since the explanation of q that Boyle actually provided does not in fact satisfy I' is it faulty? Must Boyle's explanation of q be criticized on the ground that it fails to provide a micro-theory, when such a theory is called for by instructions satisfying conditions of appropriateness?

The answer is No. Given the information available to Boyle, it was

20. *Harvard Case Histories in Experimental Science* 1, James B. Conant, ed. (Cambridge, 1948), p. 57.
21. *Ibid.*, p. 58.

reasonable for him to hold these beliefs: that his audience did not understand q in a way that satisfies *macro*-instructions I; that air is compressible and that underlying the observed regularities in the mercury levels there is a macro-law relating properties of air (so that there is a correct answer to Q satisfying I); and that understanding q_1 would be of interest to, and valuable for, that audience. To be sure, the same claims can be made for satisfying micro-instructions I'. But that in no way impugns the value of satisfying I. Indeed, Boyle in all probability believed, quite plausibly, that it was important first to be able to understand q in a way that satisfies macro-instructions before turning to micro-explanations. His audience, he assumed, understood q in neither way.

Boyle's situation can usefully be contrasted with Bohr's, both of whom explained an observed regularity. Boyle did so at a macro-level in terms of the compressibility of the air and the law relating pressure and volume. If Bohr had done a similar thing—if he had explained the spectral lines simply by citing the thermal or electrical excitation of the hydrogen gas and the Balmer formula relating the lines (e.g., if he had produced simply the "Hempelian" (α) or the "Salmonian" (β))—his explanation would have been judged inadequate. The excitation phenomenon and Balmer's formula were known to Bohr's audience; the members of this audience believed that the lines were produced by events occurring within the hydrogen atom and wanted an explanation at that level. The members of Boyle's audience had not yet reached that stage. Not all of them were even aware that the pressure of the air is responsible for the mercury level. (One of Boyle's aims was to refute a hypothesis proposed by Franciscus Linus (1595–1675) that the mercury in a simple Torricellian barometer, or indeed in a J-tube, was being pulled to its level by an invisible cord or funiculus. Boyle, by contrast, followed Torricelli in explaining the mercury level in a barometer as being due entirely to the pressure of the air on the mercury in the dish surrounding the barometer.) And even those who accepted the idea of the pressure of air were not familiar with the law relating pressure and volume.

Boyle's explanation need not be evaluated from the viewpoint of an explainer or audience with such interests and information. Judged from the perspective of someone with information we now possess and with a desire to provide a micro-theory which relates the phenomenon in question to others, Boyle's explanation will be inadequate: it is not sufficiently deep or unifying for these purposes. The situation here is analogous to one in which Bohr's explanation is judged unsatisfactory from the viewpoint of modern quantum theory.

16. THE ROLE OF GENERAL METHODOLOGICAL VALUES IN THE ASSESSMENT OF SCIENTIFIC EXPLANATIONS

Let me draw some conclusions from these examples. I have been assuming that scientists have certain general methodological values. These are particularly relevant for helping to determine the satisfaction of two of the conditions for appropriate instructions given in Section 4:

(c) A way of understanding which scientists are *interested* in achieving (or would be under the conditions noted in Section 3);

(d) A way of understanding which, if it could be achieved, would be *valuable* for a scientific audience.

Contemporary physicists seek to understand in a way requiring simplicity (quantitative hypotheses, general laws, etc.); and such an understanding, if it could be achieved, would be valuable for them. An explanation may be judged to be a good one in part because it satisfies such general values.

However, these values fail to constitute either necessary or sufficient conditions for appropriate instructions for a scientist to follow in explaining q to an audience that shares such values. They are not sufficient because conditions (c) and (d) are not sufficient conditions; (a) and (b) are also necessary (see Section 4). That is to say, one must consider not only whether the scientific audience is or would be interested in the satisfaction of these values, and whether their satisfaction is desirable, but also whether or not the audience already understands q in a way that satisfies such values, and if not, whether it is possible for it to do so. Moreover, as we saw in Part I, the methodological criteria I have mentioned are by themselves too general to provide a sufficient basis for evaluations. They do not enable one to say *enough* about why Bohr's explanation of the spectral lines is better than certain "Hempelian" explanations which also invoke laws, quantitative hypotheses, and micro-entities. Nor are the general methodological values mentioned *necessary* conditions for appropriate instructions for an explainer to follow in explaining q to a scientific audience that shares such values. As the quark, Galileo, and Boyle examples show, (c) and (d) can be satisfied even if one or more of these general values is not present. A macro-level of understanding may be valuable for Boyle's audience even if that audience places great store on the micro-level; the 18-quark theory may provide a way of understanding hadrons that is valuable for contemporary physicists, even if that understanding fails to reflect some desirable standard of simplicity.

Scientists *are* interested in understanding the world in ways satisfying instructions calling for quantitative hypotheses, micro-theories, simplicity, etc., and such understanding *is* valuable. But this does not mean that they are interested in understanding the world only in these ways, or that only such understanding is valuable in science. General methodological criteria set a direction for scientific explanation. They serve as a guide for what kind of explanation the scientist should try to achieve, at some point. But they do not constitute a set each member of which is necessary and the totality sufficient for determining a way of understanding that is valuable in science, or appropriate instructions for scientists to follow for explaining.

Finally, how do we determine (c) and (d) for appropriate instructions? How do we decide which instructions I are such that (c) an understanding of q in a way that satisfies I is something that a scientific audience is or would be interested in achieving, and (d) is something that is valuable for it?

Three factors are especially relevant for determining what type of understanding a scientific audience is interested in obtaining. First, we consider general methodological criteria valued by this audience and others like it. (We observe, e.g., that such audiences generally seek an understanding in terms of micro-theories, quantitative hypotheses, laws, etc.) Second, we consider specific beliefs this audience has about the world that pertain to the answering of Q. (Does it believe, e.g., that questions about the hydrogen spectral lines have a correct answer at a subatomic level of the hydrogen atom?) Third, we consider remarks the audience has made regarding a specific way in which it seeks to understand q. (Has the audience explicitly requested that Q be answered at the subatomic level of the hydrogen atom?) Even if the audience has no interest in understanding q in a way that satisfies I it may still be reasonable to say that it would have such an interest under certain conditions (e.g., if it was made aware of Q and I, or if it accepted assumptions in I). And this, although more difficult to ascertain than its actual interest, is not impossible to determine. For example, an audience may have no interest in understanding the spectral lines at the subatomic level because it believes that such lines are produced by vibrations of the atom, not by subatomic events. Despite this we judge that such an audience would have an interest in a subatomic understanding of the spectral lines if it thought such an understanding possible. Our judgment might be based on the fact that the audience has a general interest in micro-theoretical explanations and that it believes that other phenomena can be explained at the subatomic level.

We determine what type of understanding of q it would be *valuable* for a scientific audience to have by considering general methodological criteria valued in science, as well as specific beliefs about the world pertaining to the answering of Q. We determine that it would be valuable for a scientific audience to understand the hydrogen spectral lines at the subatomic level of the hydrogen atom by combining the general methodological principle that understanding phenomena at a micro-level is of scientific value, with a specific empirical belief that the spectral lines of hydrogen are produced by events within the hydrogen atom. The former, without the latter, would yield instructions that are too unspecific for evaluating explanations. The latter, without the former, would yield instructions whose general value for science would be undefined. General methodological criteria are important because they set a direction. Understanding in a way that fails to reflect them can be valuable in science; but once achieved it can be enhanced by understanding in ways incorporating these values. At least that is a goal.

At the beginning of this chapter I emphasized that an explanation can be evaluated from the viewpoint of ends not identical with that of rendering q understandable by producing the knowledge, of the answer one gives, that it is a correct answer to Q. In such non-illocutionary evaluations other ends (possibly related, possibly not) are selected as desirable; and the explanation is given high marks solely because it achieves those ends. Nothing I have said precludes the use of general methodological criteria in non-illocutionary evaluations. Just as an explanation can be evaluated highly because it inspired scientists to think about a new problem, so it can be considered good because of its simplicity, or the unification it achieves. (Think of the Biblical explanation of the origin of the world and its inhabitants.) In a given context it may be clear that an evaluation is being made solely from the viewpoint of one or more of these ends. But if an explanation is rated as good or bad using only these standards, nothing whatever follows about whether it is a good or a bad one for *explaining q*. Boyle's explanation is completely inadequate when evaluated using a standard requiring a micro-theoretical level. Moreover, such a level is one that is typically valued in science. But we are not justified in concluding from this that Boyle's explanation is a bad one to give, or for Boyle to have given, in explaining the observed regularity.

17. UNIVERSALIST RESPONSES

In Chapter 1, a view of the evaluation of explanations was promised which relates such evaluations to the concept of an (illocutionary) act of explaining. Such a view, for general evaluations, is expressed in (1) of Section 5. The latter makes the goodness of an explanation dependent on instructions whose appropriateness is tied to achieving, with respect to a given audience, a state which is the end of an illocutionary act of explaining. Throughout the remainder of this chapter I have been trying to show how *scientific* evaluations, for which certain methodological criteria are relevant, also involve instructions whose appropriateness can vary.

The account is different from the usual ones, such as those in Chapter 1, which offer analyses of the following type:

E is a good (scientific) explanation of q if and only if it satisfies conditions C.

Conditions C impose requirements on the kinds of propositions which comprise the explanation and on their relationships to one another. They are to be, in effect, universal scientific instructions which incorporate no specific empirical assumptions and are devoid of "pragmatic" considerations pertaining to explainers and audiences. In Section 11 I discussed various possible universalist responses to the theory of evaluation presented in this chapter. Additional ones can now be considered, in the light of Part III.

1. Universalists might claim that their aim in formulating universal scientific instructions is simply to construct a complete list of things valued in science (laws, deductive structure, simplicity, etc.), and that nothing I have said shows this task to be impossible. My reply is to agree that I have not shown this to be impossible. But universalists want more than simply a list (or even an explication) of general methodological values: they want to use this in evaluating scientific explanations. My claim has been that if such a list were constructed it would not supply a necessary or a sufficient basis for determining whether an explanation in science is a good one for explaining q. Furthermore, a (non-illocutionary) evaluation based solely on general methodological values will not enable us to see enough of what is good about certain scientific explanations. (Both Galileo's explanation and Boyle's will be severely criticized because they fail to satisfy quite a few of these values.) Nor is it true that an illocutionary evaluation ignores general methodological values. On the contrary, they form part of the basis for determining a general kind of understanding scientists are interested in achieving, and which, if it could be achieved, would be valuable for them.

2. Universalists may charge that I am concerned with the contextual or pragmatic aspect of explanation, whereas they are not. They seek to abstract a set of absolute scientific standards from contextual considerations, and to suggest that these standards constitute necessary but not sufficient conditions for good scientific explanations.

To the charge of contextualism I plead guilty (though I regard this as a virtue). In response, however, I reiterate that non-pragmatic standards of the sort universalists seek cannot provide even necessary conditions for E's being a good explanation for explaining q. Furthermore, it is important to stress three claims that are not part of my contextual position.

(i) I am not claiming that any explanation satisfying the instructions of the explainer or audience is thereby to be judged good; or even that any explanation satisfying instructions that are appropriate for the audience is a good one. For (p; explaining q) to be a good explanation for an explainer to give in explaining q to an audience it must be (reasonable for the explainer to believe) not only that it satisfies appropriate instructions, but also that it is capable of rendering q understandable to that audience, in a way that satisfies those instructions, by producing the knowledge of p that it is a correct answer to Q. Satisfaction of the latter condition is not guaranteed by satisfaction of the former.

(ii) I am not claiming that scientific explanations must be evaluated by invoking instructions that contain contextual references (e.g., specific or general references to an explainer or audience). Bohr's explanation can be evaluated using instructions such as B of Section 7, which contains no such references. The contextualism in my position amounts to this. In offering a non-epistemic evaluation of an explanation as a good one for explaining q we need to consider facts about the audience; and in offering an epistemic evaluation we need to consider beliefs that the explainer has about the world and the intended audience. Even if instructions such as Bohr's B do not mention an audience or an explainer, the use of such instructions will need to be justified by appeal to an audience and, in an epistemic evaluation, to an explainer as well. We employ instructions B in evaluating Bohr's explanation because we are taking into account the knowledge and interests of physicists in 1913. Moreover, without any appeal to contextual considerations (whether or not these are formulated in the instructions themselves) it will not be possible to justify the claim that the explanation he gave was a much better one for him to have given than ones such as the "Hempelian" (α) and the "Salmonian" (β). To abstract completely from contextual or pragmatic matters is to render it impossible to determine whether a scientific explanation is a good one for explaining q.

(iii) I am not denying the existence of general values in science that can be relevant for assessing explanations. Scientists do seek laws, quantitative hypotheses, micro-theories, simplicity, and so forth. And this may be among the reasons that certain instructions are, or ought to be, followed. But a commitment to such general values is not the only reason a scientist chooses or ought to choose the instructions he does for explaining q. By concentrating on these alone we will have an incomplete picture of standards of evaluation in science. And depending on his own information about the world and about the beliefs and interests of the audience, an explanation can receive high marks even if it satisfies instructions that fail to reflect one or more of these values.

3. Another possible universalist charge is that I am espousing an unacceptable form of historical relativism. Such a view, in the case of the evaluation of scientific explanations, makes the following claims:

(I) Any set of instructions which provides standards for evaluating scientific explanations incorporates some particular scientific theory.

(II) One who evaluates a scientific explanation by reference to such instructions is himself committed to the truth of this theory.

Despite the fact that historical relativists and I eschew the idea of universal scientific instructions, my position is not that of claims (I) and (II).

It is not part of my view that all instructions by reference to which it is appropriate to evaluate scientific explanations incorporate a particular scientific theory. Some, such as Bohr's instructions B of Section 7, will. But others will not. Recall instructions Cn of Section 11:

Cn: Determine the particular instructions I which the audience has (or ought to have) imposed on answers to Q, and answer Q in a way that satisfies I.

These instructions, although quite general, are perfectly appropriate ones for scientists (and others) to follow. Yet Cn presupposes no particular theory.

This can be so even for instructions more specific than Cn. Here are instructions for explaining (q) why a cannon ball dropped from a tower accelerates toward the earth in a straight line:

QL (for "quantitative law"): Formulate a quantitative general law of which this particular type of motion is an instance.

Such instructions would be followed, with different results, by Galileo and Newton. (For Galileo such motion is an instance of the natural tendency of bodies above the surface of the earth to persist in an accelerated state once this has been achieved by the lack of support; and to persist toward their natural place, the earth, by the shortest path—a straight line. The quantitative law governing this is Galileo's law $s = \frac{1}{2}gt^2$.[22] For Newton the motion in question is an instance of accelerated motion produced by the earth's gravitational force ex-

---

22. This would be Galileo's explanation in the first part of the *Dialogue* but not later when he considers the tower argument. Here the motion is the resultant of the tendency mentioned above and the natural circular motion.

erted on the body; and Newton's quantitative law of gravity yields different results from Galileo's.) The important point is that instructions QL do not incorporate the theory of one or the other of these physicists. To be sure, QL is committed to a certain broad empirical assumption about the motion in question, viz. that it is an instance of some quantitative general law. But the theoretical assumptions historical relativists have in mind are more specific than this.

Even if instructions by means of which an explanation is being evaluated incorporate some particular scientific theory, it does not follow that one who evaluates the explanation by reference to such instructions is himself committed to the truth of that theory. We can rate Bohr's explanation highly—as an excellent one for him to have given—based on instructions B. These instructions contain a commitment to Rutherford's model. We may recognize that, given the information available in 1913, it was praiseworthy of Bohr to have followed these instructions. And we may give Bohr's explanation high marks in virtue of the fact that it satisfies them. But we as evaluators need not be committed to the truth of the theory incorporated in these instructions.

4. Universalists may argue that there is an important relationship between the concept of a good explanation and the concept of evidence, which the theory I have presented cannot capture. Retroductivists such as Charles Peirce and N. R. Hanson claim that

(1) The fact that a hypothesis h, if true, would be a good explanation of why some proposition p is true counts as evidence that h is true.

Now this concept of evidence, it will be said, is objective and non-contextual. Whether some fact is evidence that h is true does not depend upon, or vary with, beliefs or desires of any explainer or audience. If there is a concept of good explanation that is related to an objective concept of evidence in the manner given by (1), then there must be a concept of good explanation which is not contextual; and it is this that universalists are trying to explore.

My answer is to admit that there is an important connection between an objective, non-contextual concept of evidence and a concept of explanation. This will be discussed in Chapter 10. It will be argued there that (1) above provides an inadequate account of this connection. Furthermore, an objective concept of evidence to be defined in that chapter will depend not on the (illocutionary) notion of a good explanation defined in Section 5, but on the (non-illocutionary) concept of a *correct* explanation defined in Section 1. In short, I am claiming, a concept of (correct) explanation has been provided that can be used to define an objective concept of evidence without adopting the universalist position.

5. A final universalist response consists in incorporating several ideas of the illocutionary account. A universalist might say that to ascertain whether E is a good explanation for an explainer to give in explaining q to an audience we must follow these rules:

(i) Determine whether the audience already understands q in the way provided by E;
(ii) Determine whether the audience is such that the citing of E will enable it to understand q;
(iii) Determine whether the audience is interested in understanding q in the way provided by E;
(iv) By considering general methodological values and specific empirical beliefs of the audience, determine whether understanding q in the way provided by E would be valuable for the audience.

Now it might be said that this view is still universalism in the sense that the rules (i) through (iv) are universally applicable. They are not to vary from one scientific context to another, although particular explainers, audiences, empirical assumptions, and perhaps even methodological values, may vary.

With this "universalism" the illocutionary theory can concur. Indeed, the rules (i) through (iv) above are ones expressed (more adequately) by the conditions for being appropriate instructions to follow in explaining q, and for being a good explanation of q—conditions given in Sections 4 and 5. But this picture is very different from the universalism discussed in the present chapter. For one thing, it takes into account the beliefs of the explainer and the specific knowledge and interests of the audience, as well as the general methodological values which are so important to the "old" universalist. For another, it denies that the satisfaction of these methodological values is either necessary or sufficient for being a good scientific explanation.

18. CONCLUSIONS

Let me bring together the major strands in this discussion of the evaluation of explanations. In Part I the concept of a correct answer to a content-question was defined. Using this, together with the ordered pair view, it was shown that a simple definition of a correct explanation can be given, as follows:

(1) If (p; explaining q) is an explanation, then it is correct if and only if p is true.

However, this criterion of correctness affords an insufficient basis for evaluating explanations. In Section 5 a more elaborate condition was formulated for an illocutionary evaluation of an explanation—for evaluating an explanation as a good one for explaining q—which is tied to the idea of appropriate instructions for explaining. In Part II the question was whether a set of universal instructions is possible which will provide a sufficient basis for illocutionary evaluations of explanations in science. The answer offered is that universal instructions which eschew contextual references and particular empirical commitments will not provide such a basis. The issue then becomes whether universal instructions can constitute necessary conditions for good scientific explanations. This was explored in Part III. The conclusion was that the general methodological values associated with universal instructions do not provide even necessary conditions, though such values can be important in illocutionary evaluations and do serve as a guide for the kinds of explanations scientists should try to achieve at some point.

Numerous universalist responses were considered in Sections 11 and 17. There is, however, one additional response, which will lead directly to the discussion in the next chapter. The universalist might claim that he is not (or will no longer be) concerned with evaluating an explanation as a *good* one (for any purpose, including illocutionary purposes). Rather, he is or will be concerned only with the concept of a *correct* explanation. However, he wants more than the standard of correctness supplied by (1) above. He wants what he might call a "minimally scientifically acceptable" standard of correctness. (1) does not supply such a standard, and the conditions for being a good explanation given in Section 5 go well beyond such a standard. He wants something in between. What this is will be examined in the chapter that follows.

CHAPTER 5
Can There Be a Model of Scientific Explanation?

1. INTRODUCTION

Since 1948, when Hempel and Oppenheim published their pioneering piece,[1] various models of scientific explanation have appeared. But each has had its difficulties, and observers of the philosophical scene may wonder whether models of the kind sought are really possible. Are their proponents engaged in a fruitless task of inquiry?

If modelists want to supply universal scientific instructions (of the kind described in the previous chapter) that are necessary and sufficient—or even just necessary—for evaluating an explanation as a good one for explaining q in science, then, I have been arguing, they cannot be successful. However, they may have a more limited aim. Most of them want to supply at least necessary and sufficient conditions for being a *correct* explanation. In Chapter 4, using the ordered pair view and the definition of a correct answer to a content-question, we said that

(1) If (p; explaining q) is an explanation, it is correct if and only if p is true.

Modelists, however, will not be content with just this. They want the resulting explanation not only to be correct, but correct in the right sort of way. A model of explanation must impose additional requirements.

The question, then, is whether conditions can be formulated for a correct explanation which, although not guaranteeing that an explanation will be a good one for explaining q, do go beyond (1). Such conditions might be thought of as defining a concept of correctness

1. Carl G. Hempel and Paul Oppenheim, "Studies in the Logic of Explanation," *Philosophy of Science* 15 (1948), pp. 135–75.

that satisfies certain minimal demands scientists might make as well as certain basic standards of illumination philosophers might impose. Let me refer to this as "minimal scientific correctness." Because of the nature of the requirements for this sort of correctness that modelists propose, I shall first formulate their views using their own terminology. Later this will be related to the ordered pair view.

Hempel and other modelists are particularly concerned with explanations that answer what they call explanation-seeking why-questions, which have the form

(2) Why is it the case that p?[2]

The sentence replacing "p" in (2) Hempel calls the *explanandum*. It describes the phenomenon, or event, or fact, to be explained. The answer to an explanation-seeking why-question of form (2) Hempel calls the *explanans*. It is the sentence or set of sentences that provides the explanation. It will be convenient to follow modelists in speaking of the explanans as explaining the explanandum. And we can also say that an explanans *potentially* explains an explanandum when, if the sentences of the explanans were true, the explanans would correctly explain the explanandum.

For example, if the explanation-seeking why-question is

Q: Why is it the case that this metal expanded?

then the explanandum is

(3) This metal expanded.

If in reply to Q, an explainer utters

(4) This metal was heated; and all metals that are heated expand,

then (4) is the explanans for the explanandum (3). And (4) potentially explains (3) if, given the truth of (4), (4) would correctly explain (3).

A *model* of explanation is a set of conditions that determine whether the explanans correctly explains the explanandum (where the explanation-seeking question is of form (2)). It can also be described as a set of conditions that determine whether the explanans potentially explains the explanandum. If the latter conditions are satisfied by a given explanans and explanandum, then the explanans correctly explains the explanandum, provided that the explanans is true. These conditions will be subject to certain requirements which guarantee not just correctness (as does (1) above), but correctness of a sort desired by modelists.

In what follows, my concern will be with models as sets of sufficient (rather than necessary) conditions for correct explanations; and as providing such conditions for explanations of particular events or facts

2. Hempel, *Aspects of Scientific Explanation*, p. 334.

rather than of general laws. Most of the counterexamples in the literature have been raised against models construed in this way. I shall argue that one important reason for the failure of these models is that their proponents want to impose requirements which in effect destroy the efficacy of their models.

2. THE NES REQUIREMENT

The first requirement is that no singular sentence in the explanans (no sentence describing *particular* events), and no conjunction of such sentences, can entail the explanandum.[3] I will call this the No-Entailment-by-Singular-Sentence requirement, or NES for short. What are the reasons for it?

There are, I think, three. The first two are explicit in the views of the modelists I have in mind; the third is an important implicit component of such views.

First, NES precludes certain "self-explanations" and "partial self-explanations." Suppose that our explanation-seeking why-question is

Why is it the case that this metal expanded?

for which the explanandum is

(1) This metal expanded.

The NES requirement precludes (1) itself from being, or being part of, an explanans for (1). It also precludes from an explanans for (1) sentences such as "This metal was heated and expanded," and "This metal expanded, and all metals that are heated expand," which would be regarded as partial self-explanations of (1).

Second, modelists emphasize the importance of general laws in a scientific explanans. Such laws provide an essential link between the singular sentences of the explanans and the singular sentence that

---

3. Or, in a tighter formulation, where the explanans is a set or conjunction of sentences: (i) no subset of these all members of which are singular sentences entails the explanandum; (ii) the explanandum is not entailed by a singular conjunct in a conjunctive equivalent of the explanans. To explicate the latter let S be a sentence containing within it one or more occurrences of some singular sentence (or some compound of singular sentences) P. Delete from S one or more occurrences of P, obtaining a sentence S*. Form a conjunction of S* with P. If this conjunction is equivalent to S it is a conjunctive equivalent of S containing P as a singular conjunct. E.g., (a) "(Ga v Ja) & -Ja & (x) (Fx ⊃ Gx)" is a conjunctive equivalent of (b) "(x) ((Ga v Ja) & -Ja & Fx ⊃ Gx)," which contains "(Ga v Ja) & -Ja" as a singular conjunct. Since this singular conjunct entails "Ga," explanans (b) violates NES for explanandum "Ga." The notion of a singular sentence which Hempel and others use in characterizing their models is often employed in a more or less intuitive way. And when there is an attempt to make this notion more precise difficulties emerge, as Hempel himself is aware (see *ibid.*, p. 356). Hempel suggests that the concept can be adequately defined for a formalized language containing quantificational notation; but there are problems even here (see my *Law and Explanation*, pp. 36-37). However, the kinds of cases I will be concerned with are quite simple and would, I think, be classified by modelists both as singular sentences and as ones to be excluded from the explanations in question.

constitutes the explanandum. Intuitively, to explain a particular event requires relating it to other particular events *via* a law; if the singular sentences of the explanans themselves entail the explanandum, laws become unnecessary, on this view.

Third, the NES requirement in effect removes from an explanans certain sentences which involve explanatory connectives like "explains," "reason," "because," "on account of," "due to," and "causes." Let me call sentences in which such terms connect phrases or other sentences *explanation-sentences*. Here are some examples:

(2) This metal's being heated explains why it expanded;
The reason this metal expanded is that it was heated;
This metal expanded because it was heated;
This metal's expanding is due to its being heated;
The fact that this metal was heated caused it to expand.

NES precludes any of the explanation-sentences in (2) from being, or being part of, an explanans for (1), since each of them is a singular sentence that entails (1).[4] (More generally, if Q is a question of the form "Why is it the case that p?", where p is a singular sentence, then NES excludes from an explanans any sentence expressing a complete content-giving proposition with respect to such a Q.) Without NES one could simply require, e.g., that an explanans for an explanandum p be a singular explanation-sentence of the form "E explains (why) p," or "the reason that p is that E," or "p because E." Any such explanans, if it were true, would correctly explain the explanandum p. Of course, each of the five sentences in (2), if true, could be cited in correctly explaining why this metal expanded. Modelists need not deny this. Their claim is that the sentences in (2) do not correctly explain (1) in the right sort of way. They would exclude sentences in (2) from an explanans for (1) because they think that an adequate explanans for (1) must reconstruct the sentences in (2) so that the explanatory connectives in the latter are, in effect, analyzed in non-explanatory terms. One of the purposes of a model of explanation is to define terms like "explains," "reason," "because," and "cause," and not to allow them to be used as "primitives" within an explanans. By providing such definitions modelists want to show why it is that this metal's being heated explains why it expanded.

Recall that on the ordered pair theory, if (p; explaining q) is an

---

4. To classify them as singular, of course, is not to deny that they have certain implicitly general features which further analysis might separate from the singular ones. Some might claim that NES should be applied only to completely singular sentences—ones with no generality present at all. The sentences in (2) must first be reduced. For example, in the case of the last the reduction might be "This metal was heated; this metal expanded; and whenever a metal is heated it expands." If the latter is now taken to be the reduced form of the explanans, then the reduced explanans contains a singular sentence—"This metal expanded"—which entails the explanandum (1).

explanation of q, then p is a complete content-giving proposition with respect to Q. Among complete content-giving propositions are ones expressible by sentences of the form

The reason that _____ is that _____ ,

or

_____ because _____ .

In Chapter 2, I argued that the fact that one might explain q by uttering sentences expressing such propositions does not render circular the illocutionary account of explaining. Still, modelists may urge, explanations with constituent propositions of these types are not sufficiently illuminating. It ought to be possible to define a concept of a correct explanation which excludes such propositions. Since the ordered pair theory's truth-conditions for "E is a correct explanation of q" permits these as constituent propositions in an explanation, modelists will not accept this theory's account of correctness.

NES does not exclude all singular explanation-sentences (or all explanatory connectives) from an explanans.[5] But by precluding those that entail the explanandum it does eliminate ones that, from the viewpoint of the modelists, most seriously reduce the possibility of philosophical illumination from the resulting explanation. (Such modelists would probably advocate an even broader requirement eschewing all explanation-sentences from an explanans; but this possibility I shall not discuss.)

NES also excludes certain sentences from an explanans that do not explicitly invoke explanatory connectives but are importantly like those that do which are excluded. Suppose we want to explain why the motion of this particle was accelerated. Our explanandum is

(3) The motion of this particle was accelerated.

Consider the explanans

(4) An electrical force accelerated the motion of this particle.

Although (4) itself contains no explicit explanatory connective such as "explains," "because," or "causes," it nevertheless carries a causal implication concerning the event to be explained. It is roughly equivalent to the following explanation-sentences which do have such connectives:

(5) An electrical force caused the motion of this particle to be accelerated;

---

5. For example, it permits the following. Explanandum: "An event of type C occurred." Explanans: "An event of type A caused one of type B; whenever an event of type B occurs so does one of type C."

The motion of this particle was accelerated because of the presence of an electrical force.

NES precludes (4) as well as (5) from an explanans for (3), since (4) is a singular sentence that entails (3). Those who support NES would emphasize that (4), no less than (5), invokes an essentially explanatory connection between an explanans-event and the explanandum-event which it is the task of a model of explanation to explicate.

3. THE A PRIORI REQUIREMENT

The second requirement imposed by modelists, which I shall call the a priori requirement, is that the only *empirical* consideration in determining whether the explanans correctly explains the explanandum is the truth of the explanans; all other considerations are a priori. Accordingly, whether an explanans potentially explains an explanandum is a matter that can be settled by a priori means (e.g., by appeal to the meanings of words, and to deductive relationships between sentences). A model must thus impose conditions on potential explanations the satisfaction of which can be determined non-empirically. A condition such as the following would therefore be precluded:

An explanans potentially explains an explanandum only if there is a (true) universal or statistical law relating the explanans and explanandum.

Whether there is such a law is not an a priori matter.

The idea is that a model of explanation should require that sufficient information be incorporated into the explanans that it becomes an a priori question whether the explanans, if true, would correctly explain the explanandum. There is an analogy between this and what various logicians and philosophers say about the concepts of *proof* and *evidence*.

Often a scientist will claim that a proposition q can be proved from a proposition p, or that e is evidence that h is true, even though the scientist is tacitly making additional empirical background assumptions which have a bearing on the validity of the proof or the truth of the evidence claim. If all these assumptions are made explicit as additional premises in the proof, or as additional conjuncts to the evidence, then whether such and such is a proof, or is evidence for a hypothesis, is settleable a priori. Similarly, often a scientist who claims that a certain explanans correctly explains an explanandum will be making relevant empirical background assumptions not incorporated into the explanans; if the latter are made explicit and added to the explanans, it becomes an a priori question whether the explanans, if true, would correctly explain the explanandum.

There is an additional similarity alleged between these concepts. A scientist would not regard a proof as correct—i.e., as proving what it purports to prove—unless its premises are true. Nor would he regard e as evidence that h (or as confirming or supporting h) unless e is true. (That John has those spots is not evidence that he has measles unless he does have those spots.)[6] And whether the premises of the proof, or the evidence report, is true is, in the empirical sciences at least, not an a priori question. Nevertheless, logicians in the Carnapian tradition believe that they can isolate an a priori aspect of proof and evidence such that the only empirical consideration in determining whether a proof or a statement of the form "e is evidence that h" is correct is the truth of the premises of the proof or of the e-statement in the evidence claim; all other considerations are a priori. What I have been calling the a priori requirement makes the corresponding claim about the concept of explanation: the only empirical consideration in determining whether an explanation is correct is the truth of the explanans.

Given the a priori and NES requirements, we can now see why

If (p; explaining q) is an explanation, it is correct if and only if p is true

does not constitute an adequate model of explanation, according to modelists. Since we are restricting our attention to why-questions about particular events and not laws, Q will be a question expressible by an interrogative of the form "Why did event e occur?" Let us say that the explanandum in such a case has the form

(1) Event e occurred.

We may identify the explanans of the explanation (p; explaining q) as the proposition p. Since p is a complete content-giving proposition with respect to Q, and since Q has the form "Why did event e occur?," the explanans is expressible by a sentence of the form

(2) The reason that event e occurred is ———.

The a priori requirement is satisfied. The only empirical consideration in determining whether a complete content-giving explanans of form (2) correctly explains an explanandum of form (1) is the truth of the explanans. However, the NES requirement is violated. Any sentence of form (2) is a singular sentence that entails the explanandum

6. See Chapter 10.

(1). Indeed, an explanans of form (2) is an explanation-sentence containing an explanatory connective that modelists seek to analyze by means of their model.[7]

Are there models that do satisfy both requirements?

4. MODELS PURPORTING TO SATISFY THE NES AND A PRIORI REQUIREMENTS

a. Hempel's basic D-N model. Consider this model as providing a set of sufficient conditions for correct explanations of particular events, facts, etc. The explanation-seeking question is of the form "Why is it the case that p?" where p, the explanandum, is a sentence describing the event to be explained. The explanans is a set containing sentences of two sorts. One sort purports to describe particular conditions that obtained prior to, or at the same time as, the event to be explained. The other are lawlike sentences (sentences that if true are laws). The model requires that the explanans entail the explanandum and that the explanans be true.

No singular sentence (or conjunction of such sentences) that entails the explanandum will appear in the explanans.[8] (Hence, NES is satisfied.) Any such sentence which is an explanation-sentence that someone might utter in explaining something will itself be analyzed as a D-N explanation, i.e., as a deductive argument in which no premises that are singular sentences entail the conclusion. For example, if someone utters the explanation-sentence

(1) This metal's being heated explains why it expanded (what caused it to expand, etc.)

in explaining why the metal expanded, a D-N theorist will restructure (1) as an argument such as this:

(2) This metal was heated
Any metal that is heated expands
Therefore,
This metal expanded.

---

7. Could the ordered pair view be reformulated so as to satisfy NES? Where the constituent question is expressible by the interrogative "Why did event e occur?" the constituent proposition will not be one expressible by a sentence of form (2) but by a sentence of form (2)' "the reason is _____," which does not entail the explanandum (1). The problem is that (2)' uttered as a response to "Why did event e occur?" is elliptical for (2), which does entail (1). Moreover, the present proposal will still require the use of explanatory connectives such as "reason" which modelists eschew in an explanans.

8. This is required in Hempel's informal and formal characterizations of his model. See his *Aspects of Scientific Explanation*, pp. 248, 273, 277.

And he will identify the premises of (2) as the explanans of the explanation and the conclusion as the explanandum. The premise in (2) that is a singular sentence does not entail the conclusion.

The a priori requirement also seems to be satisfied by this model. The only empirical consideration in determining whether the explanans correctly explains the explanandum is the truth of the explanans, i.e., the truth of the premises of a D-N argument; all other considerations are a priori. What are these other considerations? They are whether the explanans (but not the conjunction of singular sentences in it) deductively entails the explanandum, and whether it contains at least some sentences that are lawlike. The former is not an empirical question, nor is the latter, as construed by Hempel, since whether a sentence is lawlike depends only on its syntactical form and the semantical interpretation of its terms.[9]

The D-N model as a set of sufficient conditions for particular events is very broad, and one might seek to add further restrictions. Three more limited versions will be noted.

b. The D-N dispositional model.[10] Here the explanandum is a sentence with a form such as

(3) x manifested P.

And the explanans contains sentences of the form

(4) x has F, and conditions of type C obtained;

(5) Anything with F manifests P when conditions of type C obtain.

For an explanans consisting of (4) and (5) to provide a correct *D-N dispositional* explanation of (3) the model requires that F be a dispoposition-term, that (5) be lawlike, that (4) and (5) entail (3), and that (4) and (5) be true. The singular sentence in the explanans is not to entail the explanandum. And the satisfaction of all the conditions of the model, save for the truth of the explanans, is determined a priori. The only condition in addition to those of the basic D-N model is that F be a disposition-term, something settleable syntactically and/or semantically.

c. The D-N motivation model.[11] Here the explanandum is a sentence saying that some agent acted in a certain way. The explanans contains a singular sentence attributing a desire (motive, end) to that agent, a

---

9. See Hempel's discussion of lawlikeness, *ibid.*, pp. 271–2, 292, 340. The considerations mentioned above apply as well to the more formal account given by Hempel and Oppenheim in section 7 of their paper.
10. See Hempel, *ibid.*, p. 462.
11. See *ibid.*, p. 254.

singular sentence attributing the belief to that agent that performing the act described in the explanandum is in the circumstances a (the best, the only) way to satisfy that desire, and a lawlike sentence relating desires, beliefs, and actions of the kind in question. For example, the explanandum might be a sentence of the form

(6) Agent x performed act A,

and the explanans might contain sentences of the form

(7) x desired G;

(8) x believed that doing A is, in the circumstances, a (the best, the only) way to obtain G;

(9) Whenever an agent desires something and believes that the performance of a certain act is, in the circumstances, a (etc.) way to satisfy his desire he performs that act.[12]

For an explanans consisting of (7) through (9) to provide a correct D-N motivational explanation of (6) the model requires that (9) be lawlike, that (7) through (9) entail (6), and that (7) through (9) be true. In this model, like the others, the singular sentences in the explanans are not to entail the explanandum, and the satisfaction of all the requirements of the model, save for the truth of the explanans, is settleable a priori.

d. Woodward's functional interdependence model. James Woodward proposes adding to the basic D-N conditions the following additional necessary condition:

(10) *Condition of functional interdependence:* the law occurring in the explanans for the explanandum p must be stated in terms of variables or parameters variations in the values of which will permit the derivation of other explananda which are appropriately different from p.[13]

Suppose that the explanation-seeking question is "Why is it the case that this pendulum has a period of 2.03 seconds?," for which the explanandum is

(11) This pendulum has a period of 2.03 seconds.

Consider the following D-N argument:

---

12. I do not for a moment believe that (9) is true. But this is not the problem I want to deal with here.
13. James Woodward, "Scientific Explanation," *British Journal for the Philosophy of Science* 30 (1979), p. 46.

(12) This pendulum is a simple pendulum
The length of this pendulum is 100 cm
The period T of a simple pendulum is related to the length L by the formula $T = 2\pi \sqrt{L/g}$, where $g = 980$ cm/sec^2
Therefore,
This pendulum has a period of 2.03 seconds.

The third premise in (12) is a law satisfying Woodward's condition (10). Its variables are the period T and the length L. And variations in the values of these variables will permit the derivation of explananda which Woodward regards as appropriately different from (11). For example, if we change the explanandum (11) to

(13) This pendulum has a period of 3.14 seconds,

the law in (12) allows the derivation of (13) if the value of L is changed to 245 cm. Woodward is impressed by the fact that explanations, particularly in science, permit a variety of possible phenomena to be explained. He writes:

> The laws in examples [of this sort] formulate a systematic relation between . . . variables. They show us how a range of different changes in certain of these variables will be linked to changes in others of these variables. In consequence, these generalizations are such that when the variables in them assume one set of values (when we make certain assumptions about boundary and initial conditions) the explananda in the . . . explanations are derivable, and when the variables in them assume other sets of values, a range of other explananda are derivable.[14]

The satisfaction of condition (10) is settleable a priori. If this condition is the only one to be added to those of the basic D-N model, then the resulting model appears to satisfy the a priori requirement and NES.[15]

5. VIOLATION OF THE A PRIORI REQUIREMENT

Despite the claims of these models, the a priori requirement is not in fact satisfied (or else we will have to call certain explanations correct which are clearly not so). In order to show this I shall make use of some of the many counterexamples that have been employed against the D-N model. In these examples, the explanans is true and the other D-N conditions are satisfied. Yet the explanandum-event did

---

14. *Ibid.*, p. 46.
15. Woodward does not claim that (10) plus the basic D-N model provides a set of sufficient conditions for correct explanations—a point to which I will return later. But for the moment I want to treat his model as if it is supposed to provide such a set.

not occur because of the explanans-event, but for some other reason; and this can only be known empirically.

Consider this example:

(a) Jones ate a pound of arsenic at time t
Anyone who eats a pound of arsenic dies within 24 hours
Therefore,
Jones died within 24 hours of t.

Assume that the premises of (a) are true. Then it is supposed to be settleable a priori whether these premises correctly explain the conclusion. According to the D-N model all we need to determine is whether the second premise is lawlike (let us assume that it is), and whether the conjunction of premises (but not the first premise alone) entails the conclusion (it does). Since these D-N conditions are satisfied, the explanans should correctly explain the explanandum; and assuming the truth of the explanans, this matter is settleable on a priori grounds alone, no matter what other empirical propositions are true. However, the matter is not settleable a priori, since Jones could have died within 24 hours of t for some unrelated reason. For example, he might have died in a car accident not brought on by his arsenic feast, which, given the information in the explanans, could only be determined empirically. Suppose he did die from being hit by a car within 24 hours of t. Then the explanans in (a) does not correctly explain the explanandum, even though all the conditions of the D-N model are satisfied. Assuming the truth of the premises in (a) it is not settleable a priori whether these premises correctly explain the explanandum.

A similar problem besets all the more specialized versions of the D-N model cited above. Thus consider these D-N arguments:

(b) Disposition example:
That bar is magnetic, and a small piece of iron was placed near it
Any magnetic bar is such that when a small piece of iron is placed near it the iron moves toward the bar
Therefore,
This small piece of iron moved toward the bar.

Suppose, however, that a much more powerful contact force had been exerted on the small piece of iron, and that it moved toward the bar because of this force, not because the bar is magnetic. (Assume that the magnetic force is negligible by comparison with the mechanical force.)

(c) Motivational example:

> Smith desired to buy eggs and he believed that going to the store is the only way to buy eggs
> Whenever, etc. (law relating beliefs and desires to actions)
>   Therefore,
> Smith went to the store.

But suppose Smith went to the store because he wanted to see his girlfriend who works in the store, not because he wanted to buy eggs.

(d) Functional interdependence example:

> This pendulum is a simple pendulum
> The period of this pendulum is 2.03 seconds
> The period T of a simple pendulum is related to the length L by the formula $T = 2\pi\sqrt{L/g}$, where $g = 980$ cm/sec^2
>   Therefore,
> This pendulum has a length of 100 cm.

A pendulum has the period it has because of its length, but not vice versa. (This type of counterexample is like the others in so far as it invokes an explanans-fact that is inoperative with respect to the explanandum-fact; but the case is also different because there is no intervening cause here, although there is in the others.)

With each of the D-N models considered, whether a particular example satisfies the *requirements of the model* (other than the truth-requirement for the explanans) is settleable a priori. Yet in all of these cases, given the truth of the explanans, whether the latter correctly explains the explanandum is not settleable a priori. Thus in example (c), even if Smith desired to buy eggs, and he believed that going to the store is the only way to do so, and the lawlike sentence relating beliefs and desires to actions is true, it does not follow that

> (1) Smith went to the store *because* he desired to buy eggs and believed that going to the store is the only way to buy eggs.

The explanans of (c) correctly explains the explanandum only if (1) is true. Yet given the truth of the explanans of (c) it is not settleable a priori, but only empirically, whether (1) is true.

We could, of course, see to it that the matter is settleable a priori by changing the motivational model so as to incorporate (1) into the explanans of (c). Assuming that this enlarged explanans is true, whether the latter correctly explains the explanandum is a priori—indeed, trivially so. But now, of course, the NES requirement is violated, since (1) is a singular sentence that entails the explanandum of (c).

In example (d), even if the explanans is true and Woodward's condition of functional interdependence is satisfied, it does not follow that

> (2) This pendulum has a length of 100 cm *because* it is a simple pendulum with a period of 2.03 seconds and the law of the simple pendulum holds.[16]

The explanans of (d) correctly explains the explanandum only if (2) is true. Yet assuming the truth of the explanans of (d) the truth-value of (2) is not settleable a priori, but only empirically. Given that there is a lawlike connection between the period and length of a simple pendulum, whether the pendulum has the period it does because of its length, or whether it has the length it does because of its period, or whether neither of these is true, is not knowable a priori.[17]

More generally, in the explanans in each of these models some factors are cited, together with a lawlike sentence relating these factors to the type of event or fact to be explained. But given that the factors were present and that the lawlike sentence is true, there is no a priori guarantee that the event in question occurred because of those factors. Whether it did is an empirical question whose answer even the truth of the lawlike sentence does not completely determine. And if we include in the explanans a sentence to the effect that the event in question did occur because of those factors we violate the NES requirement.

6. D-N REPLIES

It might be replied that we should tighten the conditions on the lawlike sentence in the explanans by requiring not simply that it relate the factors cited in the explanans to the type of event in the explanandum, but that it do so in an explicitly explanatory way. Thus in example

> (a) Jones ate a pound of arsenic at time t
> Anyone who eats a pound of arsenic dies within 24 hours
> Therefore,
> Jones died within 24 hours of t,

---

16. We might be willing to say that this pendulum *must* have a length of 100 cm because it is a simple pendulum with a period of 2.03 seconds, etc. But here the explanandum is different.

17. Woodward recognizes the pendulum example (d) as a genuine counterexample to the D-N model even when supplemented by his functional interdependence condition (p. 55). He believes that a causal condition, which he does not formulate, will need to be added to the D-N model in addition to his condition. This proposal will be examined in Section 11 when Brody's causal model is discussed.

we might require the lawlike sentence not to be simply "Anyone who eats a pound of arsenic dies within 24 hours," but

(1) Anyone who eats a pound of arsenic dies within 24 hours because he has done so.

It is settleable a priori whether an explanans consisting of (1) together with "Jones ate a pound of arsenic at time t" is such that, if true, it would correctly explain the explanandum in (a).

This solution, however, would not be an attractive one for D-N theorists, since (1) is just a generalized explanation-sentence containing an explanatory connective that such theorists are trying to analyze by means of their model. Moreover, tightening the lawlike sentence in this way will produce many false explanations, since such tightened sentences will often be false even though their looser counterparts are true. For example, (1), construed as lawlike, is false since people who eat a pound of arsenic can die from unrelated causes. And if we weaken (1)—still keeping the explanatory clause—by writing

(2) Anyone who eats a pound of arsenic *can* die within 24 hours because he has done so,

we obtain a sentence that is true but not powerful enough for the job. It is not settleable a priori whether an explanans consisting of (2) together with "Jones ate a pound of arsenic at t," if true, would correctly explain why Jones died within t + 24, since he could have died for a different reason even though this explanans is true.

Another possible way of tightening the conditions on the lawlike sentence in the explanans is to require that it relate spatio-temporally contiguous events.[18] (This would mean that the explanans would have to describe an event—or chain of events—that is spatio-temporally contiguous with the explanandum-event.) Jaegwon Kim has discussed laws of this sort, and he provides schemas for them which are roughly equivalent to the following:

(3) $(x)(t)(t')(x$ has P at t and loc(x,t) is spatially contiguous with loc (x,t') and t is temporally contiguous with $t' \rightarrow x$ has Q at $t')$;

(4) $(x)(y)(t)(t')(x$ has P at t and loc(x,t) is spatially contiguous with loc(y,t') and t is temporally contiguous with $t' \rightarrow y$ has Q at $t')$.[19]

By "loc(x,t)" is meant the spatial location of x at time t. Kim does not specify a precise meaning for the arrow in (3) and (4) except that it is

---

18. This suggestion was made to me by Richard Creath.
19. Jaegwon Kim, "Causation, Nomic Subsumption, and the Concept of Event," *Journal of Philosophy* 70 (1973), pp. 217–36; see p. 232. This is a modification of Kim, whose formulations require a slight repair.

to convey the idea of "causal or nomological implication."[20] Under the present proposal, the arsenic explanation (a) would be precluded, since the only law invoked in (a) is not of forms (3) or (4). It does not express a relationship between types of events that are spatio-temporally contiguous. And, indeed, the explanans-event in (a) is not spatio-temporally contiguous with the explanandum-event.

This solution, like the previous one, may succeed in excluding intervening cause counterexamples such as (a). But it would not, I think, be welcomed by D-N theorists. If the arrow in (3) and (4) is to be construed causally as meaning (something like) "causes it to be the case that," then, as with (1), the laws in D-N explanations will be generalized explanation-sentences containing an explanatory connective that D-N theorists seek to define by means of their model. Furthermore, requiring laws of forms (3) or (4) in an explanans will disallow explanations that D-N theorists, and many others, find perfectly acceptable. For example, it will not permit an explanation of a particle's acceleration due to the gravitational force of another body acting over a spatial distance. It will not permit explaining why a certain amount of a chemical compound was formed by appeal simply to (macro-)laws governing chemical reactions—where the formation of that amount of the compound takes time and is not temporally contiguous with the mixing of the reactants. Nor will the present proposal suffice to preclude all the counterexamples in Section 5. In particular, the pendulum example (d)—in which the pendulum's length is explained by reference to its period—is not disallowed. Assuming that the arrow in (3) and (4) represents nomological but not causal implication, we can express the following "law":

> $(x)(t)(t')(x$ is a simple pendulum with a period T at time t, and loc(x,t) is spatially contiguous with loc(x,t'), and t is temporally contiguous with $t' \rightarrow x$ has a length L at t' which is related to T by the formula $T = 2\pi \sqrt{L/g}$).

This, being of form (3), can be used in the explanans in (d), which, when suitably modified, will permit an explanation of the pendulum's length by reference to its period. For these reasons the present proposal does not seem promising.

Our observations regarding the various D-N models can be generalized. Assume that the explanans satisfies the NES requirement. In the explanans we can describe an event of a type always associated with an event of the sort described in the explanandum. We can include a law saying that such events are invariably and necessarily related. The truth of the explanans event-description and of the law is

---

20. *Ibid.*, p. 229, ftn. 19.

no guarantee that the explanandum-event occurred because of the explanans-event. It could have occurred because some event unrelated to the one in the explanans was operative whereas the explanans-event was not. And this cannot be known a priori from the explanans. We can make it a priori by including in the explanans an appropriate singular sentence that entails the explanandum (e.g., an explanation-sentence that says in effect that the explanandum is true because the explanans is, or that the explanans-event caused the explanandum-event). But then the NES requirement would be violated. Or we can make it a priori by using a generalized explanation-sentence. But since this is contrary to the philosophical spirit of such models and will, in any case, tend to produce false explanations, it will not be considered a viable solution. We can also make it a priori by requiring laws of forms (3) or (4) and construing the arrow causally. But this too does not satisfy the intent of such models, and, in addition, will not permit wanted explanations. On the other hand, if the arrow is understood nomologically but not causally, then whether the explanans, if true, correctly explains the explanandum is, in general, not knowable a priori.

To avoid these problems we can say that it is an empirical, not an a priori, question whether a true explanans describing events and containing laws relating these types of events to the explanandum-event correctly explains the latter. Or we can include in the explanans some singular sentence—either an explanation-sentence or something like it (e.g., (4) in Section 2)—that entails the explanandum. In the first case the a priori requirement is violated, in the second, NES. For this reason, I suggest, D-N models which attempt to provide sufficient conditions for correct explanations in such a way as to satisfy both of these requirements will not be successful.

7. SALMON'S STATISTICAL-RELEVANCE (S-R) MODEL

The models I shall mention in this and the next two sections satisfy NES but overtly violate the a priori requirement. Their proponents seem to recognize that if the former requirement is to be satisfied it is not an a priori but an empirical question whether the explanans if true would correctly explain the explanandum. However, I shall argue, the empirical considerations they introduce are not of the right sort to avoid the problem discussed above.

As we recall from Chapter 1, Salmon's S-R model is concerned with explanations of particular events. Salmon construes such explanations as answering a question of the form

(1) Why is X, which is a member of class A, a member of class B?

Although Salmon does not do so, I shall say that the explanandum in such a case has the form

(2) X, which is a member of class A, is a member of class B,

which is presupposed by (1). The explanans consists of a set of empirical probability laws relating classes A and B, together with a class inclusion sentence for X, as follows:

(3) $p(B, A \& C_1) = p_1$
$p(B, A \& C_2) = p_2$
.
.
.
$p(B, A \& C_n) = p_n$
$X \epsilon C_k$ $(1 \leq k \leq n)$.

Two conditions are imposed on the explanans. One is that the probability values $p_1, \ldots, p_n$ all be different. The other is

(4) *The homogeneity condition:* $A \& C_1, A \& C_2, \ldots, A \& C_n$ is a partition of A, and each $A \& C_i$ is homogeneous with respect to B.

The definitions of "partition" and "homogeneity" are given in Chapter 1, and I will not repeat them here. If A is homogeneous with respect to B then A is a "random" class with respect to B.

I shall assume that for Salmon if the explanans (3) and the explanandum (2) are true, then the explanans correctly explains the explanandum, provided that Salmon's two conditions are satisfied.[21] To proceed, it will be useful to consider a simple example different from that in Chapter 1. There is a wire connected in a circuit to a live battery and a working bulb, and we want to explain why the bulb is lit, or more precisely, what, if anything, the wire does which contributes to the lighting of the bulb. Putting this in Salmon's form (1), the explanatory question becomes "Why is this wire, which is a member of the class of things connected in a circuit to a live battery and a working bulb, a member of the class of circuits containing a bulb that is lit?" The explanandum is

(5) This wire, which is a member of the class of things connected in a circuit to a live battery and a working bulb, is a member of the class of circuits containing a bulb that is lit.

Letting

---

21. In *Statistical Explanation and Statistical Relevance*, Salmon does not explicitly say this, but this seems the most reasonable interpretation of his position. See, e.g., his remarks on pp. 79–80 in which he is distinguishing homogeneity from epistemic homogeneity, and in which he compares his model with Hempel's. In private conversation he assures me that this is the correct interpretation.

A = the class of things connected in a circuit to a live battery and a working bulb
B = the class of circuits containing a bulb that is lit
C_1 = the class of things that conduct electricity
C_2 = the class of things that don't conduct electricity
X = this wire,

we can construct the following explanans for (5):

(6) $p(B, A \& C_1) = 1$
$p(B, A \& C_2) = 0$
$X \in A \& C_1$.

Salmon's two conditions are satisfied since the probability values are different, and since $A \& C_1$ and $A \& C_2$ is a partition of A, and each $A \& C_i$ is homogeneous with respect to B. Roughly, (6) explains why the bulb is lit by pointing out that the probability that the bulb in the circuit will be lit, given that the wire conducts electricity, is 1, that the probability that it will be lit, given that the wire does not conduct electricity, is 0, and that the wire does in fact conduct electricity.

Salmon's model satisfies the NES requirement since the only singular sentence in the explanans will not entail the explanandum. (Otherwise at least one of the probability laws in the explanans would be a priori, not empirical.) However, the homogeneity condition prevents his model from satisfying the a priori requirement.[22] Whether $A \& C_1, A \& C_2, \ldots, A \& C_n$ is a partition of A, i.e., whether these classes have any members in common and every member of A belongs to one of them, is not in general an a priori question (though it happens to be in the above example). Nor is the question of whether each $A \& C_i$ is homogeneous with respect to B. For example, it cannot be decided a priori whether there is some subclass of the class of electrical conductors such that the probability of the bulb being lit is different in this subclass from what it is in the class as a whole; this is an empirical issue. Accordingly, whether the explanans (6), if true, would correctly explain the explanandum (5) is not settleable a priori.

Does the inclusion of the empirical homogeneity condition avoid the kind of problem earlier discussed plaguing the D-N models? Unfortunately not. This can be seen if we change our circuit example a bit. Let A and B be the same classes as before. We now introduce

C_3 = the class of things that conduct heat
C_4 = the class of things that do not conduct heat.

---

22. It is possible to construe Salmon's model as requiring the satisfaction of the homogeneity condition to be stated in the explanans itself. If so the model would purport to satisfy the a priori requirement. However, in what follows, I shall continue to assume that only sentences of the type in (3) comprise the explanans, and thus that the model is an empirical one. On either construction the same difficulty will emerge.

I shall make the simplifying assumption that it is a law that something conducts heat if and only if it conducts electricity. Now consider this explanans for the explanandum (5):

(7) $p(B, A \& C_3) = 1$
$p(B, A \& C_4) = 0$
$X \in A \& C_3$.

Although Salmon's two conditions are satisfied, (7) ought not to be regarded as a correct explanation of (5), even if (7) is true. Intuitively, if we took (7) to correctly explain (5) we would be saying that the bulb is lit because the wire conducts heat (where the probability that it is lit, given that the wire does (not) conduct heat, is 1 (0)). But this seems incorrect. The bulb is lit because the wire conducts electricity not heat, though to be sure it does both, and that it does one if and only if it does the other is a law of nature. Admittedly, by our assumption, the class $A \& C_3$ = the class $A \& C_1$. But it is not a class which explains for Salmon, but a sentence indicating that an item is a member of a class. If the class is described in one way the explanation may be correct, while not if described in another way. In sentences of the form

(8) X's being an $A \& C_i$ (a member of the class $A \& C_i$)—together with such and such probability laws—correctly explains why X, which is a member of A, is a member of B,

the "$A \& C$" position is referentially opaque. A sentence obtained from (8) by substituting an expression referring to the same class as "the class $A \& C_i$" will not always have the same truth-value.

The kind of example here used against Salmon's model[23] is similar to those raised earlier in the following respect. In the explanans a certain fact about the wire is cited, viz. that it conducts heat, which (under the conditions of the set-up) is nomologically associated, albeit indirectly, with the fact to be explained, viz. the bulb's being lit. However, it is not the explanans-fact that is the operative one in this case but the fact that the wire conducts electricity. By invoking the homogeneity condition Salmon in effect recognizes that the question of the explanatory operativeness of the explanans is not an a priori matter. The problem is that his homogeneity condition is not sufficient to guarantee that the explanans-fact is operative with respect to the explanandum-fact.

8. BRODY'S ESSENTIAL PROPERTY MODEL

Brody construes this model as providing a set of sufficient conditions for explanations of particular events. As noted in Chapter 1, these

---

23. A variety of such examples, as well as other trenchant criticism, can be found in John B. Meixner, "Homogeneity and Explanatory Depth," *Philosophy of Science* 46 (1979), pp. 366–81.

conditions are those of the basic D-N model together with the following

Essential property condition: "The explanans contains essentially a statement attributing to a certain class of objects a property had essentially by that class of objects (even if the statement does not say that they have it essentially) and . . . at least one object involved in the event described in the explanandum is a member of that class of objects."[24]

For example, since Brody thinks that atomic numbers are essential properties of the elements, he would regard the following explanation as correct, provided its premises are true:

(1) This substance is copper
Copper has the atomic number 29
Anything with the atomic number 29 conducts electricity
Therefore,
This substance conducts electricity.

(1) satisfies the D-N conditions plus Brody's essential property condition. Brody proposes the latter condition in order to preclude certain counterexamples to the basic D-N model. Moreover, he regards the satisfaction of this condition as an empirical matter.[25] Brody's model, in cases in which the explanans does not say explicitly that the property in question is essential, does not satisfy the a priori condition.[26] To know whether (1) is a correct explanation if its premises are true is not an a priori matter, since we must know whether having the atomic number 29 is an essential property of copper; and this knowledge is empirical, according to Brody. On the other hand, the NES requirement is satisfied since the singular premises in an explanans will not entail the explanandum.

One might question Brody's notion of an essential property. However, the problem I want to raise is not this, and so I shall suppose the model is reasonably clear; indeed, I shall stick to atomic number, which is the sort of property Brody claims to be essential to the element which has it.

Consider the following argument:

(2) Jones ate a pound of the substance in that jar
The substance in that jar is arsenic
Arsenic has the atomic number 33

---

24. Brody, "Towards an Aristotelean Theory of Scientific Explanation," p. 26.
25. See *ibid.*, p. 27.
26. If the explanans explicitly says that the property is essential, then the model purports to satisfy the a priori condition. But I want here to consider a model that explicitly violates this condition. (In either case the model turns out to be unsatisfactory.)

Anyone who eats a pound of substance whose atomic number is 33 dies within 24 hours
Therefore,
Jones died within 24 hours of eating a pound of the substance in that jar.

The explanans contains essentially a statement attributing to a class of objects (those which are arsenic) a property that Brody takes to be essential to that class (having the atomic number 33). And there is an object involved in the event described in the explanandum (the substance in that jar) which is a member of that class of objects. Suppose, however, that Jones died within 24 hours of eating a pound of the substance in that jar not because he ate the substance but because of an unrelated car accident that occurred within 24 hours of his feast. Although Brody's essential property condition is satisfied, as are the conditions of the D-N model, the explanans in (2) does not, even though true, correctly explain the explanandum.

Brody may in effect recognize that it is not an a priori but an empirical question whether a D-N explanans if true correctly explains its explanandum. Nevertheless, the empirical requirement which his model invokes—the essential property condition—is not of the right sort to avoid the kind of problem plaguing this and previous models. Like Salmon's homogeneity condition, Brody's essential property condition is not sufficient to guarantee that the explanans-fact is explanatorily operative with respect to the explanandum-fact.

9. JOBE'S PRIORITY MODEL

Evan Jobe proposes adding a condition to the D-N model that is different from any of the previous modifications.[27] To formulate this condition Jobe introduces some new concepts. A sentence p is said to *theoretically imply* a sentence q if p, possibly together with definitions from some relevant scientific theory and mathematical apparatus, implies q. A D-N explanation E is said to *involve* a sentence S if and only if "the conjunction of the sentences comprising the initial conditions of E [the singular sentences in the explanans] theoretically implies S." Finally, a sentence p is said to be *explanatorily dependent* on sentence q if and only if there are D-N explanations of q that do not involve p; and every D-N explanation of p involves q. The condition which Jobe suggests adding to those of the basic D-N model is

The priority condition: A D-N explanation of a sentence q must not involve a sentence p such that p is explanatorily dependent on q.[28]

27. Evan K. Jobe, "A Puzzle Concerning D-N Explanation," *Philosophy of Science* 43 (1976), pp. 542–49.
28. Jobe goes on to generalize this by introducing a concept of dependence at different levels; but what I shall say will not be affected by this generalization.

We might put this by saying that the explanans must be explanatorily prior to the explanandum.

To see how this condition is supposed to work, consider the previous pendulum counterexample:

(1) (a) This pendulum is a simple pendulum
(b) The period of this pendulum is 2.03 seconds
(c) The period T of a simple pendulum is related to the length L by the formula $T = 2\pi\sqrt{L/g}$, where $g = 980$ cm/sec^2
Therefore,
(d) This pendulum has a length of 100 cm.

According to Jobe, the priority condition is violated. The explanation (1) involves sentence (b) that is explanatorily dependent on the explanandum (d). There are D-N explanations of why this pendulum has a length of 100 cm that do not involve the assumption that this pendulum has a period of 2.03 seconds. But, says Jobe, every D-N explanation of why this pendulum has a perod of 2.03 seconds involves the assumption that this pendulum has a length of 100 cm. "Roughly speaking," writes Jobe

> we know that we can deductively explain why a pendulum has a certain length without utilizing in any way the fact that the period is such and such. On the other hand, it would appear that any D-N explanation of the fact that a pendulum has a particular period must in some sense involve the fact that it has such and such a length. From an explanatory viewpoint, then, the period of a pendulum is dependent on its length in a way in which its length is not dependent on its period. We might say that the period is "explanatorily dependent on" the length.[29]

Although he does not say so explicitly, Jobe leaves the impression that the basic D-N model supplemented by his priority condition provides a set of sufficient conditions for correct explanations.[30] Whatever his intentions, in what follows I shall consider the model in this manner.

The satisfaction of Jobe's priority condition is an empirical matter. Whether (1) is a correct explanation if its premises are true is not settleable a priori, since we must know whether (b) is explanatorily dependent on (d). The latter depends on whether there is some (correct) D-N explanation of (d) that does not involve (b) and whether every D-N explanation of (b) involves (d). These are clearly issues that involve empirical considerations. Jobe's priority model thus violates the a priori requirement. However, NES is satisfied: the singular premises in an explanans will not entail the explanandum.

---

29. *Ibid.*, p. 543.
30. See pp. 547–48, where he seems to be claiming that he is unable to find counterexamples.

The model, nevertheless, does not avoid previous counterexamples. Consider once again

(2) (a) Jones ate a pound of arsenic at time t
    (b) Anyone who eats a pound of arsenic dies within 24 hours
        Therefore,
    (c) Jones died within 24 hours of t.

Assume that although (a) through (c) are true, Jones died in an unrelated car accident, so that (2) is an incorrect explanation. Is the priority condition satisfied? Yes, it is. Sentence (a) is not explanatorily dependent on sentence (c). There is, we may suppose, a correct D-N explanation of (c) that does not involve (a)—a D-N explanation citing the particulars of the car crash. But it is not true that every D-N explanation of (a) involves (c). Jones did not eat a pound of arsenic at t because of his death within 24 hours of t. Jobe may recognize that it is an empirical, not an a priori, question whether a D-N explanans, if true, correctly explains its explanandum. But the empirical requirement introduced by his model—the priority condition—does not suffice to avoid counterexamples.

Indeed, contrary to Jobe's claim, it is not even clear that counterexamples like (1) are eliminated. Jobe's thesis is that (1) violates the priority condition since there are D-N explanations of (d) in (1) that do not involve (b); but every D-N explanation of (b) involves (d). What D-N explanation of the former sort does Jobe have in mind? Perhaps it is one that, satisfying the D-N motivational model, appeals to the beliefs and desires of the pendulum builder. Such an explanation might take the form

(3) X desired G
    X believed that building this pendulum with a length of 100 cm is, in the circumstances, the best way to obtain G
    (law relating beliefs and desires to actions)
        Therefore,
    X built this pendulum with a length of 100 cm (which entails that this pendulum has a length of 100 cm).

Explanation (3) does not involve (b) in (1). But a D-N explanation for (b) in (1) can be produced in exactly the same fashion. Simply change the expression "with a length of 100 cm" to "with a period of 2.03 seconds" in the second premise and in the conclusion of (3). We will then obtain a D-N explanation of (b) in (1) that does not involve (d). That is, we will have a D-N explanation of the period of the pendulum that does not involve its length. If so, the priority condition is satisfied and (1) will have to be counted as a correct explanation.

Jobe, like Salmon and Brody, seems to recognize that if the NES requirement is to be satisfied it is an empirical, not an a priori, question whether an explanans if true correctly explains an explanandum. Yet none of the empirical considerations any of these modelists deploy is sufficient to insure that if satisfied an explanation will be correct if its explanans is true. Can problems of the sort generated by these models be avoided in any way other than by abandoning NES?

10. VAN FRAASSEN'S QUESTION MODEL

In the present section I turn to a very recent model whose intent, although not as clear as that of the previous three models with respect to the a priori and NES requirements, might be construed as abandoning the former while retaining the latter.

According to van Fraassen, "an explanation is an answer to a why-question. So a theory of explanation must be a theory of why-questions."[31] Van Fraassen formulates his theory using some technical notions which can best be introduced by means of an example. Consider the question

(1) Why did Jones die?

The proposition "Jones died" is the *topic* of this question. The question has a *contrast class*, which is a set of propositions containing the topic and alternatives. For example, for question (1) the contrast class could be: "Jones died," "Jones lived." (In such a case the questioner wants to know why Jones *died*, rather than lived. With another contrast class the questioner might want to know why *Jones*, rather than someone else, died.) Finally, writes van Fraassen, "there is the respect-in-which a reason is requested, which determines what shall count as a possible explanatory factor, the relation of *explanatory relevance*. In the . . . example [above], the request might be *for events 'leading up to'* [*the death*]."[32] Accordingly, he speaks of the (explanatory) *relevance relation*. A (why-)question Q, on his theory, is determined by three factors: the topic P, the contrast class X, and the relevance relation R. For example, a question expressed by (1) can be given by the topic "Jones died," the contrast class "Jones died, Jones lived," and the relevance relation "events leading up to."

Suppose now that in the light of the background information K a question Q arises given by (P,X,R). Suppose further that an answer is given of the form "P because A," or more briefly, "because A," where A is a sentence expressing a proposition. How good is that answer? Van Fraassen indicates three things that must be determined:[33]

31. Bas van Fraassen, *The Scientific Image* (Oxford, 1980), p. 134.
32. *Ibid.*, p. 142.
33. *Ibid.*, pp. 146–47.

1. We must determine whether proposition A is "acceptable" or "likely to be true."
2. We must determine the extent to which A favors the topic P as against other members of the contrast class. (Does A shift the probability toward P more than toward other members of the contrast class?)
3. We must compare "because A" with other possible answers to Q, in three respects:
    a. Is A more probable than other answers, given the background information K?
    b. Does A favor the topic P to a greater extent than other answers do?
    c. Is A made wholly or partially irrelevant by other answers? More specifically, does some other answer "screen off" A from P? (That is, is there an answer A' such that $p(P, A' \& A) = p(P, A')$?)

Let us illustrate these considerations by reference to our example. The question is (1) above, given by the topic P ("Jones died"), the contrast class ("Jones died," "Jones lived"), and the relevance relation "events leading up to." The background information K includes the proposition "Anyone who eats a pound of arsenic dies." Consider the following answer to (1):

(2) because *Jones ate a pound of arsenic,*

in which the emphasized proposition is A. How good is the answer "because A"? Let us suppose that it is known that Jones did eat a pound of arsenic, so that A is "acceptable." Van Fraassen's first condition is satisfied. Secondly, p(Jones died, Jones ate a pound of arsenic & K) = 1; whereas p (Jones lived, Jones ate a pound of arsenic & K) = 0. (Suppose that p(Jones died, K) = .2.) Proposition A thus shifts the probability toward the topic P more than toward the other member of the contrast class, "Jones lived." Therefore, van Fraassen's second condition is satisfied. To determine whether the third condition is satisfied, let us compare the answer (2) above with the following answer:

(3) because *Jones was crushed by a car,*

in which the emphasized proposition is A'. Let us suppose that given our background information K,

$p(A, K) = .99$
$p(A', K) = .90.$

Since proposition A ("Jones ate a pound of arsenic") is more probable than proposition A' ("Jones was crushed by a car"), van Fraassen's condition 3a is satisfied with respect to these two answers. Let us also suppose that our background information K includes the proposition that 95 percent of those crushed by a car die. Then proposition A favors the topic P to a greater extent than does proposition A'. (p(Jones died, Jones was crushed by a car & K) = .95; whereas p(Jones died, Jones ate a pound of arsenic & K) = 1.) So condition 3b is satisfied with respect to these two answers. Finally, proposition A' does not screen off A from P. (p(P, A' & A) = 1; p(P, A') = .95.) Accordingly, condition 3c is also satisfied with respect to the answers given by (2) and (3). If other answers are like A' with respect to A (i.e., if A satisfies the third condition better than other answers), then we may conclude that answer (2) satisfies van Fraassen's conditions for being a good answer to question (1), and hence is a good explanation of why Jones died.

Is the satisfaction of these conditions supposed to guarantee that an explanation is correct? For the moment let us suppose they are so construed. (Later I will consider alternatives.) We may take an explanation of q, where Q is a why-question, to be an answer of the form

p because a.

And, following the procedure of this chapter, I shall call p the explanandum and a the explanans. On the present construction, if p and a satisfy van Fraassen's conditions, then "p because a" is true, and the explanation is correct. It seems reasonable to suppose that van Fraassen would reject the a priori requirement. Whether his conditions (other than the truth requirement for a) are satisfied is not an a priori matter. One must determine the truth of certain probability statements, which van Fraassen seems to be construing as empirical. (See his chapter 6.) And one must determine whether a satisfies certain conditions better than other answers—which, again, seems to be a matter not settleable a priori. Is the NES requirement to be satisfied? Van Fraassen is not explicitly committed to this, but the types of cases he mentions suggest that he might well accept NES (understood now to mean that no singular sentence or conjunction of such sentences in explanans a is to entail the explanandum p). He mentions a type of case in which the explanans a *plus the background information K* (but not the explanans alone) entails the explanandum. And he mentions the type of case in which the explanans plus K does not entail the explanandum but gives it a certain probability.

Van Fraassen's question model, if construed in the present manner, would be subject to the previous counterexamples. This can be seen by reference to the arsenic case above. Answer (2), we may assume, satisfies all of van Fraassen's conditions, even though the explanation it provides is incorrect, while that given by (3) is correct. Jones died because he was crushed by a car, not because he ate arsenic, although both events did occur—both events satisfy the relation "events leading up to" (Jones's death)—and although (2), but not (3), satisfies all of van Fraassen's probability conditions. These conditions do not suffice to guarantee the truth of sentences of the form "p because a."

Nor is this problem avoided by construing van Fraassen to be providing a set of sufficient conditions not for a "correct explanation" but for an "explanation that is *more likely* to be correct than any other," or for an "explanation that is more likely to be correct than any other with which we actually compare it."[34] In the example above, just because A ("Jones ate a pound of arsenic") is (slightly) more likely than A' ("Jones was crushed by a car"), it does not follow that A is more likely to be a correct answer to Q than A'; it does not follow that it is more likely that Jones died because he ate arsenic than because he was crushed by a car. (Remember that A and A' are compatible; and given the probabilities, both are likely to be true.)

Similarly, the fact that A shifts the probability slightly more toward P ("Jones died") than does A' does not make A more likely than A' to be a correct answer to Q. That is, just because the probability of P is raised more by A than by A', why suppose that the probability of "P because A" is higher than the probability of "P because A'"? The probability of P is raised more by P itself than by many other things, e.g., than by A'. But this does not raise the probability of "P because P" over "P because A'."

Even if we exclude cases of the latter kind, the general principle that if a proposition P_1 raises the probability of a hypothesis h more than does some other proposition P_2, then p(h because P_1) > p(h because P_2) cannot be accepted. Let h be the hypothesis that Smith is a man, let P_1 be the proposition that Smith swims at Bethany Beach, and let P_2 be the proposition that Smith swims at Rehobeth Beach. The background information K includes the fact that 60 percent of the swimmers at Bethany Beach are men, while 45 percent of the swimmers at Rehobeth Beach are men. Given the background information, the proposition P_1 raises the probability of h more than does P_2. But (where "because" is being used in an explanatory sense) we cannot conclude that

(4) p(Smith is a man because he swims at Bethany Beach) > p(Smith is a man because he swims at Rehobeth Beach).

Given background information which tells us that the reason that a person is a man has nothing to do with where that person swims, both probabilities in (4) should be 0. Moreover, (4) remains false even if we

34. In conversation, van Fraassen indicates that he favors the latter interpretation.

also suppose that p(Smith swims at Bethany Beach)>p(Smith swims at Rehobeth Beach).

To conclude, then, if van Fraassen's model is construed as providing sufficient conditions for an explanation that is correct, or for one that is more likely to be correct than any other, or for one that is more likely to be correct than any other with which it has been compared, then his conditions are subject to the same difficulties plaguing earlier models.[35]

11. TWO CAUSAL MODELS

The final two models I shall discuss seem to offer a solution. They satisfy NES, violate the a priori requirement, and yet avoid the previous problems. However, whether they provide accounts that would be welcome to most of those seeking models of explanation is dubious.

a. Brody's causal model. As in the case of his essential property model, Brody regards this as providing a set of sufficient conditions for explanations of particular events. These conditions are those of the basic D-N model together with the

Causal condition: The explanans contains "a description of the event which is the cause of the event described in the explanandum."[36]

To see how this is supposed to work let us reconsider

(1) Jones ate a pound of arsenic at time t
    Anyone who eats a pound of arsenic dies within 24 hours
        Therefore,
    Jones died within 24 hours of t.

The problem we noted with the basic D-N model occurs if both premises of (1) are true but Jones died within 24 hours of t for some unrelated reason. Brody's causal condition saves the day since in such a case the event described in the first premise of the explanans was not the cause of the event described in the explanandum. Hence, on this model we cannot conclude that the explanans if true correctly explains the explanandum.

The present model, like those in Sections 7 through 9, violates the a priori requirement. Whether the explanans if true correctly explains the explanandum is not an a priori question, since the causal condi-

---

35. As van Fraassen himself recognizes, his written account is rather guarded, and certain questions of interpretation are left unanswered. After presenting his conditions for the evaluation of explanations, he writes: "The account I am able to offer is neither as complete nor as precise as one might wish."
36. Brody, "Towards an Aristotelean Theory of Scientific Explanation," p. 23.

tion must be satisfied; and whether the explanans-event caused the explanandum-event is, in general, an empirical matter. It is not completely clear whether Brody wants to exclude from the explanans itself singular causal sentences that entail the explanandum, but I shall consider that version of his model which makes this exclusion. Like the models in Sections 7 through 9, this model, I shall assume, is to satisfy the NES requirement.

Woodward (see Section 4) is another modelist who proposes the need for a causal condition to supplement the basic D-N model and his own functional interdependence condition:

> These examples suggest that a fully acceptable model of scientific explanation will need to embody some characteristically causal notions (e.g., some notion of causal priority), or some more generalized analogue of these (e.g., some notion of explanatory priority).[37]

However, unlike Brody, he leaves open the question of how such a causal condition should be formulated.[38]

b. The causal-motivational model.[39] Here, as in the D-N motivational model (Section 4), the explanandum is a sentence saying that some agent acted in a certain way. The explanans contains a sentence attributing a desire to that agent, and a sentence attributing the belief to that agent that performing the act described in the explanandum is in the circumstances a (the best, the only) way to satisfy that desire. Thus the explanandum might be a sentence of the form

> Agent X performed act A,

and the explanans might contain sentences of the form

> X desired G;
>
> X believed that doing A is in the circumstances a (the best, the only) way to obtain G.

Unlike the D-N motivational model, however, a law in the explanans relating beliefs and desires to actions is not required. What is required is the satisfaction of a

> *Causal condition:* X's desire and his belief (described in the explanans) caused X to perform act A.

37. Woodward, "Scientific Explanation," p. 53.
38. In his most recent writings Salmon too advocates supplementing his S-R model with a causal condition. See his "Why Ask, 'Why'?" *Proceedings and Addresses of the American Philosophical Association* 51 (1978), pp. 683–705.
39. See C. J. Ducasse, "Explanation, Mechanism, and Teleology," *Journal of Philosophy* 22 (1925), pp. 150–55; Donald Davidson, "Actions, Reasons, and Causes," *Journal of Philosophy* 60 (1963), pp. 685–700; Alvin I. Goldman, *A Theory of Human Action* (Englewood Cliffs, N.J., 1970), p. 78.

The counterexample cited earlier against the D-N motivational model is now avoided, since in that example it was not the agent's belief and desire mentioned in the explanans, but some other belief and desire, that caused him to act. As in the case of Brody's causal model, the a priori requirement is not satisfied, but NES is.

I shall not here try to defend or criticize these two models.[40] I shall assume for the sake of the argument that each avoids, or can be modified so as to avoid, the kind of problem I have been concerned with. However, each does so by violating NES in spirit, whereas the earlier models satisfy this requirement both in spirit and in letter. In order to apply the present causal models one must determine the truth of sentences such as these:

(2) Jones's eating a pound of arsenic at time t caused him to die within 24 hours of t;

(3) Smith's desire to buy eggs and his belief that going to the store is the only way to do so caused him to go to the store.

But these are singular explanation-sentences that entail the explanandum. To be sure, neither model requires such sentences to be in the explanans. Still in each model to determine whether the explanans if true correctly explains the explanandum one has to determine the truth of such sentences. I am not criticizing the models on these grounds. But I believe that many of those who seek models of explanation will want to do so. They will say that in order to know whether the explanans in such a model correctly explains the explanandum one has to determine, independently of the truth of the explanans, the truth of sentences of a sort these modelists want to exclude from the explanans itself. Moreover, they will point out that there is not much difference between determining the truth of (2) and (3), on the one hand, and that of

Jones's eating a pound of arsenic at t explains why he died within 24 hours of t (or Jones died within 24 hours of t because he ate a pound of arsenic at t); and

Smith's desire to buy eggs and his belief that going to the store is the only way to do so explains why he went to the store (or Smith went to the store because he had this desire and belief)

on the other. Models which impose the above causal conditions, they are likely to say, provide insufficient philosophical illumination for the

---

40. A cogent attack on Brody's model can be found in Timothy McCarthy, "On an Aristotelean Model of Scientific Explanation," *Philosophy of Science* 44 (1977), pp. 159–66. See Davidson, "Psychology as Philosophy," in J. Glover, ed., *The Philosophy of Mind* (Oxford, 1976), pp. 103–4, for criticism of the causal-motivational model (which Davidson himself once supported).

concept of explanation, even though the explanation-sentence expressing the causal relationship is not itself a part of the explanans.[41] If one of the aims of modelists is to define terms like "explains," "because," and "causes," this excludes their employment as primitive notions in the explanans, or, as in the present case, in the conditions of the model.

Regardless of whether we view such criticism as important, the present models are of interest because to avoid the sorts of problems raised in Section 5 these models, unlike their D-N ancestors, require establishing the truth of empirical sentences to determine whether the explanans if true correctly explains the explanandum. In this respect they are like Salmon's S-R model, Brody's essential property model, and Jobe's priority model. However, unlike the latter, the empirical sentences whose truth they require establishing are themselves singular sentences that entail the explanandum.

12. MODELS OF EXPLANATION AND THE ILLOCUTIONARY THEORY

It is time to draw conclusions and relate these to the illocutionary theory developed in earlier chapters. Our discussion suggests the following two points: (i) If the explanans is not to contain singular sentences that entail the explanandum, then it will be an empirical not an a priori question whether the explanans if true correctly explains the explanandum. (ii) This empirical question will involve determining the truth-values of certain singular sentences (either explanation-sentences or something akin to them) that do entail the explanandum; otherwise factors will be citable in the explanans which are not explanatorily operative with respect to the explanandum-event.

Can there be a model of explanation? Specifically, can there be a set of sufficient conditions which are such that if they are satisfied by the explanans and explanandum the former correctly explains the latter? Conditions of this sort can be formulated if we are willing to rest content with the following criterion expressed using the ordered pair view of Chapter 3:

(1) If (p; explaining q) is an explanation, then it is correct if and only if p is true.

However, modelists are unwilling to stop here, since they want explanations not only to be correct, but correct in the right sort of way

---

41. Indeed, some formulations of the causal-motivational model (e.g., Ducasse's and Goldman's) seem to allow as an explicit part of the explanans a singular explanation-sentence that entails the explanandum. In such cases the NES requirement is violated in letter as well as in spirit.

("minimal scientific correctness"). They want further requirements to be satisfied. The explanans p in (1) can be a proposition expressible by a sentence of the form

(2) The reason that e occurred is that f occurred,

where the explanandum is

(3) e occurred.

In this case, the a priori requirement is satisfied, since it is an a priori question whether the explanans (2), if true, correctly explains the explanandum (3). But NES is violated, since (2) is a singular sentence that entails the explanandum (3). It is an explanation-sentence containing the term "reason"—a term of the kind modelists want to eliminate from an explanans.

Suppose, then, that the following explanans, which contains no such explanation-terms, is substituted for (2):

(4) f occurred.

It will then be an empirical, not an a priori, question whether the explanans, if true, correctly explains the explanandum. But if the conclusions suggested previously are right, the catch is that this empirical question will involve determining the truth-value of an explanation-sentence like (2). It cannot be determined simply by appeal to the existence of a law relating e and f (even one involving spatio-temporal contiguity), or by appeals to homogeneity, essential properties, or explanatory dependence. If we want a set of conditions whose satisfaction guarantees that an explanans will correctly explain the explanandum, then we are saddled with explanation-sentences (or more generally, with content-giving sentences) such as (2).

Modelists exclude sentences of form (2) from an explanans, since such sentences violate NES. But this, I urge, is not a good enough reason for doing so. First, considerations in the present chapter suggest that if we want a set of conditions which, if satisfied, guarantees that an explanans will correctly explain the explanandum, then we are not going to be able to satisfy NES. Second, in Chapter 3 we saw the advantages in taking the constituent proposition of an explanation of q to be a complete content-giving proposition with respect to Q: doing so will avoid the emphasis problem. But on this strategy, where Q is a question of the form "Why is it the case that x?," the constituent proposition in an explanation will be expressible by a sentence of the form

The reason that x is that _____,

which violates NES.

Third, we can retain much of the essence of the modelist's view within the ordered pair theory, *even if NES is violated.* For example, as was shown in Chapter 3, Hempel's D–N theory can be reformulated as follows, within the ordered pair view, in such a way that laws and a deductive relationship are still required:

> (5) If Q is a content-question expressible by a sentence of the form "Why is it the case that x?" and (p; explaining q) is an explanation of q in which proposition p is expressible by a sentence of the form "The reason that x is that y," then (p; explaining q) is a good explanation of q ("minimally scientifically correct") if and only if
> (i) y is a true sentence;
> (ii) y is a conjunction at least one of whose conjuncts is a law;
> (iii) y entails x.

The following explanation satisfies this reformulation:

> (6) (The reason that this metal expanded is that this metal was heated and all metals expand when heated; explaining why this metal expanded).

Taking the explanandum in (6) to be "This metal expanded" and the explanans to be "The reason that this metal expanded is that, etc.," the NES requirement is violated. However, (6) does satisfy the D–N requirement for laws, and there is a deductive relationship between "This metal was heated, and all metals expand when heated" and "This metal expanded."

Why, then, not accept the reformulated D–N view (5) as an adequate account of "minimal scientific correctness" even though NES is violated? For one thing, it is too restrictive, since it confines explanatory questions to ones of the form "Why is it the case that x?" Content-questions with forms other than this are certainly answered in scientific explanations. For another thing, unlike (1), (5) does not provide a sufficient condition for correctness. The following explanation satisfies (5):

> (The reason that Jones died within 24 hours of t is that Jones ate a pound of arsenic at t, and anyone who does so dies within 24 hours; explaining why Jones died within 24 hours of t).

But if Jones died from unrelated causes, then the explanation is incorrect. (Whatever else "minimal scientific correctness" entails it is supposed to entail correctness.)

More importantly, there is a certain arbitrariness about this view. If we want to supply conditions simply for correctness in an explanation, then (1) above (unlike (5)) suffices. The D–N insistence on laws and a deductive relationship is unnecessary. If we want an *illocutionary* evaluation, then, as we saw in Chapter 4, D-N requirements will sometimes be too restrictive, sometimes too lenient, depending on information about the world, the explainer, and the audience. As indicated in Chapter 4, it is also possible to give various non-illocutionary evaluations of an explanation based on one or more of the general methodological criteria valued in science. If we want such an evaluation, then D-N requirements will sometimes be reasonable, sometimes not, depending on which methodological values are chosen. Is there, then, a fourth type of evaluation, based on a standard of "minimal scientific correctness"? Such a standard is to guarantee that an explanation provides scientific and philosophical illumination; it is to furnish the basis for a non-illocutionary evaluation more stringent than that supplied by (1) and less variable than the diverse non-illocutionary evaluations based on one or the other methodological values.

Modelists might well be construed as attempting to formulate such a standard. On this construction, what they do, in effect, is choose from among the general methodological criteria of value in science certain ones to count as necessary for "minimal scientific correctness" (for example, the presence of laws and a deductive structure). But why choose these values rather than others? Why not require that in a "minimally correct scientific explanation" all hypotheses be formulated quantitatively, or that micro-phenomena always be postulated, or that unification of diverse phenomena always be achievable by the explanans, or that a certain measure of simplicity be satisfied, and so forth? What I am questioning is whether in addition to

(i) a non-illocutionary standard of correctness of the sort supplied by (1), which invokes no general methodological values,

(ii) an illocutionary evaluation of goodness, which incorporates such values, among other considerations,

(iii) various non-illocutionary evaluations of goodness, based solely on one or more of these values,

there is also

(iv) a non-illocutionary standard of "minimal scientific correctness" which is restricted to some particular subset of these values.

The choice of some methodological values over others (in addition to mere correctness) seems arbitrary, or at best, needs arguing. The choice of all of them would defeat the idea of "minimality."

These, then, are my conclusions:

(a) There can be no model of scientific explanation if, like D-N theorists, modelists insist that both the a priori and NES requirements be satisfied.

(b) Modelists will not be successful in discovering a model in which the NES requirement is satisfied and in which it is not an explicit condition of the model that some singular sentence be true that entails the explanandum (e.g., a complete content-giving sentence).

(c) We can retain complete content-giving sentences with words like "reason" and "because" and still capture much of the essence of the modelists' view (e.g., the D–N conditions or those of Salmon); there is no need to exclude such sentences from an explanans, and there are good reasons for retaining them.

(d) However, there is no particular reason to want to capture this essence. The model that results (e.g., (5) above) will not guarantee correctness. And there seems to be no non-arbitrary standard of minimal scientific correctness based on some particular subset of methodological values.

CHAPTER 6
The Causal Relation

So far in this book I have characterized illocutionary acts of explaining, and I have defined the concept of explanation and considered the evaluation of explanations by reference to such acts. Can the illocutionary theory be used to explicate particular kinds of explanations found in the sciences? Specifically, using this theory, I want to define a concept of causal explanation and determine whether various types of explanations in the sciences, which can also be characterized on the basis of the illocutionary theory, are causal. This will be the subject of Chapter 7. In the present chapter I propose to explore an ontological question that should help illuminate causal statements: is causation a relation? More precisely, is a singular causal sentence to be construed as expressing a relationship between events in the world? Or are such sentences non-extensional? The argument I shall consider makes essential use of emphasis, which also formed the basis of one of the two objections raised in Chapter 3 against standard views of explanation products. The present discussion will buttress that objection by clarifying the concept of emphasis upon which it rests.

1. IS CAUSATION A RELATION?

The most prevalent view is that causation is a relation between events. Thus, it is held, a singular causal sentence such as

(1) Socrates' drinking hemlock at dusk caused his death

expresses a relation between two events, referred to respectively by the expressions "Socrates' drinking hemlock at dusk" and "his death."[1]

---

1. On some views, it expresses a relation between a *fact*, referred to by the expression "Socrates' drinking hemlock at dusk," and an event. See Zeno Vendler, "Causal Relations," *Journal of Philosophy* 64 (1967), pp. 704–13. In what follows I will speak of events, but everything said will be applicable as well to facts.

It is part of such a claim that singular causal sentences are *referentially transparent* in the cause- and effect-positions. This means that if

(2) Socrates' drinking hemlock at dusk = Xantippe's husband's drinking hemlock at dusk,

i.e., if these events are identical, then from (1) and (2) we may infer

(3) Xantippe's husband's drinking hemlock at dusk caused his death.

And if

(4) His (i.e., Socrates') death = Xantippe's husband's death,

then, from (1), via (2), (3), and (4), we may infer

Xantippe's husband's drinking hemlock at dusk caused Xantippe's husband's death.

I propose in this chapter to argue that singular causal sentences are not relational because they are not, in general, referentially transparent in cause- and effect-positions.

We begin by supposing that

(5) Socrates' *drinking hemlock* at dusk caused his death

is true. The emphasis in (5) indicates that some feature of the situation—the hemlock-drinking—was causally efficacious. Let us assume that "Socrates' drinking hemlock at dusk" refers to a particular event, and that it refers to the same event no matter which words, if any, are emphasized within it. That is,

(6) Socrates' drinking hemlock at dusk = Socrates' *drinking hemlock* at dusk = Socrates' drinking hemlock *at dusk*.

If singular causal sentences are referentially transparent in cause-positions, then from (5) and (6) we may infer

(7) Socrates' drinking hemlock *at dusk* caused his death.

But (7) is false, since it falsely selects the time of the drinking as causally efficacious; it states that the event's being at dusk is what caused Socrates to die. Since we can infer a false sentence from a true one by substituting expressions referring to the same event, I conclude that singular causal sentences are referentially opaque, i.e., not transparent, in the cause-position. A similar argument, to be given in Section 9, shows that they are referentially opaque in the effect-position as well. Hence singular causal sentences such as (1) and (5) are not relational.

Since this conclusion will no doubt bring rage to the hearts of many who have expressed views about causation, in what follows I consider various objections to the argument. These objections will serve as a basis for clarifying and defending the argument.

2. "CAUSATION IS NOT MIND-DEPENDENT"

The argument above cannot be right, it will be said, since if it were it would prove that causation is intensional rather than extensional—that causal contexts are like belief-contexts. The latter are intensional because they are mind-dependent. But surely causal contexts are not mind-dependent. Whether one thing caused another does not depend upon what anyone believes or intends. The gravitational force of the moon would cause the tides even if there existed no minds at all.

I agree that (many) causal sentences are not mind-dependent. However, "mentalistic" terms are not the only ones that can produce opaque contexts. Modal operators have a similar effect; so do quotation marks. And neither of these contexts need be mentalistic. How, then, does emphasis yield opacity in causal contexts? It does so via the phenomenon of *semantical aspect-selectivity*. An emphasized word or phrase in a nominalized cause-term of a causal sentence is selected as expressing an aspect of the situation that is causally operative. A shift in emphasis shifts the causally operative aspect and thus changes the meaning, and hence possibly the truth-value, of the sentence.

Two broad uses of emphasis can be distinguished: semantical and non-semantical. If I say that

(1) Socrates' drinking hemlock at dusk occurred in prison,

and you, who are hard of hearing, think I said that his drinking hemlock at dawn occurred in prison, I might reply that

(2) Socrates' drinking hemlock *at dusk* occurred in prison.

This is a "denial" use of emphasis. I utter (2) and emphasize "at dusk" because I believe that some related statement has been (or might be) made or assumed which is false; and the aspect responsible for its falsity has been replaced by the emphasized expression. Or suppose that you know that Socrates' drinking hemlock occurred in prison, but you don't know at what time of day this happened. I might utter (2) and emphasize "at dusk" because I believe that you are unfamiliar with that aspect of the situation but familiar with the rest. The emphasized words provide the new information for you. Or I might utter (2) and emphasize "at dusk" simply to call attention to something I regard as particularly interesting, important, or surprising. In uttering (2) and employing a "denial," "new information," or "interest" use of emphasis, I am not uttering a sentence whose meaning is different from that of (1). I am making the same statement, though in a different way, by the use of emphasis. These are non-semantical uses of emphasis. The meanings, and hence the truth-values, of (1) and (2) are identical.

The same is not so with

(3) Socrates' *drinking hemlock* at dusk caused his death

and

(4) Socrates' drinking hemlock *at dusk* caused his death,

even though (3) and (4) differ only in emphasis. This difference in emphasis makes a difference in the meanings and also the truth-values of these sentences. The reason is that the emphasized words become selected or captured by the word "caused," indicating that a particular aspect is causally operative. This is a semantical use of emphasis; its use affects the meaning and the truth-value of what is asserted. In non-semantical uses, the speaker selects a word to emphasize because of certain beliefs he has about the sentence, or aspects of it, or related sentences, e.g., the belief that a contrasting sentence has been uttered that is false, or that some aspect of the situation reported in the sentence is important. A shift in emphasis will not alter the meaning or the truth-value of the sentence, though it may not adequately reflect the accompanying belief of the speaker. By contrast, in semantical uses such as (3) and (4) the emphasized word or phrase becomes captured by the emphasis-selective word "caused," indicating the causal operativeness of the aspect selected by these words. A shift in emphasis will select a different aspect as causally operative. In both the semantical and non-semantical cases, use of emphasis can focus upon some aspect of the situation; in both we can have aspect-selectivity. But in semantical cases, because of an emphasis-selecting term in the sentence, the sentence itself makes a claim about that aspect.[2]

The term "cause" is one of a number of emphasis-selecting words that generate opaque contexts. Others include "explain," "reason," "advise," and "know." Each of these, to be sure, might be claimed to be "mentalistic," and so it should not be surprising that each generates opaque contexts. But there are non-mentalistic words in addition to "cause" that produce such contexts as well, e.g., "dangerous," "important," "mistake," and "illegal." Thus,

---

2. This is not the only semantical use of emphasis. Emphasis can also be used to disambiguate phrases and references and to give words special meanings. For examples, see my "Causation, Transparency, and Emphasis," *Canadian Journal of Philosophy* 5 (1975), pp. 1–23.

(5) Alice's *turning* suddenly was dangerous (important, a mistake, illegal)

might be true, while

(6) Alice's turning *suddenly* was dangerous (etc.)

is false, even though

(8) Alice's *turning* suddenly = Alice's turning *suddenly*.

The emphasized word is selected by "was dangerous," indicating the danger of a particular aspect of the situation. (6) asserts that it was the suddenness of her turning that was dangerous, while (5) asserts that what was dangerous was her turning, which was done suddenly. There should be little temptation to trace the opacity here to mind-dependence. The truth-values of (5) and (6) are unaffected by anyone's beliefs or intentions.

Although emphasis selects an aspect of the situation as being causally operative, dangerous, important, etc., this does not mean that it also *excludes* other aspects. Both (5) and (6) might be true. The emphasis in (5) selects the turning as being dangerous without excluding the possibility that the suddenness of the turning is also dangerous. My claim is not that if (5) is true then (6) must be false, but only that this is possible.

It will be noted that among the emphasis-selective words are many content-nouns (e.g., "explanation," "cause," "reason," "desire," "importance"), as well as cognate verbs and adjectives (e.g., "explains," "causes," "important"). Moreover, content-nouns that are emphasis-selective are so in content-giving sentences. Thus

The explanation of Socrates' death is that he *drank hemlock* at dusk

is a content-giving sentence for the content-noun "explanation"; the latter captures the emphasized expression. By contrast,

The explanation of Socrates' death is difficult for *contemporary* scholars to understand

is not a content-giving sentence for "explanation." Here there is no semantical capturing of the emphasized term; a shift in emphasis would not change the meaning of the sentence. However, it must not be supposed that all content-nouns are emphasis-selective, even in content-giving sentences (e.g., "event," "fact"). Nor can we assume that all emphasis-selective words are content-nouns or cognate verbs and adjectives (e.g., "knowledge," "illegal").

To conclude, then, an opaque context is one in which the substitution of co-referring terms does not always preserve truth-value. Opac-

3. "EMPHASIZED EVENT-EXPRESSIONS ARE NOT REFERRING EXPRESSIONS"

The next objection will be that I have falsely supposed that event-expressions containing emphasized words are referring expressions. This charge is leveled by Kim, who claims that the sentence

(1) Susan's *stealing* the bike occurred at 3 P.M.

does not seem to make any sense; the reason, he suggests, is that the expression "Susan's *stealing* the bike" is not referential.[3] I agree that if no context is supplied I will be puzzled by (1). My puzzlement, however, will not be over what it means, but over why the word "stealing" is being emphasized. (I would have the same kind of puzzlement if "stealing" had been written in extra small letters or upside down.) The use of emphasis in (1) is non-semantical, and to understand the particular non-semantical use to which it is being put a context will need to be provided. Thus suppose you claim that

Susan's borrowing the bike occurred at 3 P.M.

In response to your innocence, or lack of morality, or whatever, I utter (1)—thus employing a denial use of emphasis. Does what I have said make no sense? Is the sentence I have used meaningless? This seems too strong a claim to make.

An analogous point holds for referring expressions themselves. Suppose I am referring to Susan's stealing the bike, and you mistake me to be referring to her borrowing the bike. I reply that I am referring to

(2) Susan's *stealing* the bike.

When I use the emphasized expression (2) in saying what I am referring to, have I suddenly ceased to refer to anything? Again, this seems unwarranted. When I use the expression (2) I have indeed referred, though I have done so using emphasis in a non-semantical way. In using the emphasized expression (2) in the context imagined I have performed two feats: I have referred to something, viz. Susan's stealing the bike; and, by using emphasis, I have corrected a mistake you have made about the event I am referring to.

---

3. Jaegwon Kim, "Causation, Emphasis, and Events," *Midwest Studies in Philosophy* 2 (1977), pp. 100–103.

4. "EMPHASIZED EVENT-EXPRESSIONS ARE REFERRING EXPRESSIONS, BUT THEY DO NOT REFER TO EVENTS"

This claim is made by Fred Dretske, whose work on emphasis has been of seminal importance.[4] According to Dretske, a proposition such as

(1) Socrates drank hemlock at dusk

can be given different "embodiments," which he calls *propositional allomorphs,* depending on differences in emphasis. Thus,

Socrates *drank hemlock* at dusk

and

Socrates drank hemlock *at dusk*

are different propositional allomorphs of (1). When we nominalize a proposition such as (1) we obtain a noun phrase referring to an event. But, Dretske asks, what happens when we nominalize a propositional allomorph? To what does that refer? We should, he claims, be referring to something different when we refer to

(2) Socrates' *drinking hemlock* at dusk

from when we refer to

(3) Socrates' drinking hemlock *at dusk.*

Dretske's answer is that what these expressions refer to are not events but what he calls "event allomorphs"—aspects or features of events. Moreover, they refer to different aspects—the former to the hemlock-drinking aspect, the latter to the temporal aspect. This allows Dretske to agree that

(4) Socrates' *drinking hemlock* at dusk caused his death

is true, and

(5) Socrates' drinking hemlock *at dusk* caused his death

is false, without having to abandon the doctrine that causal contexts are referential. He can do this because, he maintains, although the cause-terms in (4) and (5) denote entities, they do not denote the same entities; they denote different allomorphic events or event-aspects. Hence,

---

4. Fred Dretske, "Referring to Events," *Midwest Studies in Philosophy* 2 (1977), pp. 90–99. For an earlier study of emphasis, see his "Contrastive Statements," *Philosophical Review* 81 (1972), pp. 411–37.

Socrates' *drinking hemlock* at dusk = Socrates' drinking hemlock *at dusk*

is false, and (5) cannot be inferred from (4).

How does Dretske argue for his claim that expressions such as (2) and (3) denote different things? He does so by assuming that the a- and b-positions in sentences of the form "a caused b" are transparent, i.e., that causal contexts are extensional. Then, since a sentence such as (4) is true, but (5) false, the emphasized expressions (2) and (3) cannot denote the same thing—otherwise both (4) and (5) would be true. Thus, Dretske argues, the transparency of causal contexts shows that differently emphasized event-expressions (different nominalized propositional allomorphs) are not co-referential. He writes:

> When we nominalize different propositional allomorphs we obtain noun phrases that refer to quite different things, allomorphic events, and the difference in what is being referred to manifests itself when we begin to talk about the causal relatedness of these items.[5]

Dretske's argument that expressions such as (2) and (3) do not denote the same things should carry no weight for our purposes, since whether causal contexts are extensional is just what is being questioned. My claims in Section 1 are not refuted by an argument which depends crucially on the assumption that causal contexts are extensional. Nor does Dretske offer any argument in favor of this assumption beyond noting that it is a "prevalent view."[6] I would certainly agree with him that if causal contexts are extensional, and if the cause-terms in (4) and (5) are referential, then these cause-terms denote different things, since (4) and (5) have different truth-values. But the question is whether the first clause of this conditional is true. Until it is established, Dretske's argument that expressions (2) and (3) denote different entities is not decisive.

Putting aside Dretske's argument, let us ask, independently, whether it is reasonable to suppose that expressions (2) and (3) denote different entities. Suppose I refer to

(6) Socrates' drinking hemlock at dusk,

and you mistake me to be referring to Socrates' eating dinner at dusk, while someone else mistakes me to be referring to Socrates' drinking hemlock at dawn. To you I reply that I am referring to

(2) Socrates' *drinking hemlock* at dusk;

to the other person, that I am referring to

---

5. "Referring to Events," p. 97.
6. *Ibid.*, p. 94.

(3) Socrates' drinking hemlock *at dusk.*

Am I therefore referring to different things in (6), (2), and (3)? If so this would preclude the denial use of emphasis as a means of correcting a mistake about which entity is being referred to. I could not use expressions (2) and (3) to correct the mistakes in question and to indicate what it is that I am referring to, because in each case I would be referring to something else.

Or suppose I am referring to Socrates' drinking hemlock at dusk, and you know that I am referring to Socrates' doing something or other at dusk, but you don't know what, while the second person knows that I am referring to Socrates' drinking hemlock at some time of the day, though he doesn't know what time. To you I reply that I am referring to

(2) Socrates' *drinking hemlock* at dusk,

and to the other person that I am referring to

(3) Socrates' drinking hemlock *at dusk,*

where the emphasized words provide the new information for you and the other person about what I am referring to. If the introduction and shifting of emphasis in my referring expression entails a shift in reference, then emphasis is precluded as a means of providing new information about what is being referred to. In general, Dretske's position here commits him to what strikes me as an unwarranted abandonment of non-semantical uses of emphasis in referring expressions.

A more reasonable approach would seem to be that when I use expressions (2) and (3) I am referring to the same thing, and I am employing emphasis (on the above occasions) to correct mistakes of others about what I am referring to, or to fill in their informational lacunae about this, or to focus upon some interesting aspect of the referent. I do not change the referent in these cases simply by adding or changing emphasis. Interestingly, in his first paper on emphasis, "Contrastive Statements," Dretske himself accepts this very conclusion. He argues that there is no difference in meaning, and hence reference, between expressions such as (2) and (3).

5. "IF EVENT-ASPECTS ARE CAUSALLY OPERATIVE, THEN CAUSAL SENTENCES MUST RELATE EVENT-ASPECTS"

Since I claim that

(1) Socrates' *drinking hemlock* at dusk caused his death

is true in virtue of the fact that the hemlock drinking aspect was causally operative, am I not committed to the claim that (1) reports a causal relationship between an event-aspect of Socrates' drinking hemlock at dusk and his death? Am I not saying that event-aspects are causal entities? (This claim is essentially Dretske's, though his argument for it depends crucially on the assumption that causal contexts are referential.) The answer to the question is no. To say that (1) is true in virtue of the causal efficacy of the hemlock drinking aspect of the event is not necessarily to say that the cause-term in (1) denotes an event-aspect. By analogy to say that

Socrates' *drinking hemlock* at dusk explains his death,

or that

Socrates' *drinking hemlock* at dusk is evidence for the fact that he died,

or that

Socrates' *drinking hemlock* at dusk was dangerous (important, required by law)

is true in virtue of the explanatory or evidential efficacy, or of the danger, importance, or legal requirement, of the hemlock drinking aspect of the event is not necessarily to be committed to the claim that the expression "Socrates' *drinking hemlock* at dusk" in these sentences denotes an event-aspect. The position taken by this expression in each of these sentences is referentially opaque. One need not postulate the existence of event-aspects as a new ontological species in order to make any of the claims above. Each of them can be true even though there exists no *entity* which caused, explains, or is evidence for, Socrates' death, or which was dangerous, important, or required by law.

There is, however, a related objection. I have been speaking of focusing upon, or selecting, event-aspects. But how can one do this without assuming that they are entities (i.e., without *referring* to them as entities)? Consider the event consisting of

Socrates' drinking hemlock at dusk in prison.

A number of questions about this event can be answered simply by examining the event-description itself, viz.

When was Socrates drinking hemlock in prison?

Where was Socrates drinking hemlock at dusk?

Who was drinking hemlock at dusk in prison?

What was Socrates doing at dusk in prison?

To provide answers to these and other questions is to focus upon (and provide information about) aspects of the event. When we answer the last question by saying drinking hemlock,

it was drinking hemlock that Socrates was doing at dusk,

Socrates was *drinking hemlock* at dusk in prison,

we are focusing upon a particular aspect of the event. But we need not be referring to any *entity* in so doing. One can use the expression "drinking hemlock" by itself, or in a prominent position in the sentence, or with emphasis, to focus upon a particular aspect of the event without thereby referring to an entity, which is that aspect. By analogy, the term "bald" can be used by itself or in

It was bald that Socrates was,

Socrates was *bald,*

to focus upon a particular aspect of Socrates, without thereby referring to an entity denoted by "bald."

Granted that we can focus upon aspects of events without assuming that they are entities, how can we make claims about them without supposing that they are? In the *de re* sense of "about," we cannot say anything about aspects of events if these are not entities. But we can in the *de dicto* sense. Just as we say that the non-extensional

What Jones fears is terrifying

makes a claim about what Jones fears, without supposing that "what Jones fears" denotes some entity, so we can say that

(1) Socrates' *drinking hemlock* at dusk caused his death

makes a (causal) claim about an aspect of an event without supposing that "Socrates' *drinking hemlock* at dusk" in (1) denotes an entity.

6. "EMPHASIZED CAUSAL SENTENCES ARE ELLIPTICAL"

A number of extensionalists have urged this thesis. The claim is that when an event-expression with emphasized words appears in a causal sentence it does not refer. Instead the causal sentence is to be understood as elliptical for something else. Let me mention five possibilities.

a. The conjunction paraphrase. This is proposed by Kim and Levin. The idea is that the emphasized cause-term in

(1) Socrates' drinking hemlock *at dusk* caused his death

does not refer to anything; instead (1) is to be understood as elliptical for a conjunction whose first member is the *unemphasized* (1)—in which the unemphasized cause-term is referential—and whose second member is a counterfactual, or a sentence invoking a law, or another causal sentence. For example, it might be claimed that (1) is elliptical for the conjunction

> Socrates' drinking hemlock at dusk caused his death, and if he had drunk hemlock at some other time he would not have died.

Levin claims that it is elliptical for

> Socrates' drinking hemlock at dusk caused his death, and there is a law in which "drinks at dusk" appears and which subsumes Socrates' death.[7]

Kim suggests

> Socrates' drinking hemlock at dusk caused his death, but Socrates' drinking hemlock did not cause his death.[8]

In all these cases the conjuncts contain only unemphasized causal sentences whose cause-terms are referential and refer to the event of Socrates' drinking hemlock at dusk.

b. *The operator paraphrase.* There is another possibility, which Kim also notes (without exploring it for the causal case). It asserts that in

(1) Socrates' drinking hemlock *at dusk* caused his death

the cause-term does not refer, and the emphasis is to be understood as an operator that operates on the entire causal sentence without emphasis. We are to understand (1) as

(2) $E_{\text{at dusk}}$ (Socrates' drinking hemlock at dusk caused his death),

in which "$E_{\text{at dusk}}$" is an emphasis-operator that operates on the sentence inside the parentheses to produce a sentence in which "at dusk" is emphasized, i.e., (1). The cause-position in (2) is referentially transparent.

c. *The "fact that an event has a feature" paraphrase.* A third possibility is to say that the cause-term in (1) is non-referential and to paraphrase (1) into the unemphasized

(3) The fact that Socrates' drinking hemlock occurred at dusk caused his death.

In (3) the cause-term is of the form

7. Michael Levin, "Extensionality of Causation and Causal-Explanatory Contexts," *Philosophy of Science* 43 (1976), pp. 266–77.
8. "Causation, Emphasis, and Events." Kim offers this paraphrase in the context of his own particular theory of causation, according to which events are construed as things having properties at times, and causation is analyzed in Humean terms. In Section 7, Kim's paraphrase will be examined in the light of his theory of causation.

The fact that (event) e had (feature) f,

which is to be understood as referential in the e- and f-positions.

d. A second "fact" paraphrase. The cause-term in (1) is non-referential, and (1) is elliptical for the unemphasized

> The fact that it was at dusk that Socrates drank hemlock caused his death

in which the entire fact-expression is to be construed as referential.

e. The "causally explains" paraphrase. Finally, it might be said that the cause-term in (1) does not refer to an event and that (1) is elliptical for

> Socrates' drinking hemlock *at dusk* causally explains his death.

Since "causally explains" (on this view) relates sentences, not events, the above sentence is, in turn, elliptical for

> The sentence "Socrates drank hemlock *at dusk*" causally explains the sentence "Socrates died,"

in which the cause-position is transparent, taking sentences as explainers.[9] Such a view might be suggested by Donald Davidson, who holds that causal sentences like

> The collapse was caused by the fact that the bolt gave way so suddenly and unexpectedly

are to be understood as saying not that one event caused another, but that one sentence causally explains another.[10] The causal sentences to which Davidson gives this paraphrase are importantly like the emphasized causal sentences I am considering: they involve selecting some features from the cause term as causally operative.

All five of these views suppose that emphasized event-expressions in causal sentences do not refer to events; and the first four views suppose that in causal sentences they do not refer at all. Now, of course, I am in agreement with these suppositions. But my reason is quite different from theirs. My reason is that causal contexts are referentially opaque—an assumption which those defending these paraphrases do not accept. What is *their* reason for claiming that emphasized event-expressions in causal sentences do not refer (or do not

---

9. It will make no difference in what follows whether explanations are taken to relate sentences or propositions expressed by them.
10. Donald Davidson, "Causal Relations," *Journal of Philosophy* 64 (1967), pp. 691–703.

refer to events)? It can only be that emphasized event-expressions *never* refer (or never refer to events) in any linguistic contexts. And these two claims I have already criticized in Sections 3 and 4.

Even so, one might wonder whether these paraphrase views are successful. Are the paraphrases trouble-free? Is the problem of emphasis really avoided? I think not.

Let me consider all the conjunction paraphrases in (a) together. On the conjunctive proposals, a sentence such as

(4) Socrates' *drinking hemlock* at dusk caused his death

is supposed to be elliptical for a conjunction, one of whose conjuncts is

(5) Socrates' drinking hemlock at dusk caused his death.

(5) contains no emphasis at all. If so, then, I suggest, there are two possible claims we might make about it. One is that it is ambiguous and therefore without unique truth-value. Depending on where emphasis is placed (or understood) in the cause-term, (5) becomes true or false, as the case may be. If "drinking hemlock" is emphasized it is true; if "at dusk" is emphasized it is false; as is, it is neither true nor false. The second possible claim is that (5) without emphasis is not ambiguous but is to be understood as if *all* the words in the cause-term are emphasized; it is to be understood as implying that all aspects of the event (implicit in its description) were causally operative. Emphasizing one word or phrase, e.g., "drinking hemlock," would indicate that it was the hemlock drinking that caused Socrates' death, but would make no claim about the causal efficacy of other aspects. But where all aspects are being claimed to be causally operative no emphasis is needed.[11]

In either of these cases the conjunctive view is in trouble. On this view, if (4) is true, then (5) must be true also. But this is not the case if (5) is ambiguous and therefore lacking in truth-value. Nor is it the case if (5) is construed as implying that all aspects of the event were causally operative; for this would imply that the time of the drinking was causally operative, which is false. It is the essence of the conjunction paraphrase view to paraphrase a causal sentence such as (4) into a conjunction whose members lack emphasis. But without emphasis a conjunct such as (5) either lacks a truth-value or else has one that is not necessarily the same as that of the sentence being paraphrased.

11. This second interpretation seems to me more dubious in the present case if only because it is difficult to imagine someone who knows anything about the situation uttering (5) with this intention. But other examples provide somewhat more believable alternatives. Thus, someone might utter "Kennedy's suddenly dying caused great grief," without emphasis, intending to make the claim that the dying, the suddenness of it, and the fact that it was Kennedy were all causally operative.

The "operator" (b) and "fact that an event has a feature" (c) paraphrases of emphasized causal sentences lead to similar difficulties. On both views, emphasized causal sentences generate unemphasized ones which should, but may not, have the same truth-values as the originals.[12]

The second "fact" paraphrase view (d) also has problems. On this view

(4) Socrates' *drinking hemlock* at dusk caused his death

is elliptical for

The fact that it was drinking hemlock that Socrates did at dusk caused his death,

in which the fact-expression is supposed to be referential. Assume, as we have done previously, that emphasizing words in this fact expression will not change the fact referred to; i.e.,

The fact that it was drinking hemlock that Socrates did at dusk = the fact that it was drinking hemlock that Socrates did *at dusk*.

We derive

The fact that it was drinking hemlock that Socrates did *at dusk* caused his death.

But this, unlike (4), implies the causal operativeness of the time of the drinking, which is false.[13]

The problem with all such views is a general one. When an event-expression of the form

(6) x's A-ing y B-ly (or a fact-expression of the form "the fact that x A-ed y B-ly),

in which "x" and "y" are nouns or noun phrases, "A" is a verb, and "b-ly" is an adverb or prepositional phrase (or several such), is embedded in certain larger contexts, including

_____ caused e,
_____ was dangerous (important, evidence that h),

the terms following the blank become aspect-selective. The meaning, and truth-value, of the resulting sentence depends on which aspect(s) of x's A-ing y B-ly they select. Emphasis provides a key to selectivity. An emphasized phrase is captured by the emphasis-selecting word and is understood as expressing the causally operative (dangerous, etc.)

---

12. See my "Causation, Transparency, and Emphasis," sections 5 and 6, for the details.
13. Either that, or the sentence has no clear meaning and hence no truth-value.

feature. A shift of emphasis will shift the feature selected, and thus the meaning, and possibly the truth-value, of the sentence. Without emphasis, either explicit or understood, sentences of the form

x's A-ing y B-ly caused e (was dangerous, important, etc.)

will be ambiguous or (possibly) construed as asserting that all aspects are operative. The first four paraphrase views recommend paraphrasing an emphasized causal claim of the above form into a sentence containing an unemphasized one. But without emphasis—without an indication of which aspect is being claimed to be causally efficacious— we get ambiguity or the implication that all are. A causal claim is thus generated which either lacks a unique truth-value or has one that need not be the same as the original's.

Nor, finally, do I find the Davidsonian "causally explains" paraphrase a particularly attractive alternative. According to this view,

(1) Socrates' *drinking hemlock* at dusk caused his death

is elliptical for

(7) The sentence "Socrates *drank hemlock* at dusk" causally explains the sentence "Socrates died,"

in which cause- and effect-positions are referential. However, Davidson provides no criterion for deciding whether a given causal sentence is to be construed as relating events or sentences. Even if he did, there are major objections against taking sentences as explainers and as objects of explanation.[14] Moreover, if Davidson wants to avoid even further difficulties he will have to choose e-sentences rather than sentences as relata (see Chapter 3, Section 3). The introduction and shifting of emphasis in the expression

The sentence "Socrates drank hemlock at dusk"

should not change the sentence being referred to. That is,

(8) The sentence "Socrates drank hemlock at dusk" = the sentence "Socrates drank hemlock *at dusk*" = the sentence "Socrates *drank hemlock* at dusk."

But if the cause-position in (7) is referentially transparent with respect to expressions denoting sentences, then from (7) and (8) we obtain

The sentence "Socrates drank hemlock *at dusk*" causally explains the sentence "Socrates died."

But the latter is what

14. See Chapter 3, and my "The Object of Explanation," in S. Körner, ed., *Explanation* (Oxford, 1975), pp. 1–45.

Socrates' drinking hemlock *at dusk* caused his death

is supposed to express, and this sentence is false. In short, the shift from "cause" as a relation between events to "causally explains" as a relation between sentences will not prevent inferences from truths to falsehoods due to shifts in emphasis.[15]

One final question about the "causally explains" view. How are we to construe *unemphasized* causal sentences whose cause-terms are nominals? Presumably at least some of them are to be construed as relating events, not sentences. This is Davidson's view. Thus he tells us that

(9) Flora's drying herself with a towel on the beach at noon caused her splotches

relates events, not sentences. But (9) *unemphasized* generates the very same difficulties as the unemphasized hemlock statement (5). Either it is ambiguous, or else (possibly, though not plausibly in this case) it is to be understood as implying the causal efficacy of all aspects of the event. Use of such causal sentences, then, will be severely restricted. If we want to avoid ambiguity, we will be able to use them only when we are prepared to make unusually strong causal claims. I am not saying that when someone asserts (9) he is making such a claim. (When someone utters (9) usually emphasis on some words in the cause-term is understood, at least implicitly.) I am saying only that if no emphasis is to be understood at all, and if ambiguity is to be avoided, then an unusually strong causal claim will have to be intended.

7. "EMPHASIZED CAUSAL SENTENCES MUST BE UNDERSTOOD IN CONJUNCTION WITH A PARTICULAR ANALYSIS OF CAUSATION"

This idea, which is an extension of the paraphrase view, has been suggested to me by Kim, especially in connection with his own theory of causation.[16] Kim's theory asserts that singular causal sentences relate events, where the latter are construed as things having properties at times. The event of Socrates' drinking hemlock would thus be rendered as

Socrates' having the property of drinking hemlock at time T,

---

15. It is, of course, open to the present paraphrase theorist to say that the cause- and effect-positions in "causally explains" statements are not referential. Conceivably, this is Davidson's view, although it does not seem to be, since he explicitly says that "explanations typically relate statements"; if causal-explanatory statements are genuinely relational, then the cause- and effect-positions must be referential.

16. See his "Causation, Nomic Subsumption, and the Concept of Event," *Journal of Philosophy* 70 (1973), pp. 217–36.

which Kim symbolizes as "[Socrates, drinking hemlock, T]." The causal sentence

Socrates' drinking hemlock caused his death

is rendered by Kim as

(1) [Socrates, drinking hemlock, T] caused [Socrates, dying, T'].

More generally, singular causal sentences involving reference to one individual and one property (though not necessarily the same ones) in both cause- and effect-positions have the form

(2) [a, P, T] caused [b, Q, T']

in which a and b denote individuals, P and Q properties, and T and T' times. Now Kim proposes to give a Humean analysis of sentences of form (2), the most important feature of which involves a constant conjunction requirement for the properties P and Q. Kim considers three such analyses, of which he seems to favor two. However, I will mention only one of those favored, since what I will say will be applicable to the other as well.

Using "loc(a,T)" for "the spatial location of a at time T," one of Kim's analyses can be expressed as follows:

(3) [a, P, T] is a (direct contiguous) cause of [b, Q, T'] provided that
 (i) [a, P, T] and [b, Q, T'] both exist;
 (ii) loc(a,T) is contiguous with loc(b,T'), and T is temporally contiguous with T';
 (iii) If a = b:
  $(x)(t)(t')([x,P,T]$ exists, and loc(x,t) is contiguous with loc(x,t'), and t is temporally contiguous with $t' \rightarrow [x,Q,t']$ exists);
  If $a \neq b$:
  $(x)(y)(t)(t')([x,P,t]$ exists, and loc(x,t) is contiguous with loc(y,t'), and t is temporally contiguous with $t' \rightarrow [y,Q,t']$ exists).[17]

Kim then defines "contiguous cause" in terms of the ancestral of direct contiguous causation. In what follows, however, I will focus on "direct contiguous causation" which I will simply call "causation." Kim's condition (i) requires the occurrence (existence) of both cause- and effect-events. ([a,P,T] exists, according to Kim, if and only if a has P at T.) Condition (ii) is a spatio-temporal contiguity requirement. Condition (iii) expresses a constant conjunction requirement involving the constitutive properties P and Q of the cause- and effect-events. Kim

17. In Chapter 5, Section 6, we noted the possibility of invoking laws of the type in (iii) in connection with the D-N model of explanation.

does not specify any precise meaning for the arrow in (iii) except to say that it is to "denote whatever type of implication the reader deems appropriate for stating laws."[18] It is to convey the idea of "causal or nomological implication."[19]

We are now in a position to consider the present proposal. The emphasized sentence

(4) Socrates' drinking hemlock *at dusk* caused his death

is to be rendered as a conjunction, viz.

(5) [Socrates, drinking hemlock at dusk, T] caused [Socrates, dying, T'], and [Socrates, drinking hemlock, T] did not cause [Socrates, dying, T'],

where T denotes the particular time at which Socrates drank the hemlock, and T' the particular time at which he died. Assuming, for the sake of argument, that the causation in (5) is "directly contiguous," (5) is to be understood in terms of the analysis provided by (3). Thus the first conjunct of (5) would be analyzed as:

(a) [Socrates, drinking hemlock at dusk, T] and [Socrates, dying, T'] both exist; and

(b) loc(Socrates, T) is contiguous with loc(Socrates, T') and T is temporally contiguous with T'; and

(c) $(x)(t)(t')([x, $ drinking hemlock at dusk, $t]$ exists, and loc(x,t) is contiguous with loc(x,t'), and t is temporally contiguous with t', $\rightarrow [x,$ dying, $t']$ exists).

Kim's strategy for handling the emphasis in (4) thus amounts to this: capture the emphasis (i.e., the semantical aspect-selectivity) in (4) by formulating the conjunction (5); but construe the latter as a Humean causal claim involving the idea of constant conjunction. The Humean translation once obtained, Kim would claim, contains no ambiguities and is not subject to semantical uses of emphasis. In Section 6, when I considered the conjunctive paraphrase view, (4) was formulated as

Socrates' drinking hemlock at dusk caused his death, but Socrates' drinking hemlock did not cause his death.

The complaint was that the first conjunct here is subject to various semantical interpretations depending on where emphasis is to be understood. The present proposal avoids this embarrassment, since the first conjunct of (5) is to be understood as the conjunction of sen-

---

18. "Causation, Nomic Subsumption, and the Concept of Event," p. 229, ftn. 19.
19. *Ibid.*

tences (a), (b), and (c), which are not subject to various semantical interpretations depending on emphasis. The truth-value of any member of this conjunction is not affected by the introduction and shifting of emphasis. The present proposal, then, purports to capture the semantical aspect-selectivity introduced by emphasis while retaining the extensionality of causal sentences.

Does this invocation of Humean causation save the day? I am not convinced that it does. To begin with, it saddles us with a theory of causation subject to well-known difficulties (of which Kim is aware). If the arrow in the constant conjunction condition is construed quite broadly, then, as Kim himself notes (p. 235), there are counterexamples to the analysis which involve properties exhibiting a lawlike, but not a causal, connection. Kim's analysis also seems overly restrictive, since it requires a nomological connection between the constitutive properties of (Kim's) events. Many causal sentences which we ordinarily accept as true would thus be rendered false, including

(6) Socrates' drinking hemlock caused his death,

or Kim's (1), since there is no true law relating the properties of drinking hemlock and dying *in the way (iii) specifies*. (Whether one dies from drinking hemlock depends upon how much one drinks, whether one's stomach is pumped immediately, etc.)

Most importantly, is Kim correct in assuming that a Humean translation of the sort he advocates is not subject to semantical uses of emphasis? That depends on the interpretation to be given to the arrow in Kim's laws of constant conjunction. Let us focus on such laws and simplify them by omitting the contiguity clauses. Kim's present solution commits him to the view that such sentences are not subject to semantical uses of emphasis, that the introduction and shifting of emphasis will not alter truth-values. Now I believe that such a view is correct for general sentences such as

(7) Voters in Maine are conservatives,

which carry no particular causal force. But there are other general sentences that do; and when they do, it is not so clear that they are (semantically) emphasis-insensitive.

Gazing at the awesome north face of the Eiger, I say

(8) Anyone who climbs *that dangerous wall* in the morning will die,

while you say

(9) Anyone who climbs that dangerous wall *in the morning* will die.

I believe that both of our claims have causal implications. Neither of us is claiming merely that climbing will be followed by death; we are saying, in addition, that death will result from, will be caused by, will occur because of, the climbing. Moreover, I suggest that our claims have different causal implications. My claim might be formulated more fully as being, or at least implying, (something like)

(10) Anyone who climbs that dangerous wall in the morning will die because it is that dangerous wall that he will be climbing.

Your claim might be expressed as

(11) Anyone who climbs that dangerous wall in the morning will die because it will be in the morning that he climbs it.

And these claims need not have the same truth-value. Sentences such as (8) and (9), by contrast with (7), would normally be understood as carrying causal implications. And the emphases here have different semantical consequences, since they focus upon different aspects of the event-type "climbing that dangerous wall in the morning" as being causally operative.

Accordingly, I suggest, Kim's position is faced with the following dilemma: Either the arrow in his laws of constant conjunction carries causal implications or it does not. (Either it permits inferences of sentences like (10) and (11) from (8) and (9), or it does not.) If not—if, e.g., it expresses a material conditional or some nomological relation which is non-causal—then there are well-known counterexamples to the resulting Humean analysis. (Many false causal claims will be generated.) On the other hand, if the arrow, whether analyzed or primitive, carries causal implications, then the laws are semantically emphasis-sensitive, and emphasis is not eliminated from the final analysis of causal sentences, as Kim requires.

8. "EMPHASIS DOES NOT AFFECT THE MEANING OR THE TRUTH-VALUE OF CAUSAL SENTENCES"

So far all of my real and imagined extensionalist critics agree with me on at least one point, viz. that emphasis does affect the meaning and truth-value of causal sentences. Dretske obviously holds this view; so do the paraphrase theorists of Section 6, since differently emphasized causal sentences like

(1) Socrates' *drinking hemlock* at dusk caused his death

and

(2) Socrates' drinking hemlock *at dusk* caused his death

will have different paraphrases and truth-values. However, there are other critics who would deny the truth of this common assumption.

Peter Unger, e.g., argues that emphasis never affects the meaning or truth-value of any sentence.[20] Suppose I utter (1), and you mishear "dawn" for "dusk." I reply:

(3) I didn't say anything about dawn, I said that Socrates' drinking hemlock *at dusk* caused his death.

According to Unger, in (3) I would be making the same causal claim as in (1), although the emphasis has shifted. Unger concludes

> So, emphasis does not alter the content of what is said. The reason for this is that it doesn't alter the meaning of the (unemphasized) sentence which standardly serves to express what is said. . . . Unsurprisingly, then, what emphasis does is just to emphasize. It helps get us to focus on what might be thought important to think about.[21]

I would have no quarrel with Unger here if his claim were only that there are uses of emphasis that do not affect the meaning or truth-value of the sentence asserted. These I have called non-semantical uses of emphasis, and Unger's "focusing on what is thought important" is an example of what I have called the interest use. What Unger is denying, and what other extensionalists are accepting, is that emphasis has semantical uses as well.[22] No doubt when one person asserts (1) and another asserts (2) they are focusing on different things. But in a causal context this difference of focus has different semantical implications. The speakers here are not simply focusing on something of particular importance to them; they are focusing on something in such a way as to be making different causal claims. This is why I and the earlier extensionalists agree that (1) and (2) have different meanings. Where we disagree is over what this semantical focusing amounts to. Dretske wants to claim that it involves making a reference to entities, viz. event-allomorphs; some other extensionalists that it involves making an unemphasized conjunctive claim one of whose conjuncts is a causal sentence referring to events and the other conjunct a counterfactual, lawlike, or causal statement; Davidson (perhaps) that it involves making a claim about causal explanation and a reference to sentences, not events, as relata; and I, that this focusing in causal and other emphasis-selecting contexts is a semantical phenomenon that does not involve reference to any entity focused upon.

Turning, then, to Unger's particular argument, am I making different causal claims when I utter (1) and (3)? I am prepared to agree with him that I am not. But this is because the linguistic context—in

---

20. Peter Unger, *Ignorance* (Oxford, 1975), pp. 78–79.
21. *Ibid.*, p. 79.
22. It might be noted that while Unger refuses to countenance semantical uses of emphasis, Dretske's position, by contrast, commits *him* to abandoning non-semantical uses.

particular the first clause of (3)—makes it clear that emphasis is being employed here solely in its denial use, which is non-semantical. Although I have said that emphasized words in a causal sentence such as (1) are selected as expressing a causally operative aspect, it is possible to cancel this type of selectivity in favor of a non-semantical one, if a suitable preamble or context is made explicit. But without this linguistic context—in the absence of the first clause of (3) or something akin to it—the most natural construal of emphasis in (3) is semantical. In the absence of such contextual semantical canceling, it seems most natural to take people who assert (1) and (2) to be making different claims.

At this point a critic may agree that people who assert (1) and (2) are making different claims, but deny that this is true of sentences (1) and (2) themselves. Such a critic will distinguish speaker-meaning from sentence-meaning. Indeed, this line is taken by Steven E. Boër, who argues that emphasis concerns the former not the latter.[23] Although he does not deal explicitly with sentences (1) and (2), his view commits him to saying that these sentences have the same meaning, even though what speakers would normally mean when they utter these sentences (the proposition each speaker would express) is different. Borrowing a term from Grice, Boër suggests that emphasis is involved with what a speaker either conventionally or non-conventionally "implicates" when he utters a sentence, rather than with the meaning of the sentence uttered (which Grice characterizes as "what is said (in a favored sense)").[24] Indeed, Grice explicitly discusses emphasis, using as one of his examples (p. 122)

(4) *Jones* didn't pay the bill, *Smith* did.

His claim is that a speaker who utters (4) "implicates" that

(5) Someone thinks or might think that Jones did pay the bill.

But (5) does not give the meaning of the sentence (4), nor is it entailed by that sentence.

My reply is to agree with Grice's conclusion about (4) and (5). Sentence (4), uttered in the context Grice imagines, illustrates what I earlier called the "denial" use of emphasis: a speaker utters (4) and places the emphasis where he does because he believes that some related sentence has been, or might be, asserted which is false (in this case, "Jones paid the bill"). The denial use of emphasis is but one of the non-semantical uses. And in such cases, when a speaker utters a sen-

---

23. Steven E. Boër, "Meaning and Contrastive Stress," *Philosophical Review* 78 (1979), pp. 263-98.
24. See H. P. Grice, "Further Notes on Logic and Conversation," *Syntax and Semantics*, vol. 9.

tence emphasizing some word or phrase he "implicates" something which is not part of the meaning of the sentence itself. However, my claim is that, by contrast, there are semantical uses of emphasis as well, illustrated in sentences (1) and (2). When a speaker utters (2) he is not "implicating" that

> (6) The fact that it was at dusk that Socrates drank hemlock is what caused his death,

in the way that an utterer of (4) "implicates" that (5). To be sure, when a speaker utters (2) *he* means to be claiming that (6). But, on my view (and on that of all of my previous extensionalist critics), what makes it possible for the speaker to mean to be claiming that (6) when he utters (2) is that (2) and (6) are equivalent in meaning, or at least that (2) entails (6).

The major problem I find with Boër's position is that he provides no account of the difference between speaker-meaning and sentence-meaning that will establish the thesis that emphasis always pertains to the former but not the latter. He simply cites Grice's discussion of this distinction without showing how such a thesis follows. Nor is it clear that he could show this.

Grice's basic notion is that of speaker-meaning. He defines a locution of the form "By uttering x utterer U meant that p" in terms of U's intentions. He then defines various notions of *sentence*-meaning by reference to speaker-meaning. For example, one definition he offers (in simplified form) is this:

> For U (within U's idiolect) the sentence S means (has as one of its meanings) "p" = (def.) U has a resultant procedure for S, viz. to utter S if, for some audience A, U wants A to believe that U believes that p.[25]

This definition uses the notion of a "resultant" procedure, concerning which Grice says:

> As a first approximation, one might say that a procedure for an utterance type X will be a resultant procedure if it is determined by (its existence is inferable from) a knowledge of procedures
> (a) for particular utterance-types which are elements in X, and
> (b) for any sequence of utterance types which exemplifies a particular ordering of syntactical categories (a particular syntactical form).[26]

Grice's subsequent remarks about "resultant" procedures are, by his own admission, quite sketchy and limited in scope, and it is by no

---

25. H. P. Grice, "Utterer's Meaning, Sentence Meaning, and Word Meaning," *Foundations of Language* 4 (1968), pp. 225–42. See particularly pp. 233, 236.
26. *Ibid.*, p. 235.

means obvious, e.g., that the following procedure is precluded from being a "resultant" one: to utter

(2) Socrates' drinking hemlock *at dusk* caused his death,

if, for some audience A, the speaker U wants A to believe that U believes that the fact that it was at dusk that Socrates drank hemlock is what caused his death. If this procedure is "resultant," then, by Grice's definition, for the speaker U the sentence (2) means the same as (6); i.e., these *sentences* have the same meaning.

9. THE EFFECT SIDE OF CAUSATION

I have been concentrating on the cause-side of singular causal sentences. An argument that such sentences are also referentially opaque in the effect-position is readily constructable.

I have a special roulette wheel on which each number appears twice, once below a black rectangular space and once below a red one. But this is a crooked wheel since two secret buttons are attached. If I push button 1 the ball will land on a red number; if I push button 2 it will land on one of the Sevens. I now push both buttons, as a result of which the following sentences are true:

(1) My pushing button 1 caused the ball's landing on the *red* Seven;

(2) My pushing button 2 caused the ball's landing on the red *Seven*.

We may assume that "the ball's landing on the red Seven" refers to a particular event, and that it refers to the same event no matter which words, if any, are emphasized. That is,

(3) The ball's landing on the red Seven = the ball's landing on the *red* Seven = the ball's landing on the red *Seven*.

If singular causal sentences are referentially transparent in the effect-position, then from (1), (2), and (3) we may infer

My pushing button 1 caused the ball's landing on the red *Seven*

and

My pushing button 2 caused the ball's landing on the *red* Seven,

both of which are false.

If the conclusion of the previous sections is accepted, then singular causal sentences do not express a relation between events and what is caused. The present argument gives us a reason to suppose that they do not express a relation between causes and events.

CHAPTER 7
Causal Explanation

1. INTRODUCTION

Appeals are often made to the concept of a causal explanation. Typically those who characterize such explanations do so by utilizing some particular model of explanation such as one of those discussed in Chapter 5. Thus, according to Hempel, a causal explanation is a D-N explanation whose explanans describes events that occur prior to the explanandum-event and which contains laws of succession relating types of events that do not occur simultaneously.[1] According to Brody, a causal explanation is one that satisfies the D-N model and whose explanans describes an event that caused the explanandum-event. Very recently, Salmon has proposed that a causal explanation is an S-R explanation in which the statistical regularities introduced are causal.[2]

Some who characterize causal explanations by appeal to a model of explanation are committed to a particular philosophical analysis of causation. Hempel, in effect, adopts a Humean regularity analysis which requires the cause to precede the effect and to be related to the effect via a general law. Salmon proposes an analysis of causation based on certain technical notions he introduces such as the "intersection of processes" and "interactive forks."[3] Others, e.g., Brody, characterize causal explanations in terms of some favorite model of explanation without invoking or presupposing some particular analysis of cause.

Suppose that one rejects the models of explanation on which these characterizations of causal explanation depend. Assume, further, that one is sympathetic to the illocutionary theory, and in particular, to the

---

1. Carl G. Hempel, *Aspects of Scientific Explanation*, p. 352.
2. Wesley C. Salmon, "Why Ask, 'Why'?" *Proceedings and Addresses of the American Philosophical Association* 51 (1978), pp. 683–705.
3. *Ibid.*

ordered pair view developed in Chapter 3. Suppose, finally, that one wishes to characterize causal explanations utilizing the ordered pair view without, however, committing oneself to any particular analysis of causation. (Perhaps causation is fundamental and cannot be usefully and non-circularly defined in non-causal terms.) Is such a task possible, and if so, will the result be of value?

One potential benefit is in helping to settle the issue of whether all explanations in the sciences are, or ought to be, causal. Hempel, by reference to his own models of explanation and "Humean" analysis of causation, answers: No. Salmon, in his most recent writings, disagrees. He regards the claim that scientific explanations are causal as one of two "quite sound" intuitions about scientific explanations (the other being that they involve subsumption under laws).[4] And he supports his intuitions by appeal to his own S-R model supplemented with a causal condition for whose concept of causation he proposes an analysis.

Nancy Cartwright believes that a good explanation "must increase the probability of the fact to be explained" and that "what counts as a good explanation is an objective, person-independent matter."[5] She argues that both of these criteria are satisfied if and (I take her to be saying) only if the explanation is causal. (For her this means that it must use causal laws—laws which state that some type of thing causes some other type of thing, e.g., "Force causes motion.")

Robert Causey, who adopts Hempel's D-N model of explanation as a set of necessary conditions, recognizes that "some D-N derivations are not explanatory, i.e., some do not represent causal explanations."[6] He proposes no sufficient set of conditions for a causal explanation, and suggests, indeed, that "the conditions for causal explanations are in part theory dependent."[7] Nevertheless, he assumes that scientific explanations generally are, and must be, causal.

David Lewis, in his recent Howison Lectures, goes even further than Salmon, Cartwright, or Causey, appears to go. All explanations of occurrences, he claims, whether scientific or otherwise, provide information about the causal history of those occurrences.[8] His thesis is presented along with his own counterfactual analysis of causation, but it is not supposed to depend for its truth on the adequacy of that analysis.

Even if we reject analyses of explanation or causation defended by these writers, we can still examine the particular thesis about scientific

---

4. *Ibid.*, p. 685.
5. Nancy Cartwright, "Causal Laws and Effective Strategies," *Noûs* 13 (1979), p. 424.
6. Robert Causey, *Unity of Science* (Dordrecht, 1977), p. 25.
7. *Ibid.*, p. 27.
8. David Lewis, "Causal Explanation" (mimeograph), p. 6.

explanation they are supporting or attacking. Some illumination concerning the claim that scientific explanations are causal may be forthcoming by utilizing the ordered pair view of explanation, even in the absence of a definition of causation.

2. A PRELIMINARY DEFINITION OF CAUSAL EXPLANATION

It might be thought that the task of defining "causal explanation" using the ordered pair view is possible but trivial. We can simply say that

(1) (p; explaining q) is a causal explanation of q if and only if it is an explanation of q and Q is a causal question.

But what is a causal question? Perhaps it is one that has among its presuppositions that something caused (or causes, or will cause) something else. This is too broad, for it will entail that

What is the meaning of the sentence Jane was reading when the gas leak caused the explosion? and

When did the gas leak cause the explosion?

are both causal questions. (Both presuppose that the gas leak caused the explosion.) We need a narrower concept.

Let us begin not with questions, but with interrogative sentences that express them, in particular, with interrogatives of the form

(2) What (_____) caused (causes, will cause) _____?

Sentences of this form will be called what-causal interrogatives. The first parentheses in (2) need not be filled. If they are, then Q will be a what-causal interrogative only if they are filled with a noun or noun phrase (including nominals). (Thus, "What is the meaning of the sentence Jane was reading when the gas leak caused the explosion?" is not a what-causal interrogative.) Examples of what-causal interrogatives are

What caused the explosion?
What leak caused the explosion?
What causes bodies to accelerate?

There are interrogatives with forms other than (2)—indeed, ones not containing the word "cause"—which are equivalent to interrogatives of form (2). Thus,

Who accelerated the car?

is equivalent to

What person caused the car to be accelerated?

I shall say that Q is a *what-causal question* if it is expressible by means of a what-causal interrogative. All of the interrogatives just given express what-causal questions.

We might now modify (1) by formulating the following sufficient condition:

(3) (p; explaining q) is a causal explanation of q if it is an explanation of q, and Q is a what-causal question.

By this condition, assuming that

(4) (The gas leak caused the explosion; explaining what caused the explosion)

is an explanation, it is a causal explanation, since "What caused the explosion?" is a what-causal question. By contrast, assuming that

(The meaning of the sentence Jane was reading when the gas leak caused the explosion is "All Gaul is divided into 3 parts"; explaining what the meaning of that sentence is)

is an explanation, (3) would not count it as a causal explanation, since its constituent question is not a what-causal question. Moreover, (3) counts as causal explanations not only ones, such as (4), whose constituent propositions are singular, but also ones whose constituent propositions are general laws, e.g.,

(The presence of unbalanced forces causes bodies to accelerate; explaining what causes bodies to accelerate).

Nevertheless, (3) needs broadening, as can be seen by considering the question

(5) Why did the bell ring?

and the following answer to (5):

(6) The reason the bell rang is that the sexton pulled the rope attached to the bell.

Intuitively, we would want to say that the ordered pair

(7) (The reason the bell rang is that the sexton pulled the rope attached to the bell; explaining why the bell rang)

is a causal explanation of why the bell rang. Yet (5) is not a what-causal question. It is a more general question which asks for a reason for the bell's ringing. One way of answering this more general question is by supplying a cause. But another way is by supplying an end, or purpose, or function, e.g.

(8) (The reason the bell rang is to summon the congregation; explaining why the bell rang).

Although some have argued that teleological explanations are reducible to causal ones, I shall make no such assumption here. I will assume that (7) is a causal explanation, but will leave it open whether (8) is too.

If (7) is a causal explanation, and its constituent question is not a what-causal question, then (3) does not supply a necessary condition for causal explanations. (p; explaining q) can be a causal explanation even if Q is not a what-causal question. Can (3) be expanded so that (7) is counted as a causal explanation?

3. A BROADER DEFINITION

The question

(1) What caused the bell to ring?

will be said to be a *what-causal transform* of question

(2) Why did the bell ring?

In general, Q_2 will be said to be a what-causal transform of Q_1 under these conditions:

(i) Q_2 is a what-causal question;
(ii) Every presupposition of Q_1 is a presupposition of Q_2;
(iii) There is some sentence expressing a complete presupposition of Q_1 in which there is a ϕ-existential term, such that a cause is a kind of ϕ.[9]

To illustrate these conditions, consider questions (1) and (2). The sentence

The bell rang for some reason

expresses a complete presupposition of (2) in which there is a reason-existential term, viz. "for some reason." Since a cause is a kind of reason, condition (iii) is satisfied.[10] So are conditions (i) and (ii), since (1) is a what-causal question, and every presupposition of (2) is a presupposition of (1). Therefore, (1) is a what-causal transform of (2).

By contrast, (1) is not a what-causal transform of

(3) Where was Jones when the bell was caused to ring?

Condition (i) is satisfied (since (1) is a what-causal question), but not (ii) or (iii). There is a presupposition of (3)—"Jones was some place

9. See Chapter 2, Section 4, for the notion of a ϕ-existential term.

when the bell was caused to ring"—that is not a presupposition of (1). Moreover, a cause is not a kind of place.

In order to decide how to broaden our definition of a causal explanation, consider the following propositions:

(4) The reason the bell rang is that the sexton pulled the rope attached to the bell;

(5) The cause of the bell's ringing is that the sexton pulled the rope attached to the bell.

I shall assume that what makes the sexton's pulling the rope the reason the bell rang is that it was the cause; and if the cause of the bell's ringing is that the sexton pulled the rope, then this is also the reason the bell rang. In short, I shall assume that (4) and (5) are equivalent in the sense that one is true if and only if the other is. Now consider the following explanations, in which (4) and (5), respectively, are constituent propositions:

(6) (The reason the bell rang is that the sexton pulled the rope attached to the bell; explaining why the bell rang);

(7) (The cause of the bell's ringing is that the sexton pulled the rope attached to the bell; explaining what caused the bell to ring).

In view of the fact that the constituent propositions in (6) and (7) are equivalent, and that the constituent question in (7) is a what-causal transform of the constituent question in (6), I shall say that the explanation (7) is a *what-causal equivalent* of explanation (6).

More generally, (p'; explaining q') will be said to be a *what-causal equivalent* of (p; explaining q) if and only if

(i) (p; explaining q) is an explanation of q, and (p'; explaining q') is an explanation of q';

(ii) Q' is a what-causal transform of Q;

(iii) p' is true if and only if p is.

We can now expand definition (3) of Section 2 by supplying the following sufficient condition for causal explanations:

(A) (p; explaining q) is a causal explanation of q if it is an explanation of q, and either Q is a what-causal question or (p; explaining q) has a what-causal equivalent.

By this condition both

---

10. Some might question the first part of this claim on the grounds that causal contexts are extensional whereas reason-contexts are intensional. But I have tried to show in Chapter 6 that such grounds are nonexistent: causal contexts are not extensional.

(The gas leak caused the explosion; explaining what caused the explosion)

and

(6) (The reason the bell rang is that the sexton pulled the rope attached to the bell; explaining why the bell rang)

are causal explanations. What about the following?

(The reason the bell rang is to summon the congregation; explaining why the bell rang).

Since the explanatory question here is not a what-causal question, this explanation must have a what-causal equivalent if it is to be a causal explanation by the lights of (A). The constituent question in such an equivalent would be

What caused the bell to ring?

And the constituent proposition would be one expressible by filling the blank in

The cause of the bell's ringing is _____,

where such a proposition is true if and only if

The reason the bell rang is to summon the congregation.

Whether there is such an equivalent proposition is an issue I shall not pursue here. Teleological reductionists ("mechanists") claim there is such an equivalent, anti-reductionists deny this. If there is, then, by (A), the explanation in question is causal.[11]

4. A FINAL DEFINITION

Does

(A) (p; explaining q) is a causal explanation of q if it is an explanation of q, and either Q is a what-causal question or (p; explaining q) has a what-causal equivalent

provide necessary conditions for causal explanations? Consider the explanation

(1) (The gas leak caused the explosion by allowing the gas to come in contact with a nearby flame; explaining how the gas leak caused the explosion).

This does not satisfy condition (A). Its explanatory question

11. In Chapter 8 this issue will be explored for the case of functions.

(2) How did the gas leak cause the explosion?

is not a what-causal question. Nor does (1) have a what-causal equivalent.[12] To be sure, the boundaries of the class of causal explanations are not very sharp. Still I am inclined to suppose that (1) ought to count as a causal explanation of how the gas leak caused the explosion. If so, then although (A) may provide a sufficient condition for causal explanation it does not provide a necessary one.

Let us call explanations satisfying (A) *what-causal explanations*. Although there are various types of causal explanations, I suggest that what-causal explanations are particularly central and that a number of other types can be understood, at least in part, by reference to these. Thus, consider explanation (1) whose explanatory question is (2). The answer to (2) provided by (1) is

(3) The gas leak caused the explosion by *allowing the gas to come in contact with a nearby flame*.

If (3) is true then so is

(4) Allowing the gas to come in contact with a nearby flame caused the explosion,

in which the cause-expression consists of the emphasized words in (3). Now by (A),

(5) (Allowing the gas to come in contact with a nearby flame caused the explosion; explaining what caused the explosion)

is a what-causal explanation. Under these conditions I shall say that explanation (1) has a *counterpart* what-causal explanation, viz. (5).

More generally, consider questions which presuppose causal statements of the form

x caused (causes, will cause) y.

Such questions may be expressed by interrogatives such as these, among others:

Q_1 How did x cause y?
Q_2 In what manner did x cause y?
Q_3 By what means did x cause y?
Q_4 Why did x cause y?

---

12. For example, "What caused the explosion?" and "What gas leak caused the explosion?" are not what-causal transforms of (2), since not every presupposition of (2) is a presupposition of these. Therefore, these could not be constituent questions in what-causal equivalents of (1).

Consider answers to these questions which are expressible by complete content-giving sentences of the form

(6) x caused y by _____ ,

or of the form

(7) The reason that x caused y is that p,

in which "p" is replaced by a sentence. Let us suppose that if the resulting sentence (6) is true, then so is

(8) _____ caused y,

where the blank in (8) is filled with the same expression that filled the blank in (6). Similarly, let us suppose that if the resulting sentence (7) is true, then so is

The fact that p caused y,

where "p" is replaced by the sentence replacing "p" in (7). (In the example above, (3) is a sentence of form (6), and (4) is of form (8); and if (3) is true, so is (4).) If Q is a question that has an answer satisfying these conditions, I shall say that an explanation of the form

(x caused y by _____ ; explaining q)

has a *counterpart what-causal explanation,* viz. one of the form

(_____ caused y; explaining what caused y);

and an explanation of the form

(The reason that x caused y is that p; explaining why x caused y)

has a counterpart what-causal explanation of the form

(The fact that p caused y; explaining what caused y).

In the example above, explanation (1) has (5) as a counterpart what-causal explanation. And

(9) (The reason the gas leak caused the explosion is that the gas leak allowed the gas to come in contact with a nearby flame; explaining why the gas leak caused the explosion)

has the following counterpart what-causal explanation:

(The fact that the gas leak allowed the gas to come in contact with a nearby flame caused the explosion; explaining what caused the explosion).

Using (A) as a definition of *what-causal explanation,* we can now expand the idea of a causal explanation, as follows:

(B) (p; explaining q) is a causal explanation of q if it is an explanation of q and either it is a what-causal explanation of q or it has a counterpart what-causal explanation.

In accordance with this, both (1) and (9) are causal explanations, even though neither is a what-causal explanation.

I shall not speculate whether more changes are required. At the least, (B) defines a concept of causal explanation that is broader than (3) of Section 2, or than (A). It covers a variety of explanations whose constituent questions are not necessarily what-causal questions (see Q_1 through Q_4 above), and whose constituent propositions can be expressed by sentences of various forms that do not necessarily contain the word "cause." Moreover, definition (B) covers explanations of general regularities as well as of particular events. It should enable us to raise questions about various types of explanations in the sciences. (Objections to the effect that this concept is still not broad enough for our purposes will be considered in Section 12.) However, before turning to scientific cases an elementary but important question can be posed.

5. WHAT MAKES IT POSSIBLE FOR CAUSAL EXPLANATIONS TO EXPLAIN ANYTHING?

An answer is immediately forthcoming if we examine the constituents of a causal explanation, as given by definition (B), and invoke the discussion in Chapter 2 (particularly Section 15). In accordance with the illocutionary theory of explaining, it is persons who (in the primary sense) explain. So we might ask: what makes it possible for a person to explain something by means of a causal explanation? And why do people so often explain by means of causal explanations? As noted in Chapter 2, the sorts of answers I reject are these: because the constituent propositions in causal explanations are "intrinsic explainers" the mere expressing of which is necessarily to explain, or because "to explain" means "to utter a sentence expressing a causal proposition."

What makes it possible for a person to explain something by means of a causal explanation? In accordance with definition (B), the constituent questions in causal explanations are questions with forms such as "What caused ———?" "Why is it the case that p?" "How did x cause y?" "Why did x cause y?" These are all content-questions. (There are complete content-giving propositions with respect to such questions.) Moreover, questions of these forms are often sound ones; i.e., they admit of a correct answer. Furthermore, people are frequently unable to answer such questions correctly. (More precisely, they are not in a complete knowledge-state with respect to such questions, without being in an alternation-state.) Under these conditions such people are in n-states—states of non-understanding—with respect to such questions. (See Chapter 2, Section 12.) Now, in accordance with definition (B), the constituent proposition in a causal explanation is a complete content-giving proposition with respect to the constituent question Q in the explanation. (It is a proposition expressible by a sentence of the form "the cause of _____ is _____," or "the reason that p is that _____," or "the manner in which x caused y is _____," and so forth.) Since such a proposition is a complete content-giving proposition with respect to Q, producing it can get people into a complete knowledge-state; therefore it can get them to understand what the cause is (the reason that p, the manner in which x caused y). Invoking a causal explanation can accomplish this by getting people to know, of the constituent proposition in the explanation, that it is a correct answer to the constituent question Q. Therefore it is possible for a person to utter a sentence expressing this constituent proposition with the following intention: that his utterance enable others to understand q by producing the knowledge, of the proposition expressed by the sentence he utters, that it is a correct answer to Q. But according to the definition of explaining in Chapter 2, this means that where (p; explaining q) is a causal explanation a person can *explain* q by uttering a sentence that expresses p.

Why do people so often explain by means of causal explanations? The answer is that events have causes (there are causal reasons for many things, when something causes something else it does so in a certain manner, etc.). But nature and our finite minds conspire to produce in us states of non-understanding—n-states—with respect to the question of what caused many of those events (what the causal reason is, in what manner the cause operated, etc.). Explainers frequently explain by means of causal explanations because this will remove states of non-understanding with which we are frequently beset.

6. ARE ALL EXPLANATIONS IN THE SCIENCES CAUSAL?

Clearly many are, in the sense expressed by (B). Scientists offer causal explanations, in this sense, of particular events as well as of general regularities. Becquerel described his famous discovery of natural radioactivity, as follows. He placed a crystal of uranium salt on several photographic plates wrapped in thick black paper. He continues:

> Some of the preceding experiments were prepared during Wednesday the 26th and Thursday the 27th of February [1896], and since on those days the sun appeared only intermittently, I stopped all experiments and left them in readiness by placing the wrapped plates in the drawer of a

cabinet, leaving in place the uranium salts. The sun did not appear on the following days, and I developed the plates on March 1st, expecting to find only very faint images. The silhouettes [of the uranium crystals] appeared, on the contrary, with great intensity. . . . A hypothesis presents itself very naturally to explain these radiations (by noting that the effects bear a great resemblance to the effects caused by the radiations studied by M. Lenard and M. Roentgen), as invisible radiation emitted by phosphorescence, but whose persistence is infinitely greater than the duration of the luminous radiation emitted by these bodies.[13]

Becquerel's explanation can be formulated as follows:

(What caused the uranium crystal silhouettes on the photographic plates in the experiments of February 1896 was invisible radiation from these crystals whose persistence is greater than the luminous radiation; explaining what caused the uranium crystal silhouettes, etc.).

This is readily seen to be a causal explanation, in the sense given by (B), of why certain particular events occurred.

Becquerel performed other experiments, and concluded:

All the uranium salts I have studied, whether phosphorescent or not under light, whether in crystal form or in solution, gave me corresponding results. I have thus been led to the conclusion that the effect is due to the presence of the element uranium in these salts. . . .[14]

Accordingly, Becquerel offers this *general* causal explanation:

(What causes the darkening of photographic plates containing uranium salts is radiation emitted by the element uranium; explaining what causes darkening of photographic plates containing uranium salts).

Again, we have a causal explanation, this time, of a general regularity rather than of a particular event or set of events.

None of this should be controversial. What is of special interest is whether all scientific explanations are causal in the sense expressed by (B)? Does (B) provide a standard that should be demanded in science? One of the problems in answering is that (B) employs an undefined notion of cause. In the course of the discussion in the following sections, although "cause" will not be defined, certain general claims about causation will be made (C_1, C_2, and C_3 below) which I regard as reasonable. Examples will be chosen which are typical of three types of explanations found in the sciences. It will be argued that these fail to satisfy (B) when "cause" is understood in such a way that C_1 through

13. Morris H. Shamos, ed., *Great Experiments in Physics* (New York, 1959), pp. 213–14.
14. *Ibid.*, p. 215.

7. SPECIAL-CASE-OF-A-LAW EXPLANATIONS

In explaining why projectiles continue to move after being propelled Descartes writes as follows:

> The reason why projectiles persist in motion for some time after leaving the hand that throws them is simply that when they once move they go on moving, until their motion is retarded by bodies that get in the way.[15]

The idea that bodies continue to move except when prevented by other bodies Descartes takes to be a special case of what he calls his first law:

> Every reality, in so far as it is simple and undivided, always remains in the same condition so far as it can, and never changes except through external causes.[16]

For the case of motion this law becomes:

> A moving body, so far as it can, goes on moving.[17]

Accordingly, we might identify Descartes's explanation (at least one of them) as the ordered pair

(1) (The reason that a projectile continues in motion once it has left the hand is that it is a law that any moving body will continue to move unless its motion is retarded by other bodies; explaining why a projectile will continue in motion once it has left the hand).

The constituent question in (1), viz.

(2) Why does a projectile continue in motion once it has left the hand?

is not a what-causal question. Nor does (1) have a what-causal equivalent. To be sure, there is a what-causal transform of (2), viz.

(3) What causes a projectile to continue in motion once it has left the hand?

And using this what-causal transform, we might formulate the ordered pair

---

15. Elizabeth Anscombe and Peter Thomas Geach, eds., *Descartes' Philosophical Writings* (Indianapolis, 1971), p. 216.
16. *Ibid.*
17. *Ibid.*

Causal Explanation   231

(4) (The fact that (it is a law that) any moving body will continue to move unless its motion is retarded by other bodies causes a projectile to continue in its motion once it has left the hand; explaining what causes a projectile to continue in motion once it has left the hand).

But the first members of (1) and (4) are not equivalent. While the first member of (1) is true, the first member of (4) either makes questionable sense or is false. (Why this is so I will take up in a moment.) Descartes, indeed, seems committed to saying that nothing (other than perhaps God) *causes* a projectile to continue to move. Hence, for him, the first member of (4) could not be true, although the first member of (1) is true. Since these members are not equivalent, (4) is not a what-causal equivalent of (1). As noted, Descartes would perhaps be willing to construct the following causal explanation for (3):

(5) (God causes a projectile to continue in motion once it has left the hand; explaining what causes a projectile to continue in motion once it has left the hand).[18]

However, (5) is not a what-causal equivalent of (1), since their first members are not equivalent.

Does (1) have a counterpart what-causal explanation? No, since for this to be the case the constituent question of (1) must presuppose something that it does not presuppose, viz. that something causes a projectile to continue in motion once it has left the hand. Since the constituent question in (1) is not a what-causal question, and (1) has no what-causal equivalent, and (1) has no counterpart what-causal explanation, we may conclude that (1) is not a causal explanation in the sense provided by (B).

What type of explanation is (1)? Its constituent question presupposes that there is a reason that a projectile will continue in motion once it has left the hand. And its constituent answer provides such a reason by invoking a more general law of which projectile motion is a special case. It can be represented as having the form

(6) (The reason that p is that it is a law that L; explaining why p),

in which the sentence replacing "p" describes some regularity that is a special case of law L.[19]

18. Descartes believed that God preserves the totality of motion in the universe. Perhaps he believed that God is an indirect cause of individual motions.
19. Using the simple criterion for correct explanations of Chapter 4, Section 1, we know that an explanation of form (6) is correct if it is true that the reason that p is that it is a law that L. Now, I am not here assuming that if p is a special case of law L then it must be true that the reason that p is that it is a law that L. (See Section 13.) But I do suppose that this could be true in a given case; i.e., I assume that there are correct explanations of form (6).

Science is rife with such "special-case-of-a-law" explanations, e.g.,

(7) (The reason the moon exerts a gravitational force on the waters of the earth is that it is a law that all bodies exert gravitational forces on all other bodies; explaining why the moon exerts a gravitational force on the waters of the earth).

Thus Roger Cotes might have offered this explanation, even though he believed that gravitation (like inertia) is an ultimate and simple disposition of bodies, and therefore that nothing causes the moon to exert a gravitational force on the waters of the earth.[20] Nevertheless, there is a reason why it does so, which is given in explanation (7).

Now a general feature of causation—one on which claims in the previous paragraphs depend—might be held to be that

C_1: The existence of a regularity is not caused by the existence of a law of which that regularity is a special case.

Particular definitions or analyses of causation may attempt to confront the question of whether, and if so, why, C_1 holds. Is C_1 a special case of some broader principle? On some accounts, cause and effect must be *distinct* in a sense that precludes the effect from being "included in" the cause. That is,

(8) If Y is included in X then X does not cause Y.

It is up to such accounts to provide criteria of inclusion. Principle C_1 embodies one such criterion. If Y is the existence of a regularity, and X is the existence of a law of which that regularity is a special case, then Y is included in X (so X does not cause Y). But can some more general principle of inclusion be formulated that justifies this assertion?

Consider two sentences of the form "All A's are B's" and "All C's are D's" (or "It is a law that all C's are D's"). If all A's are C's, and if the property of being a B is identical with the property of being a D, I shall say that the state of affairs described by "All A's are B's" is included in that described by "(It is a law that) all C's are D's."[21] (This is intended as a sufficient but not a necessary condition for inclusion.) By this condition, the state of affairs described by

All viscous liquids at high temperatures boil

is included in that described by

---

20. See his Preface to the second edition of Newton's *Principia* I (Berkeley, 1962), p. xxvii.
21. This condition invokes the idea of property identity. For a discussion of this see my "The Identity of Properties," *American Philosophical Quarterly* 11 (1974), pp. 257–75; and Sydney Shoemaker, "Identity, Properties, and Causality," *Midwest Studies in Philosophy* 4 (1979), pp. 321–42.

All liquids at high temperatures boil,

as well as in that described by

All liquids at high temperatures are in a state in which the vapor pressure of the liquid is equal to the external pressure acting on the surface of the liquid.

(In the latter case it is being assumed that the property of boiling is identical with that of reaching a state in which vapor pressure equals external pressure.) Similarly, the state of affairs described by

All projectiles released from the hand and not retarded by other bodies continue to move

is included in that described by

It is a law that all moving bodies not retarded by other bodies continue to move.

This simple criterion needs to be generalized to handle all cases in which the state of affairs of a regularity's obtaining is included in the state of affairs consisting of a certain law's holding (of which that regularity is a special case). But it will adequately capture at least some such cases. Using this criterion, together with (8), we may conclude, e.g., that the fact that liquids at high temperatures boil is not what causes viscous liquids at high temperatures to boil. Similarly, the fact that it is a law that all moving bodies not retarded by other bodies continue to move is not what causes projectiles released from the hand and not retarded by other bodies to continue to move. I suggest that the present criterion for inclusion, when suitably extended, together with (8), will provide a sufficient basis for asserting C_1.

Given C_1, then, criterion (B) for causal explanations (Section 4) suffices to show why "special-case-of-a-law" explanations such as (1) and (7) are not causal explanations. (C_1 by itself is not sufficient for this purpose.)

If "special-case-of-a-law" explanations of form (6) are not causal— if, in accordance with C_1, the fact that L's obtaining does not cause p to obtain—in virtue of what can L be a reason that p obtains? The answer I am suggesting is simply this. For a particular p and L, the fact that it is a law that L can be the reason that p obtains *in virtue of the fact that p is a special case of law L.* That fact by itself can constitute a reason for p. We need not, and if C_1 is right, cannot, invoke causation to further ground this reason.

8. CLASSIFICATION EXPLANATIONS

Examining the periodic table of the elements, one notes that the elements are arranged in a certain order by atomic numbers. And one might ask, "Why do the elements have the atomic numbers they do?" or more specifically, e.g.,

(1) Why does iron have the atomic number 26?

Here is one explanation:

(2) (The reason that iron has the atomic number 26 is that atomic numbers, i.e., the numbers representing the positions of the elements in the periodic table, are determined by the number of the protons in the atomic nucleus of the element, and the nucleus of iron has 26 protons; explaining why iron has the atomic number 26).

The constituent question in (2), viz. (1), is not a what-causal question. Does (2) have a what-causal equivalent? There is a what-causal transform of (1), viz.

What causes iron to have the atomic number 26?

But in response it seems plausible to say that although iron has this atomic number in virtue of the number of its nuclear protons, the fact that it has this number of protons does not cause it to have the atomic number it has. If so, (2) has neither a what-causal equivalent nor a counterpart what-causal explanation. What (2) does is to explain by indicating not that the having of 26 protons causes iron to have the atomic number 26 but that the having of this number of protons, in virtue of a certain system of classification, means that iron is *classifiable* as being something with the atomic number 26.

Let me call explanations of type (2) *classification* explanations. There are many others in this category. Thus, one may ask why H_2SO_4 is an acid, and receive either of the following explanations:

(3) (The reason H_2SO_4 is an acid is that it is a proton donor; explaining why H_2SO_4 is an acid);

(4) (The reason H_2SO_4 is an acid is that it is one of a class of substances which typically are soluble in water, sour in taste, turn litmus red, neutralize bases, etc.; explaining why H_2SO_4 is an acid).

H_2SO_4 is not caused to be an acid by its having the properties mentioned. Rather, its having these properties, in virtue of a certain definition, or criterion, or system of classification, means that it is an acid.

More generally, classification explanations such as (3) and (4) have the form

(5) (The reason x is a P_1 is that x is a P_2; explaining why x is a P_1),

where there is some definition, or criterion, or system of classification according to which something is a P_1 if and only if it is a P_2. Something's being a P_2 is a reason that it is a P_1 in virtue of the fact that there is such a definition (not in virtue of the fact that something's being a P_2 causes it to be a P_1). The definition, or criterion, or system of classification may even be included within the constituent proposition of the explanation. In such a case the explanation could have the form

(6) (The reason x is a P_1 is that x is a P_2 and according to such and such a definition, or criterion, or system of classification, something is a P_1 if and only if it is a P_2; explaining why x is a P_1).

Now I take it to be a general truth about causation—one on which claims in recent paragraphs are based—that

C_2: The presence of some property or state is not caused by the satisfaction of a set of defining criteria for that property or state.

C_2, like C_1 above, can be defended as a special case of the broader principle

(7) If Y is included in X, then X does not cause Y,

which we introduced in Section 7. Indeed, in the case of classification explanations we can use the simple criterion of inclusion also introduced in that section, provided that we assume that the property P is identical with Q if, in accordance with a definition or system of classification, something is P if and only if it is Q. By the latter assumption, the property of being a bachelor is identical with that of being an unmarried man; the property of being an acid is identical with being a proton donor. Using the criterion of inclusion of Section 7, together with principle (7) above, we may conclude that priests' being bachelors is not caused by their being unmarried men; that H_2SO_4's being an acid is not caused by its being a proton donor; and more generally, that the presence of some property or state is not caused by the satisfaction of a set of defining criteria for that property or state, i.e., C_2 above.

9. IDENTITY EXPLANATIONS

Our observations about classification explanations can be generalized. Such explanations are special cases of what I will call *identity explanations*. The latter have the form

(1) (The reason x is a P_1 is that x is a P_2; explaining why x is a P_1)

where the property of being a P_1 is identical with the property of being a P_2. Classification explanations are simply those in which the properties are identical in virtue of a definition. But there are empirical property identities that can serve as a basis for (identity) explanations such as

(2) (The reason that ice is water is that ice is composed of H_2O molecules; explaining why ice is water),

where the property of being water = the property of being composed of H_2O molecules; or

(3) (The reason this gas has the temperature T is that it has the mean molecular kinetic energy K; explaining why this gas has the temperature T),

where the property of having the temperature T = the property of having the mean molecular kinetic energy K.

In the case of identity explanations of type (1), using the criterion of "inclusion" introduced in Section 7, we can say that the state of affairs of x's being a P_1 is included in that of x's being a P_2, since the properties are identical. Therefore, by the general causal principle

If Y is included in X, then X does not cause Y,

x's being a P_2 does not cause x to be a P_1. Using the definition of causal explanation developed earlier, we may conclude that identity explanations such as (2) and (3) above are not causal explanations.

Indeed, it is possible to reverse the "explanans" and "explanandum" (and hence the properties P_1 and P_2) in an identity explanation of form (1). Thus we might not only want to explain why ice is water, by formulating (2); we might also (on some different occasion) want to explain why ice is composed of H_2O molecules, by formulating

(4) (The reason that ice is composed of H_2O molecules is that ice is water; explaining why ice is composed of H_2O molecules).

(This might be a good explanation to give to someone who would be surprised to learn that ice is frozen water.) If an identity explanation such as (2) were causal, this reversal would be ruled impossible, on the grounds that if A causes B then B does not cause A. But there is nothing intrinsically wrong with (4), or with a comparable reversal of (3). It is possible to explain the presence of a macro-property by appeal to the presence of an identical micro-property; or vice versa. If so, identity explanations are not causal.

It is not my claim that all identity explanations (i.e., all those of form (1) in which $P_1 = P_2$) are correct. For example,

(The reason that ice is water is that ice is water; explaining why ice is water)

is not a correct explanation, since its constituent proposition is not true. Yet the relevant properties are clearly identical. Nor is it my view that an identity explanation, if correct, is necessarily a good one for explaining q. My claim is only that there are correct identity explanations, that some of them are good, and that none is causal. Such explanations include classification explanations in which the properties are identical by definition. They also include non-classification explanations, such as (2) and (3), where identical properties are picked out by "macro-" and "micro-"descriptions.

An important conclusion can now be drawn from our discussion in this and the previous two sections. Let us call an explanation of the form

(5) (The reason that p is that r; explaining why p)

a *reason-giving* explanation. There are explanations falling under the categories "causal," "special-case-of-a-law," and "classification" (or more generally, "identity"), that are reason-giving explanations. That is, there are explanations of each of these types with the form (5) in which r presents a reason for its being the case that p. Now, if we make assumptions

C_1: The existence of a regularity is not caused by the existence of a law of which that regularity is a special case,

C_2: The presence of some property or state is not caused by the satisfaction of a set of defining criteria for that property or state,

and if we employ definition (B) for causal explanations, and grant that there are "special-case-of-a-law" and "classification" explanations that are reason-giving explanations, then it follows that *not all reason-giving explanations are causal:* r may be a reason that p, but not in virtue of the fact that r's being the case causes it to be the case that p (and not in virtue of some teleological fact). It may be a reason in virtue of the fact that r gives a law of which the fact that p is a special case; or in virtue of the fact that r gives a condition which, according to a system of classification, is necessary and sufficient for its being the case that p; or in virtue of the fact that r describes a micro-property that is identical to a macro-property referred to in a sentence describing the fact that p (or vice versa).

10. DERIVATION EXPLANATIONS

It will be useful once again to invoke Bohr's explanation of the spectral lines. Using the ordered pair theory, which contains criteria for individuating explanations, we may distinguish a number of explanations here, depending on the question being raised. Some of these explanations are definitely causal by our criterion, e.g.,

(1) (Excitation of hydrogen atoms causes the spectral lines by causing the electron to jump from its normal energy state to an excited state and then to radiate energy when it falls from the latter to a state of lower energy; explaining how the excitation of the hydrogen atom causes the spectral lines).

By contrast, I want to note an explanation pertaining to the spectral lines which Bohr gives in section 2 of his paper:

(2) a. The amount of energy radiated by the hydrogen atom in the formation of a stationary state t in which the electron has a definite orbit is
$$w_t = \frac{2\pi^2 me^4}{h^2 t^2},$$ where m is the mass of the electron, e its charge, and h Planck's constant;

b. From which it follows that the amount of energy emitted by passing the system from a state t_1 to state t_2 is
$$w_{t_1} - w_{t_2} = \frac{2\pi^2 me^4}{h^2}\left(\frac{1}{t_2^2} - \frac{1}{t_1^2}\right);$$

c. But since the amount of energy emitted equals $h\nu$ (ν = frequency), we get
$$w_{t_2} - w_{t_1} = h\nu;$$

d. From which it follows that
$$\nu = \frac{2\pi^2 me^4}{h^3}\left(\frac{1}{t_2^2} - \frac{1}{t_1^2}\right),$$

which yields Balmer's formula when $t_2 = 2$ and t_1 varies.

The question Bohr is raising here is this:

(3) How is Balmer's formula, which relates the lines in the hydrogen spectrum, derivable from the assumptions Bohr makes about the hydrogen atom?

With respect to this question, Bohr's explanation is this:

(4) (Balmer's formula, which relates the lines in the hydrogen spectrum, is derivable from the assumptions Bohr makes about the hydrogen atom in the following manner (here follows (2)); explaining how Balmer's formula is derivable from the assumptions Bohr makes about the hydrogen atom).

The constituent question in (4) is not a what-causal question. Nor does (4) have a what-causal equivalent. Nothing causes Balmer's formula to be derivable from Bohr's assumptions. More generally, indeed,

C₃: Nothing causes one proposition to be derivable from others.[22]

If not, then (4) has no what-causal counterpart either. Explanation (4) is an example of what I shall call a *derivation-explanation*. It can be represented as being of the form

(5) (Proposition P_1 is derivable from a set of propositions P_2 in such and such a way; explaining how P_1 is derivable from P_2).

Derivation-explanations are not causal in the sense expressed by (B). Yet they are frequent in the sciences, especially the quantitative ones. Their point is not simply to state the assumptions from which a conclusion is derivable. Such explanations are not simply *arguments,* in the logician's sense—i.e., sets containing premises and a conclusion which follows from the premises. They are derivations. *They explain how a conclusion follows from a set of premises.*

11. COMPLEX DERIVATION EXPLANATIONS

Scientists often produce a type of explanation that contains a derivation, but is somewhat different from what I have been calling derivation explanations. The explanatory question is not "How is proposition p_1 derivable from p_2?" but "Why is it the case that p_1?" Let me discuss a simple example, which involves an appeal to conservation principles.

When one particle strikes a stationary particle of equal mass the first stops and the second moves off with the velocity of the first. (Think of billiard balls.) Why is this so? A standard explanation proceeds by appeal to the principles of conservation of momentum and energy. According to these principles, the total momentum of an (isolated) system of particles is constant, as is the total energy. In the system consisting of the two particles, the total momentum of the system before the collision $= m_1 v_0$, where m_1 is the mass of the first particle and v_0 is its velocity before collision. The total momentum of the system after collision $= m_1 v_1 + m_2 v_2$, where v_1 and v_2 are the velocities of the two particles after collision, and m_2 is the mass of the second particle. Applying the law of conservation of momentum to this case, we get

(1) $m_1 v_0 = m_1 v_1 + m_2 v_2$.

22. As with C_1 and C_2 one would like some more general principle of causation on which to ground this one. Both C_1 and C_2 might be defended, we recall, by appeal to the idea that an effect cannot be included in a cause. But this idea, as explicated earlier, is not applicable to C_3. Perhaps the latter can be defended as a special case of the principle that if p is "analytic" then nothing causes it to be the case that $\dot{p}$ is true (where the claim that one proposition is derivable from certain others is "analytic"). However, I feel more confident of C_3 than I do of this generalization of it.

The total (kinetic) energy of the system before collision = $(1/2)m_1v_0^2$; the total energy afterward = $(1/2)m_1v_1^2 + (1/2)m_2v_2^2$. Applying the principle of conservation of energy, we get

(2) $(1/2)m_1v_0^2 = (1/2)m_1v_1^2 + (1/2)m_2v_2^2$.

Eliminating v_2 from (2) by using (1), we obtain

$(m_1 + m_2)v_1^2 - 2m_1v_0v_1 + (m_1 - m_2)v_0^2 = 0$.

The roots of this are

(3) $v_1 = v_0$ and $v_1 = \dfrac{m_1 - m_2}{m_1 + m_2}v_0$.

Similarly, eliminating v_1 from (2) by using (1) will yield

(4) $v_2 = 0$ and $v_2 = \dfrac{2m_1}{m_1 + m_2}v_0$.

The first solution in each case is precluded, since it corresponds to the case in which there is no collision (i.e., where initial and final velocities of the first particle are the same, and the final velocity of the second particle is 0). The second solutions describe the result of the collision. If the masses of the particles are the same, then $m_1 - m_2 = 0$, and from (3) we get

(5) $v_1 = 0$,

i.e., the velocity of the first particle after collision is 0. Also, if the masses are the same, then $m_1 + m_2 = 2m_1$. So from (4) we obtain

(6) $v_2 = v_0$,

i.e., the velocity of the second particle after collision is the same as the velocity of the first particle before collision. And it is the facts described by (5) and (6) which were to be explained.

Now let us try to analyze this explanation. Its constituent question is

Q: Why is it the case that *when one particle strikes a stationary particle of equal mass, the first stops and the second moves off with the velocity of the first?*

Call the emphasized part of this question "p." Then Q has the form

Why is it the case that p?

The explanation appeals to the fact that the laws of conservation of momentum and energy hold, and it includes a derivation as an integral part. I shall refer to it as a *complex derivation explanation*. It seems to have this form:

(7) (The reason that p is that it is a law that the total momentum of a system of particles remains constant and it is a law that the total energy of a system of particles remains constant, from which the fact that p is derivable in the following manner (here follows the derivation above); explaining why it is the case that p).

The problem I find with this formulation is that it does not adequately reflect the explanatory role of the derivation. The constituent proposition in (7) entails that the reason that

(a) Under the circumstances described the first particle stops and the second moves off with the velocity of the first

is that

(b) it is a law that the total momentum of a system of particles remains constant and it is a law that the total energy of a system of particles remains constant.

Does the constituent proposition in (7) also entail that the *derivability in the above manner* of the proposition (a) from the proposition (b) is the reason, or part of the reason, that (a)? It does not seem to; nor should it. By analogy, if the reason that Greeks are mortal is that Greeks are men and that men are mortal, then the fact that there is such and such a derivation of the former proposition from the latter using the apparatus of quantification theory is not the reason, or part of the reason, that Greeks are mortal. Similarly, if the reason that (a) is that (b), then the derivability of the proposition (a) from the proposition (b) in the manner given is not the reason, or part of the reason, that (a). If not, then what role does the derivation play?

My suggestion is that a complex derivation explanation ought to be construed as two explanations, not one. In our example one thing that is being explained is why the particles move the way they do. (The reason the particles move the way they do is that they are subject to the principles of conservation of momentum and energy.) But since it may not be obvious what relationship these principles bear to the motion in question, another thing that is being explained is how the proposition describing the motion of the particles is derivable from the conservation principles.

Accordingly, I would analyze the complex derivation explanation given at the beginning of this section as follows:

(8) (The reason that p is that it is a law that the total momentum of a system of particles remains constant and it is a law that the total energy of a system of particles remains constant; explaining why it is the case that p);

(9) (The proposition that p is derivable from the laws of conservation of momentum and energy in the manner given above; explaining how the proposition that p is derivable from these conservation laws).

Is either of these explanations causal? (9) is not causal, since it is an (ordinary) derivation explanation, which, in Section 10, we argued is non-causal. The constituent question in (8), viz. Q, is not a what-causal question. (8) has no counterpart what-causal explanation. And (8) does not have a what-causal equivalent. There is a what-causal transform of Q, viz.

What causes it to be the case that p (i.e., that when one particle strikes a stationary particle of equal mass the first stops and the second moves off with the velocity of the first)?

But even if there is something which causes this, it is not the fact that

It is a law that the total momentum of a system of particles remains constant, and it is a law that the total energy of a system of particles remains constant.

Why not?
Let us suppose that

(10) The fact that it is a law that the total momentum of a system of particles remains constant and it is a law that the total energy of a system of particles remains constant causes it to be the case that p (that when one particle strikes another, etc.).

How does this fact (or these facts) cause this? Reverting to the original explanation at the beginning of the section, the only plausible way to explicate how such a causal connection could work would seem to be this:

(11) The fact that it is a law that the total momentum of a system of particles remains constant causes it to be the case that the total momentum of a system of 2 particles before collision is equal to the total momentum of this system after collision; i.e., it causes it to be the case that (1) is true.

Similarly,

(12) The fact that it is a law that the total energy of a system of particles remains constant causes it to be the case that (2) is true.

Finally,

The fact that (1) and (2) hold causes it to be the case that (5) and (6) are true if there is a collision, i.e., causes it to be the case that p.

But (11) and (12)—at least—must be rejected if we accept the earlier principle of causation C_1, which precludes the existence of a regularity from being caused by the existence of a law of which that regularity is a special case. In both (11) and (12) the regularities cited are special cases of the laws invoked. Perhaps the causal connection postulated in (10) could operate in ways not requiring the truth of (11) and (12); but I am unable to say how.

I conclude that the complex derivation explanation given by (8) and (9) is not causal.

Explanations that utilize conservation principles are not all derivation explanations. Suppose that a system of particles S has a certain total momentum M at a time t. Consider the question

Q': Why does system S have total momentum M at time t?

One way of explaining this is to proceed without invoking the principle of conservation of momentum simply by indicating the momentum of each particle in the system at t and taking the sum. Another way is to invoke the principle of conservation of momentum without indicating the momentum of each particle. (Of course these strategies can be combined.) Here is an example of the latter:

(13) (The reason that system S has total momentum M at time t is that the total momentum of a system of particles does not change, and the total momentum of S prior to time t is M; explaining why system S has total momentum M at time t).

Is time-reversibility possible with conservation explanations of type (13)? Consider

(14) (The reason that system S has total momentum M at time t is that the total momentum of a system of particles does not change and the total momentum of S after time t is M; explaining why system S has total momentum M at time t).

(13) explains the fact that total momentum of S at time t is M by appeal to the principle of conservation of momentum together with the fact that its total momentum *before* t is M; (14) explains it by appeal to the same principle together with the fact that its total momentum *after* t is M. If one is inclined to suppose that (13) and (14) could both be correct explanations, then, again, we would have examples of correct non-causal explanations. Or at least (14) would have to be viewed as non-causal. This is so, since, if (14) were causal, then an effect could precede a cause.[23]

---

23. It may be of interest to note that Hempel allows explanations such as (14), though he too denies they are causal. Employing an analogous example that uses Fermat's principle of least time, he argues that "sometimes a particular event can be satisfactorily explained by reference to subsequent occurrences," *Aspects of Scientific Explanation*, p. 353.

12. RESPONSES

Utilizing definition (B) of causal explanation given in Section 4, and certain general claims about causation (C_1 through C_3), I have argued that various types of scientific explanations are not causal. These include the explanation of a regularity as being a special case of a general law, the explanation of a fact by reference to a system of classification (or more generally, identity explanations), and simple and complex derivation explanations. Let me now consider four responses to my claim that not all scientific explanations are causal.

The first is that those who defend the causal thesis have in mind a narrower position than I have been discussing. Their view is that all scientific explanations of *particular events, or occurrences, or phenomena* are causal. They are not committed to the broader thesis that no matter what is explained the explanation must be causal. They would then point out that the "special-case-of-a-law," "classification," and "derivation" explanations I have introduced are not explanations of particular events.

Even this narrower thesis can be shown to be false if we use the concept of causal explanation supplied earlier, together with the general principles of causation introduced. Suppose that a particular projectile is continuing to move once it has left the hand. We have here, I assume, a particular event, or occurrence, or phenomenon, for which the following explanation is possible:

(1) (The reason that projectile is continuing to move after it has left the hand is that it is a law that any moving body will continue to move unless its motion is retarded by other bodies; explaining why that projectile is continuing to move after it has left the hand).

(1) is a "special-case-of-a-law" explanation of a particular event which, like (1) of Section 7, can be shown to be non-causal.

Or, to take an example of a different kind, suppose that the pH value of a particular solution is changing. *One* type of explanation of this occurrence is as follows:

(2) (The reason the pH value of that solution is changing is that the concentration of hydrogen ions which that solution contains is changing; explaining why the pH value of that solution is changing).

Here we have a classification explanation, since the pH value of a solution is defined as the log of the reciprocal of the concentration of hydrogen ions in the solution. And (2), like the classification explanations of Section 8, can be shown to be non-causal using our definition of a causal explanation together with principle C_2 governing causation. No doubt causal explanations can be given for why the pH value is changing (e.g., because an acid is being poured into the solution). But this does not preclude the possibility of a non-causal explanation of type (2).

Even non-causal (complex) derivation explanations of particular occurrences are possible, as can be seen if we change the conservation explanation given by (8) and (9) of the previous section to be an explanation of why, when one particular particle struck a stationary particle of equal mass, the first stopped and the second moved off with the velocity of the first.

Let me turn to a second response. When it is said that explanations in science are causal all that is meant is that scientific explanations appeal to laws. On this view, a causal explanation can be thought of as having the form

(3) (The reason that p is that r (where r cites laws); explaining why it is the case that p).

The claim that all explanations in science are causal is to be understood simply as the claim that they all have this form.

Even someone like Salmon who accepts the idea that scientific explanations are causal would not agree that a causal explanation is just any explanation of form (3). He would allow that there are non-causal explanations of form (3) that satisfy his S-R model. Salmon (like Hempel and Brody) would say that an explanation of form (3) is a causal explanation only if the laws cited are of certain types.

More importantly, the above criterion is much too broad. There are good reasons for denying that the following explanations are causal, despite the fact that they are of form (3):

(4) (The reason the moon exerts a gravitational force on the waters of the earth is that it is a law that all bodies exert gravitational forces on all other bodies; explaining why the moon exerts a gravitational force on the waters of the earth);

(5) (The reason that H_2SO_4 is an acid is that it is a proton donor; explaining why H_2SO_4 is an acid);[24]

(6) (The reason that when one particle strikes a stationary particle of equal mass the first stops and the second moves off with the velocity of the first is that it is a law that the total momentum and total energy of a system of particles remain constant; explaining why when one particle, etc.).

---

24. That H_2SO_4 is a proton donor is a law, according to the liberal notion of law utilized by Hempel and others. What is not a law, but a definitional fact, is that an acid is a proton donor.

That these explanations are non-causal can be established by invoking definition (B) of causal explanation together with the principles of causation C_1 and C_2; the latter in turn can be justified by appeal to the idea that effects are not "included in" causes, in a sense defined earlier.

If an explanation such as (4) were to be counted as a causal explanation, then Newton's claim that he did not know the cause of gravity would have to be construed in a manner different from the one he intended. In a well-known passage from Book III of the *Principia*, Newton writes:

> Hitherto we have explained the phenomena of the heavens and of our sea by the power of gravity, but have not yet assigned the cause of this power. This is certain, that it must proceed from a cause that penetrates to the very centers of the sun and planets, without suffering the least diminution of its force; that operates not according to the quantity of the surfaces of the particles upon which it acts (as mechanical causes used to do), but according to the quantity of the solid matter which they contain, and propagates its virtue on all sides to immense distances, decreasing always as the inverse square of the distances. . . . But hitherto I have not been able to discover the cause of those properties of gravity from phenomena, and I frame no hypotheses.[25]

To be sure, Newton can be said to have given a causal explanation of the tides, as follows:

> (The reason the tides occur is that the moon exerts a gravitational force on the waters of the earth; explaining why the tides occur).

(The explanation is causal in virtue of the fact that it has a what-causal equivalent.) Newton invoked gravity as a cause of various "phenomena of the heavens and of our sea." What Newton claimed he could not give is a cause of gravity itself. I suggest that this entails that he did not know what causes a body, such as the moon, to exert a gravitational force on another body, such as the waters of the earth. And from this I conclude that Newton could not give a causal explanation of why the moon exerts a gravitational force on the waters of the earth. If (4) is a causal explanation, then he could easily have given a causal explanation of why the moon exerts such a gravitational force.

In view of this, perhaps a defender of the thesis that all scientific explanations are causal will say that what he means is not simply that they all invoke laws, but that they all invoke causal laws. They are of the form

> (7) (The reason that p is that r (where r cites *causal* laws); explaining why it is the case that p).

25. Pp. 546–47.

One who makes such a claim owes us some account of causal laws. Is Newton's law of gravitation to be counted as a causal law? If it is, then by the present proposal, explanation (4) would have to be classified as a causal explanation of why the moon exerts a gravitational force upon the waters of the earth—Newton's protestations to the contrary. But even if we were to ignore such protestations and accept the present suggestion about causal explanation, it is simply false to suppose that all explanations in science are of form (7). For one thing, explanatory questions are not always why-questions, and constituent propositions are not all expressible by sentences of the form "the reason that p is that r." For another, even if we restrict the thesis to explanations whose constituent questions and propositions are of this form, the thesis that all such explanations in science cite causal laws cannot be supported. Think of the classification explanation (2) above. No laws are cited, and the only general truth presupposed—one relating pH value to hydrogen ion concentration—is a definition and not, I take it, a proposition that anyone would classify as a causal law. Or think of the explanation (6), which is part of a complex derivation explanation given earlier. If causal laws state or at least entail that something causes something else,[26] then conservation laws which state that the total quantity of, say, energy or momentum of an isolated system always remains constant ought not to be classified as causal laws. They do not state, or entail, or even presuppose, that something causes something else.

The last response I shall note is that even if not all explanations offered in the sciences are causal, a scientific explanation is a good one only if it is causal.

Our discussion in Chapter 4 of the role of general methodological values in the assessment of scientific explanations is relevant here. (See Section 16.) Such values, we recall, are particularly important for helping to determine the satisfaction of two of the four conditions for I to be appropriate instructions for explaining q to an audience. These conditions are that

(c) the audience is interested in understanding q in a way that satisfies I (or would be under the conditions noted in Chapter 4, Section 3);

(d) understanding q in a way that satisfies I, if it could be achieved, would be valuable for that audience.

Now, our concern is with scientific audiences, and, for the sake of argument, let us suppose that such audiences in general are interested

---

26. See Nancy Cartwright, "Causal Laws and Effective Strategies," p. 420.

in understanding things causally, which, if it could be achieved, would be valuable. However, as we saw in Chapter 4, general methodological values fail to provide sufficient conditions for appropriate instructions to follow in explaining q to a scientific audience that shares such values. They are not sufficient because (c) and (d) above do not suffice. What must be considered is not only whether the scientific audience is or would be interested in the satisfaction of these values, and whether their satisfaction is desirable, but also whether the audience already understands q in a way that satisfies such values, and if not, whether it is possible for it to do so. (See Chapter 4, Section 4.) Indeed, one must consider the understanding and interests of the particular scientific audience in ways more specific than simply by appeal to general methodological values. Bohr could have offered the following explanation of the spectral lines:

> (Excitation of hydrogen by electricity or heat causes the hydrogen spectral lines; explaining what causes the spectral lines of hydrogen).

The explanation is causal, by our criterion, and correct. But it would not have been a good explanation for Bohr to have given, since it ignores the specific interests and knowledge of physicists in 1913.

Nor do general methodological values provide *necessary* conditions for appropriate instructions in science. As shown in the case of Boyle's explanation of the regularities in the mercury columns, a macro-level of understanding may be valuable for a scientific audience, even if that audience also prizes micro-understanding. Similarly, a non-causal understanding of a phenomenon—e.g., an understanding provided by a special-case-of-a-law explanation, or by a macro-micro identity explanation, or by a non-causal complex derivation explanation—may be valuable for a scientific audience, even if a causal understanding is also valuable. It may be valuable (in part) because it reflects methodological values that are also sought in science, e.g., use of laws, or appeal to micro-structure, or use of derivations. There are, in short, methodological values in addition to the assignment of causes. An explanation which incorporates them can be a good one, even if it is not causal.

13. COVERING LAW AND STATISTICAL EXPLANATIONS OF PARTICULAR EVENTS

Material in the present chapter is applicable to two types of explanations widely discussed in the philosophical literature. Both are explanations of particular events or states of affairs in which other particular events or states of affairs together with laws are explicitly cited as explanatory factors. Now, as we have already seen, not all such explanations are causal. (Recall, e.g., the conservation of momentum explanation (13) of Section 11.) So that we may focus clearly on causal explanations of these sorts, let us restrict our attention to cases in which a particular event is cited, a causal law is invoked which states that an event of the first type causes one of the second, and both the event and the causal law are being invoked as explanatory factors.

Consider first an explanation of this sort involving a universal causal law (which states that any event of the first type causes one of the second). Suppose that in the course of explaining what caused the expansion of a particular piece of metal a scientist invokes the fact that the metal was heated and that it is a law that heating any metal causes it to expand. How should the explanation be formulated? One obvious proposal is this:

(1) (The fact that this metal was heated and that it is a law that heating any metal causes it to expand caused this metal to expand; explaining what caused this metal to expand).

This satisfies criterion (B) for causal explanations. However, there may be a problem with (1), depending upon certain putative features of the concept of "cause" that we have not yet discussed.

Peter Unger,[27] has suggested that causation satisfies the following uniqueness condition:

(2) If some entity caused a particular event, then no other entity caused that event.

For example, if the fact that Smith had a bacterial infection caused his death, then the fact that he had a viral infection did not cause it. To be sure, both facts together—the pair—may have caused Smith's death. But then it would be untrue to say that the first fact caused it; that fact was only part of the cause.

Applying this to (1), let us suppose—what seems plausible—that

(a) The fact that this metal was heated

caused this metal to expand. ((a) is not part of the cause, it is the cause.) Then by (2), it is not the case that

(b) The fact that this metal was heated and that it is a law that heating any metal causes it to expand

is what caused this metal to expand, since (a) and (b) are different causal entities. So if the uniqueness condition governs causation, and

---

27. Peter Unger, "The Uniqueness in Causation," *American Philosophical Quarterly* 14 (1977), pp. 177–88.

if this metal's being heated is what caused it to expand, then although (1) is a causal explanation it is an incorrect one.

Unger's (2) is a very strong condition (many will say: too strong). The following weaker principle, which also applies to (1), might be proposed:

> (2)' If the fact that this X has P caused this X to have Q, then the fact that this X has P and that it is a law that any X's having P causes it to have Q did not cause this X to have Q.

One way to justify (2)', of course, is as being a special case of the uniqueness condition (2). But there is another possibility, which does not carry a commitment to (2). It is to claim that only *particular events (or states of affairs) can be causes, or parts of causes, of other particular events—though they are so in virtue of the fact that laws obtain.* Such a claim might well be made by a Humean, who is concerned with singular causal sentences of the form "event e_1 caused event e_2." According to the Humean, sentences of this form entail general laws relating the types of events in question. But they do not entail that such laws themselves are causes, or parts of causes, of the particular event e_2. Hume himself defines a cause as "an object precedent and contiguous to another, and where all the objects resembling the former are placed in a like relation of priority and contiguity to those objects that resemble the latter." The cause is the object (the event), and it is a cause provided that there is a law governing objects of that type. But the law itself, or the fact that it obtains, is not the cause, or part of the cause. Now if the fact that this X has P and *that it is a law that any X's having P causes it to have Q* caused this X to have Q, then, contrary to what has just been said, the fact expressed by the emphasized phrase would be part of the cause of this X's having P. So (2)' is justified, on such a view.

Let us suppose that causation is subject to condition (2)', if not to (2). The justification of (2)' I have suggested is based on the principle that only particular events or states of affairs cause other particular events. Some who are willing to accept the earlier conditions C_1 through C_3 governing causation may balk at this. Nevertheless, many of those who focus on explanations of particular events are Humeans, or Humean sympathizers, who might accept (2)' for the reason suggested. How might such persons reformulate the explanation of the scientist who invokes both a particular event and a causal law in explaining what caused this metal to expand? (How can explanation (1) be reformulated in such a way that it is not incorrect?)

Taking a clue from complex derivation explanations we might say that our scientist is explaining not one thing, but two: what caused this metal to expand, and why that cause was operative. His explanations are these:

(3) (The fact that this metal was heated caused it to expand; explaining what caused this metal to expand);

(4) (The reason that the heating of this metal caused it to expand is that it is a law that heating any metal causes it to expand; explaining why the heating of this metal caused it to expand).

The first explanation (3) is, by our criterion (B), a causal explanation of what caused this metal to expand. If its constituent proposition is true, (3) is a correct explanation. By contrast, (4) does not explain what caused this metal to expand, but why the heating of this metal caused it to expand. Therefore, the truth of the constituent proposition in (4) does not violate (2)' or (2) by entailing that something in addition to the heating of this metal, viz. a law's obtaining, caused it to expand. Indeed, (4) is a special-case-of-a-law explanation, which, in Section 7, we argued must be non-causal. On the present interpretation, then, a scientist who invokes a particular event and a causal law in the course of explaining the expansion of this metal can be taken to be providing two explanations, only one of which is causal.

A second type of explanation widely discussed in the philosophical literature is an explanation of a particular event which explicitly invokes as explanatory factors another particular event together with a *statistical* causal law. (The latter states that in x percent of the cases one type of event causes another.) Suppose that in explaining what caused Smith's death, the doctor invokes the fact that Smith contracted legionnaire's disease and the statistical law that in 65 percent of the cases the contraction of this disease causes death. Shall we formulate his explanation as follows:

(5) (The fact that Smith contracted legionnaire's disease and that it is a statistical law that in 65 percent of the cases the contraction of this disease causes death caused Smith's death; explaining what caused Smith's death)?

If we do, and if we assume a version of (2)' applicable to statistical laws (or if we assume (2)), and if Smith's contracting legionnaire's disease caused his death, then, as before, we will have to regard (5) as incorrect.

An alternative is to substitute for (5) a pair of explanations, as follows:

(5a) (The fact that Smith contracted legionnaire's disease caused his death; explaining what caused Smith's death);

(5b) (The reason that it was possible for the contraction of legionnaire's disease to cause Smith's death is that it is a statistical law that in 65 percent of the cases the contraction of this disease causes death; explaining why (or how) it was possible for the contraction of legionnaire's disease to cause Smith's death).

(5a) is a causal explanation, by our criterion. (5b) does not explain the same thing as (5a). Rather, it explains why or how it was possible for the cause to cause the effect—by appeal to the fact that it is a statistical law that in x percent of the cases such causes cause such effects. (If the percentage is very high, perhaps, as in the universal cases, one can explain not only why it was possible for the cause to cause the effect, but also why it did do so; however, I shall not pursue this.) Explanation (5b) would be particularly appropriate for an audience that does not realize that legionnaire's disease can cause death, or for one that believes that it can do so only rarely. We might speak of (5b) as a special-case-of-a-*statistical*-law explanation, which, it can be argued, is not causal.

The plausibility of such reinterpretations of statistical explanations becomes especially apparent when the statistical laws involve low percentages. Consider the following example, which has been discussed in the literature.[28] Jones gets a rash; his doctor explains what caused it by saying that Jones had a penicillin injection, and that in 2 percent of the cases penicillin causes rashes. Should we construct the following explanation?

(6) (The fact that Jones had a penicillin injection and that it is a statistical law that in 2 percent of the cases penicillin injections cause rashes caused Jones's rash; explaining what caused Jones's rash).

Some philosophers, notably Hempel, object that (6) could not be a correct statistical explanation since the percentage quoted is very low. If only 2 percent of penicillin injections result in rashes, then from this fact and the fact that Jones took penicillin we could not infer that Jones got a rash. (And the satisfaction of this inference-condition, which Hempel calls the explanation-prediction symmetry requirement is, for Hempel, necessary for correct statistical and non-statistical explanations of particular events.)

All of this is avoided if we take the doctor to be formulating two explanations rather than one. First, he is offering a causal explanation of the rash:

28. The example was introduced by Arthur Collins in "The Use of Statistics in Explanation," *British Journal for the Philosophy of Science* 17 (1966), pp. 127–40, as a counterexample to Hempel's statistical model of explanation. See my discussion of this in *Law and Explanation*, pp. 105–8.

(Jones's penicillin injection caused his rash; explaining what caused Jones's rash).

Second, he is offering a non-causal statistical explanation of why it was possible for that cause to cause the rash:

(The reason that it was possible for Jones's penicillin injection to cause his rash is that it is a statistical law that in 2 percent of the cases penicillin causes rashes; explaining why it was possible for Jones's penicillin injection to cause his rash).

There is nothing untoward about saying that the penicillin injection caused the rash, while admitting that in only 2 percent of the cases it does so. There is something questionable about the claim, implied by the constituent proposition in (6), that the fact that it is a statistical law that in 2 percent of the cases penicillin causes rashes is part of the cause of Jones's rash.

Finally, we can resolve a dispute about whether events can be explained by appeal to statistical laws which assign low probabilities to such events. Salmon, on the one hand, suggests that explanations such as this—or his S-R version—can be correct, if their statistical laws are true:

(7) (The reason that the electron in a diffraction experiment hit the screen where it did is that, according to such and such statistical laws of quantum mechanics, the probability that an electron will hit the screen at that spot is such and such a (low) number; explaining why the electron hit the screen where it did).

By contrast, according to Hempel, if the probability in question is low, then (7) could not be a correct explanation.

A resolution of this dispute is to replace (7) with

(8) (The reason it was possible for the electron to hit the screen where it did is that, etc.; explaining why or how it was possible for the electron to hit the screen where it did).

(8) does not purport to explain why the electron hit the screen where it did, but why or how it was possible for it to do so. And (8) is not a causal explanation. If quantum mechanics is correct—and if there is no deeper causal theory—then nothing caused the electron to hit the screen where it did. So there is no correct causal explanation of why the electron hit the screen where it did. Still there can be a correct non-causal explanation of why or how it was possible for that event to occur.

This section began with an explanation of a particular event which invokes, as explanatory factors, another particular event and a universal causal law. If the explanation is formulated as (1) above then it is obviously causal, by the criterion (B) I have been expounding, whether or not the causal condition (2)' or (2) is accepted. However, if there is a concept of causation satisfying either of these conditions, then someone who explains a particular event by reference to another event together with a causal law may be offering a causal explanation of one thing, and a non-causal special-case-of-a-law explanation of another.

14. REASON-GIVING SENTENCES

Various types of reason-giving explanations of the form

(The reason that p is that r; explaining why p)

have been characterized in this chapter. Is it possible to use these to provide a set of truth-conditions for sentences of the form

(1) The reason that p is that r

in which p and r are sentences? The following might be suggested:

(2) A sentence of form (1) is true if and only if p is true, r is true, and a third condition obtains given by an alternation of conditions supplied by the reason-giving explanations we have distinguished.

The third condition is an alternation of the following:

(3) (Causal): The cause of its being the case that p is that r;

(4) (Special-case-of-a-law): A law of which the fact that p is a special case is that r;

(5) (Classification): p is a sentence of the form "x is a P_1" and r is of the form "x is a P_2"; and according to such and such a definition of "P_1", something is a P_1 if and only if it is a P_2;

(6) (Macro-micro): p is a sentence of the form "x is a P_1," r is of the form "x is a P_2," the property of being a P_1 = the property of being a P_2, and "P_1" is a macro-predicate and "P_2" is a micro-predicate (or conversely).

So, e.g., in view of (2) and (3), the sentence

The reason that this body accelerated is that it was subjected to unbalanced forces

is true if this body did accelerate, it was subjected to unbalanced forces, and the cause of its accelerating is that it was subjected to unbalanced forces.

Similarly, in view of (2) and (5), the sentence

The reason that H_2SO_4 is an acid is that H_2SO_4 is a proton donor

is true if H_2SO_4 is an acid, it is a proton donor, and by definition something is an acid if and only if it is a proton donor.

However, no reason has been offered to suppose that the conditions (3) through (6) are exhaustive. (There are numerous reason-giving sentences, including teleological, dispositional, and probabilistic ones which, it might well be argued, fail to fit one of these categories.) Let us ask, then, whether (2), supplemented with an alternation of (3) through (6), provides a set of sufficient (although not necessary) conditions for the truth of reason-giving sentences of form (1).

Even this does not obtain. The problem (in every case but (3)) is the same as that raised in Chapter 5 against models of explanation satisfying the a priori requirement: the law, classification, or macro-micro states may not be the "operative" ones in the case in question. Focusing on (4), consider two examples used in Chapter 5:

(7) The reason that Jones died within 24 hours of eating a pound of arsenic is that anyone who eats a pound of arsenic dies within 24 hours.

(8) The reason that this small piece of iron moved toward that magnetic bar is that any magnetic bar is such that when a small piece of iron is placed near it the iron moves toward the bar.

Suppose, however, that Jones died in an unrelated car accident, and that the iron moved toward the bar because of a much more powerful contact force that intervened (see Chapter 5, Section 5). Then neither (7) nor (8) is true, even though (2) supplemented with (4) is satisfied.

Turning to (5), consider

(9) The reason that iron has the atomic number 26 is that it has the atomic weight 55.85.

Iron does have the atomic number 26 and the atomic weight 55.85. And there is a definition of atomic number in accordance with which atomic numbers are determined by atomic weights. Yet physicists today would not regard (9) as correct, since the definition in question, though once accepted, is so no longer. That definition was offered before the discovery of isotopes of elements. (Atomic number is now defined by reference to the number of protons in the nucleus of the atom.)

Finally, in the case of (6), consider

(10) The reason that airplanes are made of aluminum is that they are made of the element whose atoms contain 13 protons.

Criterion (2) supplemented with (6) obtains. (I shall assume that "made of aluminum" and "made of the element whose atoms contain 13 protons" are macro- and micro-predicates, respectively, expressing the same property.) Yet (10) is not true under the most natural interpretation of "the reason that airplanes are made of aluminum." What is true, perhaps, is that the reason that this substance, of which airplanes are made, is aluminum is that it is the element whose atoms contain 13 protons.

To avoid these problems we need some notion of an "operative" law, definition, and macro-micro identity. For example, we need to change (4) to

(4)' An operative law of which the fact that p is a special case is that r.

What is the difference between (4) and (4)'? What can be meant by an "operative" law in (4)'? It cannot mean simply a law that holds, since that is so in both (4) and (4)'. And, if my earlier conclusions are right, it cannot mean a law, expressed by r, whose existence *causes* it to be the case that p; for this would violate principle C_1 (Section 7), according to which the existence of a regularity is not caused by the existence of a law of which that regularity is a special case. What we really intend by "operative" in (4)' is a law that r *which is such that the reason that p is that r*. But this would obviously render (4)' circular as a truth-condition for (1). Analogous remarks apply to (5) and (6) modified to include the term "operative" (definition, identity).

The present problem could be avoided by using only the causal condition (3) as a basis for (2), and saying

(2)' A sentence of the form "the reason that p is that r" is true if and only if p is true, r is true, and the cause of its being the case that p is that r.

I assume that causes are (reason-)"operative," so that if the cause of its being the case that p is that r then the reason that p is that r. With (2)', then, the "operative" problem does not arise. But (2)' belies the main conclusion of the present chapter which is that there are non-causal explanations, including non-causal reason-giving ones. Assuming that the existence of a law constitutes a reason that a certain regularity obtains which is a special case of that law, and assuming that the existence of a law does not cause the existence of a regularity which is a special case of that law, (2)' must be rejected.

Let us say that a type of reason is *necessarily operative* if and only if the truth of p and r, and the satisfaction of the condition governing a reason of that type (e.g., one of the conditions in (3) through (6)), suffices for the truth of

(1) The reason that p is that r.

We have been supposing that causal reasons are necessarily operative. By contrast, special-case-of-a-law reasons are not. This is because the truth of p and r, and the satisfaction of condition (4)—that a law of which the fact that p is a special case is that r—does not guarantee the truth of (1).

Now I have made the following claims:

(a) r can be a reason that p in virtue of being a reason of a certain type, even though it is not the case that this type of reason is necessarily operative. For example, the fact that all unsupported bodies fall with uniform acceleration is the reason that bullets when fired from a gun fall with uniform acceleration. It is a reason in virtue of the fact that the latter regularity is a special case of the former law. And this is so even though it is not true that special-case-of-a-law reasons are necessarily operative.

Note that at the end of Section 7 I claim that the fact that it is a law that L can be the reason that p obtains in virtue of the fact that p is a special case of law L. Some philosophers might suppose that unless a type of reason is necessarily operative, a particular r cannot be a reason that p in virtue of being a reason of that type. But why should that be? In a given case, r can be a reason that p in virtue of being a reason of type T, even though T is not a necessarily operative type of reason. (By analogy, in a given case eating that substance can be the cause of death in virtue of the fact that the substance is of the type "arsenic," even though eating arsenic is not a necessarily operative cause of death—an arsenic-eater may die of unrelated causes.)

(b) Non-necessarily operative reasons of the sort I have considered in this chapter do not, and indeed if I am right, cannot require causation to make them operative.

Since I have provided no general criteria that determine whether reasons are operative in a given case, the present chapter does not supply a set of truth-conditions for reason-giving sentences.

15. IMPLICATIONS FOR PROGRAMS SEEKING TO DEFINE "CORRECT EXPLANATION"

In Chapter 5 we examined various attempts to provide a set of sufficient conditions for sentences of the form

(1) E is a correct explanation of q.

According to one type of proposal, associated particularly with Hempel's D-N model, these conditions must satisfy the a priori and NES requirements. The former states that the only empirical consideration in determining whether the explanans correctly explains the explanandum is the truth of the explanans; all other considerations are a priori. The NES requirement precludes a singular explanans sentence from entailing the explanandum. If Q is a question of the form

Why is it the case that ———?

then, where ——— describes some particular event or state of affairs, NES disallows as an explanans a sentence of the form

(2) The reason that ——— is that ———.

According to a second type of proposal, associated with Salmon's S-R model, Brody's essential property model, and Jobe's priority model, the conditions for (1) must satisfy NES but not the a priori requirement. Modelists of both types, then, espouse NES. What they seek, in effect, is a set of sufficient conditions for sentences with forms such as (1) and (2)—where these conditions do not themselves invoke explanatory terms like "explanation," "reason," or "cause."

It was argued in Chapter 5 that various attempts to provide a set of conditions for (1) satisfying the NES and a priori requirements, or just NES, are unsuccessful. And in the previous section we have observed unsuccessful attempts to provide a set of sufficient conditions for sentences of form (2).

By contrast, there is a third program which can be carried out. It involves providing a set of sufficient conditions for sentences of form (1) that satisfy the a priori requirement but not NES. The conditions are those formulated at the beginning of Chapter 4 using the ordered pair theory, viz.

(3) If (p; explaining q) is an explanation, it is correct if and only if p is true.

Or equivalently,

(4) E is a correct explanation of q if Q is a content-question and E is an ordered pair whose first member is a true proposition that is a complete content-giving proposition with respect to Q and whose second member is the act-type *explaining q*.

The a priori requirement is satisfied since the only empirical consideration in determining whether the explanation is correct is the truth of the explanans p. However, NES is not satisfied, since proposition p can be one expressible by sentences with forms such as (2). Of course, if the program is simply to provide conditions for a correct explanation which do not invoke explanatory terms such as "explanation," "reason," or "cause," then (3) and (4) are successful. The notion of an explaining act, that of a content-question, and that of a complete content-giving proposition with respect to Q, are all defined in Chapter 2 without making use of explanatory terms.

The rub is that content-giving propositions, which are constituent propositions of explanations, include (within their large array) ones expressible by sentences of form (2) containing explanatory terms; and no truth-conditions for such sentences have been supplied. Does this render (3) and (4) useless as conditions for correct explanations? Several points deserve notice.

(a) As I emphasized in Chapter 4, (3) and (4) are meant to be quite weak conditions for evaluating explanations. Their satisfaction is neither necessary nor sufficient for an explanation to be a good one. This is especially so in science, where what is wanted is not just correctness (or the likelihood of correctness) but correctness "in the right sort of way" (as determined by contextual, methodological, and empirical factors of the kind described in Chapter 4).

(b) Despite their weakness, (3) and (4) are very general. They are correctness conditions for explanations whose constituent questions are not simply why-questions but any *content-questions whatever*. The constituent propositions in such explanations include not only ones expressible by sentences of form (2), but ones expressible by sentences with forms such as

The cause of ——— is that ———;

The method by which ——— is that ———;

The significance of ——— is that ———;

The excuse of ——— is that ———;

The desire of ——— is that ———.

And the theory of correct explanations embodied in (3) and (4) provides no truth-conditions for such sentences either.

(c) Theories of correct explanation that restrict their attention to why-questions and attempt to formulate truth-conditions for sentences of form (2)—theories of the sort suggested by the modelists of Chapter 5, as well as by the proposals in the previous section—are unsuccessful. They run afoul of the "inoperative" reason problem. (A strictly analogous problem plagues the many attempts since the time of Hume to supply truth-conditions for sentences of the form "The cause of ——— is that ———.") I offer the speculation that any attempt to satisfy NES—or more narrowly, any program to provide truth-conditions for sentences such as (2) which completely eschews explanatory concepts—will be subject to this problem.

(d) It will be useful to compare what I have said about "correct explanations" with Aristotle's theory of the four causes. Aristotle can be construed as providing a set of sufficient conditions for

(1) E is a correct explanation of q

in cases in which Q is restricted to why-questions. The conditions are these:

> (5) E is a correct explanation of q if Q is a question of the form "Why does X have P?" and E is a proposition of the form "X has P because r" in which r (correctly) gives one or more of Aristotle's four causes of X's having P.

Aristotle's (5), like my (4), satisfies the a priori requirement but violates NES. (We take "X has P" to be the explanandum and "X has P because r" to be the explanans.) Aristotle does provide truth-conditions for sentences of the form "X has P because r." But these invoke the explanatory notions of efficient, material, formal, and final cause, for which no definitions are given in non-explanatory terms.

Aristotle's program for correct explanations might be put like this:

> A. Its aim is to say what is common to correct explanations involving why-questions.
> B. This is done by invoking a notion of cause which is undefined and which Aristotle may be taking to be indefinable (in non-explanatory terms).
> C. Using this notion of cause Aristotle characterizes various types of causal explanations but does not explicate these in non-causal (or more broadly, non-explanatory) terms.

The program for correct explanations represented by (3) and (4) might be put (in part) like this:

> A'. Its aim is to say what is common to correct explanations of all types, not just ones involving why-questions.
> B'. This is done by invoking notions, including that of a content-giving proposition, that (unlike the case with Aristotle) are defined independently of explanatory ones. However, among such propositions are ones expressible by sentences of the form "the reason that ——— is that ———" and "the cause of ——— is that ———." And, as in the case of Aristotle, no truth-conditions for such sentences are provided in non-explanatory terms.
> C'. Using an undefined notion of cause, together with the apparatus of the ordered pair theory and certain principles governing causation, the present chapter defines a concept of causal explanation and characterizes various types of non-causal explanations.

Let us call a doctrine of correct explanations *reductionist* if

> (i) the doctrine provides a definition (or necessary and/or sufficient conditions) for "correct explanation" using only concepts defined independently of explanatory ones;

(ii) the application of the definition to particular cases—i.e., determining the correctness of particular explanations using this definition—never requires determining the truth of any explanation sentences.

Hempel's D-N model is reductionist since it satisfies (i) and (ii). Aristotle's doctrine is non-reductionist since it violates both (i) and (ii). The definitions I have proposed in terms of the ordered pair theory satisfy (i) but not (ii). They are only partially reductionistic. And I have provided reasons for doubting that one will get a successful theory of correct explanation that satisfies both reductionist conditions.

16. CONCLUSIONS

In the present chapter, utilizing the ordered pair view of the illocutionary theory, a concept of causal explanation has been defined that I take to be broad and intuitive. Certain general principles governing causation (C_1 through C_3) were also introduced and defended. Using these principles, together with the definition of causal explanation, I dispute the claim that scientific explanations are all causal. For this purpose, again utilizing the ordered pair view, a number of types of explanations frequently found in the sciences were characterized and shown to be non-causal. Among them are explanations of a regularity that cite a law of which the regularity is a special case; identity explanations (which include classification explanations as well as ones that explain why an item has a certain macro-property by appeal to a contingently identical micro-property); and simple and complex derivation explanations. Nor will the claim that scientific explanations are all causal be established if it is restricted to (non-teleological) reason-giving explanations of the form

(The reason that p is that r; explaining why p),

even to ones in which particular events are explained and laws are cited. There are "special-case-of-a-law," "identity," and "complex derivation" explanations of these types that are not causal. Moreover, someone who explains what caused a particular event by citing another event and a causal law governing such events may be offering a pair of explanations, one of which is not causal. On the other hand, if the claim that scientific explanations are causal is understood as evaluative rather than descriptive—as providing a criterion for good scientific explanations—then it provides one that is neither necessary nor sufficient.

To assert that there are various types of non-causal explanations in the sciences is not, of course, to say that science is devoid of causal explanations. That would be absurd. Nor is it to deny that causal propositions may be entailed or presupposed by the constituent propositions of non-causal explanations. The constituent proposition in the explanation

> (1) (The reason the moon exerts a gravitational force on the waters of the earth is that it is a law that all bodies exert gravitational forces on all other bodies; explaining why the moon exerts a gravitational force on the waters of the earth)

entails causal propositions. But it does not follow that (1) is a causal explanation of why the moon exerts a gravitational force on the waters of the earth. (Indeed, as I have suggested, Newton could have accepted (1) as an explanation of why the moon exerts such a force, while denying that this is a causal explanation.)

More generally, from the fact that the constituent proposition of an explanation of q entails or presupposes a law which asserts that something causes something under certain conditions, it does not follow that the explanation is a causal explanation of q.

CHAPTER 8
Functional Explanation

Talk of functions looms large in discussions of teleology. The concern has been mainly with sentences of the form "The (or a) function of x is to do y," e.g., "The function of the heart is to pump the blood." Two questions about such sentences have been of particular interest to philosophers: what do they mean, and what sort of explanation, if any, do they offer? The present chapter is devoted to these questions.

1. THREE ACCOUNTS OF FUNCTIONS

In recent years analyses of functions tend to fall into three general categories. Most popular is what I shall call the good consequence doctrine, which appears in various forms in the writings of Canfield,[1] Hempel,[2] Hull,[3] Lehman,[4] Ruse,[5] Sorabji,[6] and Woodfield.[7] The general idea is that doing y is x's function only if doing y confers some good.

Some specify what this good is, others do not. According to Sorabji, there is a sense of "function" for which a necessary condition is that "the performance by a thing of its (putative) function should confer some good."[8] But the goods can vary and particular ones should not

---

1. John Canfield, "Teleological Explanation in Biology," *British Journal for the Philosophy of Science* 14 (1963-64), pp. 285-95.
2. Carl G. Hempel, "The Logic of Functional Analysis," reprinted in Hempel, *Aspects of Scientific Explanation.*
3. David Hull, *Philosophy of Biological Science* (Englewood Cliffs, N.J., 1974).
4. Hugh Lehman, "Functional Explanation in Biology," *Philosophy of Science* 32 (1965), pp. 1-20.
5. Michael Ruse, *The Philosophy of Biology* (London, 1973).
6. Richard Sorabji, "Function," *Philosophical Quarterly* 14 (1964), pp. 289-302.
7. Andrew Woodfield, *Teleology* (Cambridge, 1976).
8. "Function," p. 291. Sorabji also recognizes a second sense of function which involves the idea of making an effort to obtain a result.

be specified as part of an analysis. Woodfield has a similar account. Hempel and Ruse, on the other hand, identify specific goods, adequate working order for the former, survival and reproduction for the latter. According to Hempel, a function of x in a system S is to y if and only if x does y and x's doing y ensures the satisfaction of certain conditions necessary for S's remaining in adequate, effective, or proper working order. For Ruse, the function of x in S is to y if and only if S does y by using x and y aids in the survival and reproduction of S.

In what follows I shall formulate the doctrine in a general way without mentioning specific goods, although this can be done to obtain particular versions.

The Good-Consequence Doctrine: The (a) function of x (in S) is to y if and only if x does y (in S) and doing y (in S) confers some good (upon S, or perhaps upon something associated with S, e.g., its user in the case of artifacts).

Thus we can say that the function of the heart in mammals is to pump the blood, because the heart does pump the blood in mammals and pumping blood confers a good upon mammals. But we cannot say that the function of a poison ivy reaction in humans is to make the skin itch, since the latter confers no good upon humans.

The second doctrine focuses upon the idea of contributing to a goal that something or someone has.

The Goal Doctrine: The (a) function of x (in S) is to y if and only if x does y (in S) and doing y (in S) is or contributes to some goal which x (or S) has, or which the user, owner, or designer of x (or S) has.

For example, a function of my hands is to grasp objects, because my hands do this, and grasping objects is, or at least contributes to, one of the goals which I, their user, have. An appeal may be made here to the notion of a *goal directed system,* roughly, a system which persists in, or tends to achieve, some state (the goal) under a variety of disturbing conditions. Those who speak this way would say that the function of the thermostat in a home heating system is to turn the heat on and off, since doing so contributes to a "goal" of the heating system, viz. to maintain a constant temperature. Anyone who refuses to attribute goals to inanimate objects can still espouse the present doctrine, provided that he attributes the goal to an animate user, owner, or designer of the object or system.

Although so far as I know no one has formulated the doctrine in exactly this way, a goal doctrine is suggested by Nagel, Adams, and Boorse. According to Nagel, a function of x in S (under environmental conditions E) is to y if and only if x is necessary for S under E to do y, and S is a goal directed system with some goal G, to the realization or maintenance of which doing y contributes.[9] On Adams's analysis, a function of x in S is to do y if and only if x does y in S, S is a goal directed system with a goal-state O, and doing y causally contributes to achieving O through the causal feedback mechanism in S.[10] And Boorse claims that x is performing the function y in the G-ing of S at time t if and only if at t, x is doing y, which is making a causal contribution to the goal G of the goal directed system S.[11] According to these authors, in the case of items such as guided missiles the goals are to be attributed to the artifacts themselves; with items such as chairs and fountain pens, according to Adams and Boorse, the goal is to be attributed to a "system" consisting of the artifact together with its user.

The good consequence and goal doctrines are quite broad, as I have stated them, and the question of their logical independence might be raised. Specific forms of one doctrine can be independent of either specific or general forms of the other. Something may contribute to a goal that I have even though it does not increase my chances of survival and reproduction (Ruse's specific good); and something may increase my chances of survival and reproduction even though survival and reproduction is not, and fails to contribute to, any goal of mine (e.g., my goal is to die quickly, without descendents). Again a good consequence theorist might refuse to attribute goals to anything but humans and higher animals while nevertheless claiming that various activities can confer a good upon things in addition to humans and higher animals. (Soaking up rainwater is beneficial for plants, he might say, even though it does not contribute to any of their goals, since they have none.)

On the other hand, if we consider these two doctrines in their most general formulations and include an appeal to goal directed systems, to goods and goals not just for S but for users, owners, and designers of S, and to cases in which doing y either is or contributes to a goal, then the doctrines may well not be independent. It is difficult to imagine a case in which doing y confers a good upon S (or upon its user, owner, or designer) but is not and does not contribute to any goal of S (or its user, etc.), or vice versa.

A third position, the explanation doctrine, which has been de-

---

9. Ernest Nagel, *Teleology Revisited and Other Essays in the Philosophy and History of Science* (New York, 1979), pp. 310, 312. Nagel regards the clause following "and" as a presupposition rather than an implication of the function statement.
10. Frederick R. Adams, "A Goal-State Theory of Function Attributions," *Canadian Journal of Philosophy* 9 (1979), pp. 493–518; see pp. 505, 508.
11. Christopher Boorse, "Wright on Functions," *Philosophical Review* 85 (1976), pp. 70–86; see p. 80.

fended by Wright,[12] Ayala,[13] Bennett,[14] and Levin,[15] seems quite different from the other two. The general idea is that function sentences provide etiological explanations of the existence or presence of the item with the function. Since Wright's account is simple and general I will use it in what follows.

> *The Explanation Doctrine:* The function of x is to y if and only if x is there because it does y (this "because" involves the idea of etiological explanation which Wright understands may include causes and agents' reasons); and y is a consequence of x's being there.

For example, the function of that switch is to turn the light on and off, since the presence of that switch in this room is etiologically explainable by reference to the fact that it turns the light on and off, and its doing this is a consequence of its being there. When it comes to natural functions Wright invokes the idea of natural selection. The function of the human heart is to pump the blood since the heart's presence in humans can be explained, via natural selection, by the fact that it does pump the blood. (Nature selected in favor of those who get their blood pumped by means of the heart.) And the blood's being pumped is a consequence of the heart's presence. Although it will often be the case, it is not required by this analysis that x's doing y confer some good or contribute to some goal of anyone or anything.

The goal and explanation doctrines are meant to apply to the functions of artifacts as well as natural objects and processes. Their proponents advocate a unified analysis. By contrast, some good-consequence theorists, e.g., Ruse, restrict their particular versions of this doctrine to the functions of natural items only, while others, such as Hempel, believe that their doctrine holds generally for all items to which functions are ascribed.

2. COUNTEREXAMPLES

It should not be surprising that counterexamples to each view have been or might readily be proposed. Whether they are completely devastating is a matter that will be taken up in Section 5.

If the good consequence doctrine is meant to apply to artifacts, the following example seems to show that the doctrine fails to provide a necessary condition. Consider a sewing machine which contains a spe-

12. Larry Wright, "Functions," *Philosophical Review* 82 (1973), pp. 139–68; *Teleological Explanations* (Berkeley, 1976).
13. Francisco Ayala, "Teleological Explanations in Evolutionary Biology," *Philosophy of Science* 37 (1970), pp. 1–15.
14. Jonathan Bennett, *Linguistic Behaviour* (Cambridge, 1976).
15. Michael E. Levin, "On the Ascription of Functions to Objects, with Special Reference to Inference in Archeology," *Philosophy of the Social Sciences* 6 (1976), pp. 227–34.

cial button designed by its designer to activate a mechanism which will blow up the machine. Activating such a mechanism, we are to suppose, will never have any good consequences whatever for the machine or its user or even its designer. This is so whether we focus on the goods usually mentioned such as survival and reproduction (in the case of user and designer), or proper working order (in the case of the machine), or indeed on any other goods it seems reasonable to consider. It's simply something which the designer designed the button to do.[16] Still, it would seem, the function of that button is to activate a mechanism which will blow up the machine.

It might be replied that if no good comes of activating such a mechanism at least its designer must have believed that it would. The present doctrine might then be weakened to require in the case of artifacts not that doing y in fact confers some good but only that x's designer believes that it will. Still, although it would be strange and pointless to design such a button if the designer did not believe that activating the exploding mechanism would confer some good, it would not be impossible. And if he did design such a button then its function would be to activate the exploding mechanism.

A second example, proposed by Larry Wright, purports to show that the present doctrine fails to provide sufficient conditions for functions. Suppose that the second hand in this watch happens to work in such a way that it sweeps the dust from the watch and doing this confers a good upon the watch and its owner since it makes the watch work more accurately. Although the good consequence doctrine is satisfied, it seems not to be the (or even a) function of the second hand in this watch to sweep dust from the watch. As Wright points out, the fact that the second hand does this is a coincidence; and, he argues, if x does y by coincidence then y cannot be the function of x, even if doing y confers a good. We might say that the second hand *functions as* a dustsweeper; but we would not identify this as its function, according to Wright.

The next example seems to impugn the claim that the good consequence doctrine provides sufficient conditions even for non-artifacts. The example, first introduced by Hempel against a simple consequence view of functions,[17] can also be used against the good conse-

---

16. To say that this button activates the exploding mechanism is to imply, of course, that it is capable of doing this, and not that the mechanism is always being activated. In such a case, however, I take the good consequence doctrine to be saying that it is the actual activation of the mechanism, and not simply the capability of doing so, that confers a good (at least on certain occasions). But even if on this doctrine the mere capability is to be taken as what confers the good the same example could be used; we can suppose that having such a capability confers no good at all.
17. That is, the view that x's function is to y if and only if y-ing is a consequence of x or of x's presence.

quence doctrine, as Frankfurt and Poole, critics of this doctrine, have noted.[18] The human heart produces heartsounds in the body, and the production of heartsounds is of benefit to the heart's owner, since it aids doctors in diagnosing and treating heart disease. In a human whose beating heart produced no sound heart disease would be more difficult to detect and treat. Yet, all these authors insist, it is not the (or a) function of the heart to produce heartsounds.

These three examples can also be used against the goal doctrine. We might imagine that neither the designer nor the user of the sewing machine with the self-destruct button has as a goal the activation of the exploding mechanism and that doing this contributes to no goal of the designer or user. We might suppose that sewing machines with such buttons are the only ones made, that people who buy them use them only for sewing, and that activating the exploding mechanism is not, and does not contribute to, one of their goals. As far as users are concerned activation of such a mechanism is completely useless and quite dangerous. Indeed, its non-activation contributes to one of their goals.

A goal-theorist might reply that it is a goal of the designer that the button on the machine be *capable* of activating the exploding mechanism. In response, however, it seems more appropriate to say that the designer's goal in such a case is *to design* (or *produce*, or *bring into existence*) a button with such a capability (see Section 6)—a goal which is not identical with activating the exploding mechanism (the putative function of the button), nor one to which such activation contributes. Moreover, on the goal theory, if it is a function of x to do y then it is the doing of y, and not merely having the capability of doing y, that is or contributes to the goal. If the function of the heart is to pump the blood then, on the present theory, pumping the blood and not merely having the capability to do so is or contributes to the goal. And, if we modify the goal theory and allow capabilities as goals we seem to generate altogether too many functions. The designers may have designed the Cadillac to be capable of traveling 120 miles per hour. Yet it seems incorrect to say that it is a function of the Cadillac to have this capability. Designers generally imbue their products with more capabilities than are functions.

A goal theorist might also respond to the sewing machine example by invoking the idea of a goal directed system. Yet if we think of the sewing machine as such a system—to use the jargon of Nagel and others—it is not a goal of this system to activate the exploding mechanism. This is not something which the machine, even when it is in use, tends to do or persists in doing under a variety of disturbing

18. H. Frankfurt and B. Poole, "Functional Analyses in Biology," *British Journal for the Philosophy of Science* 17 (1966–67), pp. 69–72.

conditions. Nor does activating the exploding mechanism contribute to some other "goal" which might be attributed to the machine, such as sewing. Should we then speak here of *conditional* persistence and say that the exploding mechanism tends to be activated under a variety of disturbing conditions *if the button is pushed?* Conditional goals, like capabilities above, will saddle us with a bevy of unwanted functions. But even if they were allowed on the goal doctrine this proposal will not work, for a reason which vitiates the previous non-conditional version as well. This sewing machine (indeed all of them) may have faulty wiring so that the exploding mechanism will not be or tend to be activated even if the button is pushed. Still the function of this button is to activate the exploding mechanism.

For these reasons the conditions of the goal doctrine do not appear to be necessary for functions. Nor are they sufficient, if we accept Wright's watch example or the heartsounds example. Let us imagine that one of my goals is always to be prompt for appointments. My watch's second hand sweeps dust from the watch and this, by making it work more accurately, contributes to the goal of promptness which I, the watch's owner, have. Despite this fact it is not the second hand's function to sweep dust from the watch. Similarly, my heart's producing heartsounds does contribute, albeit indirectly, to one of my goals, viz. good health or at least prompt diagnosis and treatment in case of bad health. Yet its producing heartsounds is not a function which my heart has.

When we turn to the explanation doctrine new examples are needed. Part of the problem with this doctrine—at least in Wright's formulation—stems from the vagueness of the "is there because" locution. Wright urges that function sentences are explanatory but that what they explain (the "object" of explanation) can vary significantly.[19] On his view a function sentence might explain how x came to exist, or how x came to be present, or why x continues to exist, or why it continues to be present, or why it exists where it does, or why it continues to exist where it does, or why it is used, and so forth. Some of his examples involve one of these explanations, others different ones. But they are not necessarily the same. (One might explain how that light switch came to exist without thereby explaining why it is present in the room.) Yet Wright seems to think that if x's doing y explains why x "is there" in any of these ways then his explanation condition for function sentences is satisfied.

Consider now an example in which the manager of the local baseball team adopts a new policy to keep a player on the first team if and only if that player continues to bat over .300. We might imagine a

---

19. See his *Teleological Explanations*, p. 81.

situation in which both of the following statements are true: (a) Jones is there (i.e., on the first team) because he continues to bat over .300; (b) Jones's continuing to bat over .300 is a consequence of his being there (i.e., on the first team, which gives him practice and confidence). In (a) we are not explaining Jones's existence but rather his presence on the first team—something allowable on Wright's analysis. In (b) we are claiming that his continuing to bat over .300 is a consequence of the same state of affairs explained in (a), viz. his presence on the first team. So Wright's conditions appear to be satisfied. Yet it seems false to say that Jones's function (or even his function on the team) is to continue to bat over .300. This may be one of his aims or goals but not his function. Similarly, suppose that a patient is being kept in a certain room in the hospital because he is sneezing a lot; but it turns out that his sneezing a lot is a consequence of his being kept in that room (which happens to be full of dust and pollens). Is his function, therefore, to sneeze a lot? In short, Wright's conditions seem not to be sufficient.

Because of the latitude in the explanation conditions it is harder, but perhaps not impossible, to show that they are not necessary either. The function of the human heart is, let us say, to pump the blood. If so Wright is committed to holding that one can explain etiologically why humans came to have hearts, or why they continue to have hearts, or why human hearts are where they are in the body, or some such, by appeal to the fact that hearts pump blood. Whether an explanation of the sort Wright envisages is possible, I shall not discuss. But let us change the case to consider just my heart. Its function, I take it, is to pump blood.[20] Yet in this case it is difficult to see how to construct a Wrightian explanation. Does the fact that my heart pumps blood causally explain how my heart came to exist, or why it continues to exist, or why it is present on the left side of my body, or indeed any of the other possibilities Wright allows? How will natural selection lead causally from the fact that my heart does pump blood to the fact that it exists or even to the fact that it continues to exist? Indeed it can continue to exist for years even after it ceases to pump blood.[21]

There is another type of example, however, which appears to impugn both of Wright's conditions as necessary conditions. Artifacts can be designed and used to serve certain functions which they are incapable of performing. The function of the divining rod is to detect the presence of water, even if such a rod is incapable of doing so. The

20. Wright might, of course, say that "the function of my heart is to pump blood" is just a figurative way of making the *general* claim that the function of the human heart is to pump blood. But he explicitly rejects such a move. See *ibid.*, p. 88.
21. This is not to deny that one can explain the presence of my heart in my body by appeal to its function. But as I shall argue in Section 7 this is not a causal or etiological explanation.

function of this paint on the faces of the savages is to ward off evil spirits, even though there are no such spirits to ward off. Now both of Wright's conditions commit him to saying that when x's function is to y, then x in fact does y (or at least that x is *able* to do y under appropriate conditions).[22] Yet in these examples x in fact does not do y nor is x able to do y. Late in his discussion Wright grants the existence of such cases,[23] but says they are non-standard or variant uses of function sentences. Whether this is so is something I shall want to take up. But at least he recognizes that such uses exist and that they fail to satisfy either of his conditions.

I will return to a number of these counterexamples for purposes of reassessment after introducing some needed distinctions.

3. THREE TYPES OF FUNCTIONS: A PRELIMINARY DISTINCTION

Suppose that magnificent chair was designed as a throne for a king, i.e., it was designed to seat the king. However, it is actually used by the king's guards to block a doorway in the palace. Finally, suppose that although the guards attempt to block the doorway by means of that chair they are unsuccessful. The chair is so beautiful that it draws crowds to the palace to view it, and people walk through the doorway all around the chair to gaze at it. But its drawing such crowds does have the beneficial effect of inducing more financial contributions for the upkeep of the palace, although this was not something intended. What is the function of the chair?

In answering this question I suggest that all of the above information would be relevant. It would be appropriate to say what the chair was designed to do, what it is used to do, and what it actually does which serves to benefit something. As this example indicates the answers to these questions need not be the same. Accordingly, one might distinguish three types (some might want to say senses) of function: design-functions, use-functions, and service-functions. The design-function of that chair is to seat the king; its use-function is to block the doorway; its service-function is to attract visitors to the palace and thus induce more financial contributions.[24]

Before trying to clarify this distinction one additional aspect of function sentences should be noted, the possibility of relativization. An item can have a function for someone or within one system or set of activities that is not its function for another. In swimming (for the

---

22. See Wright, *Teleological Explanations*, p. 81.
23. *Ibid.*, pp. 112–13.
24. Note that contrary to the explanation doctrine, the latter is a function of the chair even though the chair's existence (presence, etc.) cannot be etiologically explained by reference to its attracting visitors to the palace.

swimmer) the, or at least one, (use- or service-)function of the legs is to help keep the body horizontal in the water; in walking their function is to help keep the body vertical; in soccer it is to kick the ball. An item can also have a function at one time or occasion which it does not have at another. If the regal chair was once used to seat the President of the United States when he visited the king's palace then this was its (use-)function on that occasion even though this was not its function on any other occasion.

In pointing to various possible relativizations I am not saying what some philosophers have said about function statements, viz. that they must always be understood as relativized to some "system" and time.[25] To understand the claim that the function of that mousetrap is to catch mice one need not identify or be able to identify anyone for whom, or any system within which, or any occasion on which, this is its function. To be sure, this mousetrap may on a given occasion, or within some system, or for some particular person, serve a function other than that of catching mice, and if so we may need to identify the occasion or system. But we need not conclude from this that all function sentences must be so relativized. Indeed it is the (design-)function of that mousetrap to catch mice whether or not it serves that function for anyone, or within any system, or on any occasion.

4. CLARIFICATION

a. Design-functions. The basic idea is that x's design-function is what x was designed to do. But this is not yet quite right, since my stopwatch was designed to resist water. Yet that is not its function. Design-functions, as well as those of the other two types, involve the important idea of a *means* of doing something. The designer designs x to be or to serve as a means of doing y. This mousetrap was designed by its designer to be a means of catching mice. But my stopwatch was not designed to be a means of resisting water. (I shall have more to say about the concept of means in Section 6.) Alternatively, x may be a part of, or belong to, or otherwise be present in, a "system" S which was designed by its designer to do y by means of x. If so we say that the (design-)function of x in S is to *enable* S to do y. The function of the gasoline in this engine is to enable the engine to run. So we can write

(1) If x was designed to be or to serve as a means of doing y then x's function (at least one of them) is to do y.

(2) If S was designed to do y by means of x then x's function in (with respect to) S is to enable S to do y.

25. See, e.g., Boorse, "Wright on Functions."

Should design-functions be restricted only to cases in which x (or S) is *designed?* A chemical can be *produced,* a person *appointed,* a rock *placed where it is,* to be or to serve as a means of doing something. If so we can speak of the chemical's, person's, or rock's function, even though these items were not themselves designed. Nevertheless, design was present. Perhaps then (1) can be generalized as follows:

If x was produced (created, established, appointed, placed where it is, etc.) by design to be or to serve as a means of doing y then x's function is to do y.

A similar generalization of (2) would be possible.

However, with the inclusion of these additional activities all done by design we encounter cases in which x might have been designed for one end, produced for another, and placed where it is for a third. Suppose that the Universal Design Company designed a certain type of bolt to serve as a means of bolting a car engine to the frame. The Ford Motor Company purchases this design but produces the bolt in its own shops to serve as a means of bolting the torsion bar to the frame. The bolt is then sent unattached to the dealers who are to install it when they receive the car. The dealers, however, ignoring the previous design, affix the bolt to the wheel. What is the function of this bolt? Does it have three functions? Each of the agents noted—the designer, the producer, and the placer of the bolt—might well claim that the bolt has only one (proper) function and disagree over what this is.

In reply we might proliferate types of functions still further by talking about production-functions, placement-functions, and so forth. But the more natural way to settle this dispute is to recognize that function sentences do not always need to take the simple form "the function of x is to do y," which philosophers tend to focus upon. We also utter sentences of the form "the function x was *designed* to serve is to do y," "the function x was *produced* to serve is to do y," "the function x was *placed where it was* to serve is to do y," and so forth. Each of the three protagonists in our bolt story can then be understood as making a different type of function claim. When an agent from the Universal Design Company indicates the function of that bolt he means to indicate the function it was designed to serve; the agent from the Ford Motor Company is talking about the function it was produced to serve; the dealer is speaking of the function it was placed where it was to serve.

My solution to this dispute, then, is to recognize not different types (or senses) of function but different types of function *sentences.* Several of the latter can be grouped together under the heading "design function sentences," all of which involve the general idea of design in origin or placement. One of their basic forms is this:

(3) The function x was designed (produced, created, established, appointed, placed where it is, etc.—all by design) to serve is to do y,

which is true if and only if x was designed (produced, etc.) to be or to serve as a means of doing y. If x does not itself do y but S does we can write

(4) The function x was designed (produced, etc.) to serve is to enable S to do y,

which will be true if and only if x was designed (produced, etc.) to be or to serve as a means of enabling S to do y.

My claim now comes to this. A sentence of the form

(5) The function of x is to do y (or to enable S to do y)

is ambiguous. Someone who utters it may be making one of a number of different claims about design in origin or placement *some* of which are given by (3) and (4). But not all. The second chapter of Ayer's *Language, Truth, and Logic,* entitled "The Function of Philosophy," states that the function of philosophy is to provide analyses of certain sorts or "definitions in use." When Ayer wrote this he was aware that this is not the function that philosophy is produced or written by most philosophers to serve. Nor is his claim to be taken simply as one about the function of *analytic* philosophy. Rather he is indicating the function that (all) philosophy *ought to be* produced or written to serve. Accordingly, another type of design function statement which someone uttering (5) might mean to be asserting is

The function x ought to be designed (produced, etc.) to serve is to do y (or to enable S to do y),

which is true if and only if x ought to be designed (produced, etc.) to be or to serve as a means of doing y (or of enabling S to do y).

One can, of course, answer a question of the form "What is x's function?" by uttering sentences other than (5), for example, by saying that x is an A (a church, a mousetrap). In such a case x is not designed to be a means of being an A, although the notion of means is not irrelevant here. If that building is a church then its function is to enable Christians to worship together. And the latter claim, which can be understood as being of type (4), will be true if and only if a corresponding means statement is.

b. Use-functions. When someone utters a sentence of form (5) he may mean to be saying something not about x's origin or placement but about its use, e.g.,

The function x is used to serve is to do y (or to enable S to do y), which will be true if and only if x is used as a means of doing y (or as a means of enabling S to do y). Someone who says that the function of the human hands is to grasp objects may be saying that this is the function they are *used* to serve. And this will be so if and only if the hands are used as a means of grasping objects.

Although what function an item has been designed (or produced, etc.) to serve often coincides with the function it is used to serve, sometimes it does not. Something may have been designed or even placed where it is to serve as a means of doing y although it is never in fact used, or although it is used only as a means of doing z. The function that trough may have been designed and placed where it is to serve is to water the pigs, even though it is never used or the only function it is used to serve is to water the flowers.[26] Moreover, x may be used to serve a given function without this being so by *design*. The function a mosquito's wings are used to serve is to enable the mosquito to fly. We need not say that the mosquito uses them by design or that they were designed or created by design.

Use-function, like design-function, sentences can be prescriptive. Someone who claims that the function of a college education is to arouse intellectual curiosity and not just to get a job might mean that this is the function it ought to be used to serve, not that it is in fact used to serve this. There is a related use of function sentences of form (5). Pointing to the wings on a particular mosquito I might say that the function of those wings is to enable the mosquito to fly, even though in this case the wings are broken or for some other reason the mosquito never flies. My claim, then, could be that

(6) The function those wings are supposed to be used to serve is to enable that mosquito to fly.

And I may make such a claim because I believe something about the function of wings in mosquitos generally, viz. that

(7) The function a mosquito's wings are (generally) used to serve is to enable the mosquito to fly.

The general sentence (7) provides a norm on the basis of which the function sentence (6) about a particular x can be asserted. On the other hand, pointing to those particular broken wings I might also say that they have no function at all, since they cannot be used.

---

26. In our earlier example the function the regal chair was placed where it is to serve is the same as the function it is used to serve by the guards, viz. to block the doorway. However, if the chair remains where it is long after the guards are all dismissed by the king, then the function it was placed where it was to serve remains the same, but the function it is used to serve may not.

c. Service-functions. One who utters (5) might mean to be claiming something stronger than simply what function x was designed or is used to serve, viz.

(8) What x in fact does, the performance of which is a function that it serves, is y (or to enable S to do y).

Thus we might say that although the function the regal chair was designed to serve is to seat the king, and the function it is used to serve is to block a doorway, the only thing the chair actually does, the performance of which is a function it serves, is to draw crowds to the palace. We might say this because crowds are in fact drawn to the palace by means of the chair and this is of benefit, while the king is in fact not seated, nor is the doorway really blocked, by means of that chair.[27] More generally, a sentence of form (8) is true if and only if

(9) y is in fact done by means of x (or S is in fact enabled to do y by means of x), and either x was designed (produced, etc.) to be or to serve as a means of doing y (enabling S to do y), or x is used as a means of doing y (enabling, etc.), or y's being done confers a good.

A number of points must now be made about the last clause in (9). It would be a mistake, I think, to follow those writers who require only one specific type of good such as survival and reproduction. As Sorabji notes, a certain creature might have an organ which shuts off sensations of pain when lethal damage has been done to its body.[28] We might then identify this as the function of the organ even if shutting off the pain does not increase the creature's chances of surviving and reproducing. Analogous examples are possible against views committed to other specific goods.

Possibly, those who select one type of good such as survival and reproduction do so because they think of this as an ultimate good—as worthy of having for its own sake; other states of affairs can be beneficial only because they contribute to the ultimate good. But even if we distinguish ultimate and intermediate goods it is doubtful that there is a unique ultimate good with which all functions can be associated. Another reason for selecting survival and reproduction as the only good is that the authors that do so are concerned almost exclu-

---

27. Service-function sentences, however, need not be confined to those in which y is in fact done. Pointing to a particular defective kidney in Jones's body we might say that its function is to remove wastes, and mean that what this kidney is supposed to do, the performance of which is a function it would serve (if it did it), is to remove wastes. This claim would be based on a general service-function sentence of form (8) about what human kidneys in fact generally do.

28. Sorabji, "Function," p. 293.

sively with biology. The function of the nose, they will say, is to breathe, not to hold up eyeglasses, although both are done by means of the (human) nose and their being done confers a good. To this the correct response is that the nose serves or can serve a number of different functions not all of which are *biological*. Whether doing y is classifiable as a biological function will depend at least in part upon the type of good it confers, upon what and how; however, I shall not here try to offer criteria for such functions. If holding up eyeglasses is not a biological function of the nose it is nevertheless an important function of the nose in the eyeglass "support system."

Doing y need not be of benefit to x itself or to its owner or user but to something else. The sickness and subsequent death of various animals in a species may be of no benefit to *them*, but to other members of the species, who will now have fewer competitors for food. So the sickness and death of these animals can have a function for the species, even if not for the animals themselves.

Doing y can confer a good upon a person S even if doing y is not something which S knows about or regards as beneficial even if he knows about it. One of the functions of basic training in the army is to teach recruits to obey orders, even if the recruits don't realize that basic training is teaching them to do this and even if they don't want to be taught or regard this as beneficial. One who believes that basic training is of value in this way can make such a function statement.

By contrast, one can also make a service-function statement without committing oneself to values, if any, in that statement. Suppose that watching television dulls the mind to the problems of the world and people act as if so dulling the mind is beneficial for them. For the television viewer, one might say, what television does, the performance of which is a function it serves, is to dull the mind. One can make such a statement from the point of view of the television viewer with his set of values. On the other hand, employing one's own more intellectual values one might say that for the television viewer what television does, viz. dull the mind, serves no function at all, since it confers no benefits. The speaker can make explicit which point of view—which set of values—he is employing, or this can be clear from the context. This means that speakers with quite different values can still agree over the truth of those service-function sentences based on the conferring of a good, provided that they invoke the values of the party in the function sentence, not their own.

Finally, it should be clear, I am not maintaining that all function sentences, or even all service-function sentences, commit one to the claim that doing y confers a good. We can utter a true service-function sentence of form (8) by saying that what that button on the sewing machine in fact does, the performance of which is a function it serves, is to activate the exploding mechanism, even if activating this mechanism is of no benefit at all.

5. THE COUNTEREXAMPLES RECONSIDERED

I have claimed that a sentence of the form "The function of x is to do y" is ambiguous and can be used to make a variety of quite different functional claims. This does not mean, conversely, that whenever one explicitly utters some design, use, or service function sentence one is willing to say that the function of x is to do y. We can generally do this in the case of artifacts. With natural objects, however, such an inference may be resisted unless the identity of some system in which the object has that function, or of some user of the object, or of some item benefited is made clear either by the context or within the function sentence itself by relativization. One might be willing to utter the service-function sentence "One of the things this tree does, the performance of which is a function it serves, is to provide shade," based on the conferring of a good, but be unwilling to conclude that this tree's function is to provide shade. The latter statement should become less objectionable, however, if it is relativized to be making a claim about its function for the homeowner on whose property it lies.

By contrast to the view I have been suggesting, the good consequence, goal, and explanation doctrines say that sentences of the form "The function of x is to do y" (or in the case of some advocates of the first view, that sentences of this form in which x is not an artifact) are always used to make the same type of statement, which is subject to the same analysis. In rejecting these doctrines, however, I do not thereby accept all of the counterexamples to them in Section 2. Our discussion in the last two sections will facilitate an assessment of these counterexamples and will also illustrate some of the distinctions I have pressed.

In the case of the sewing machine button, the function it was *designed* to serve is to activate the exploding mechanism, even if this has no good consequences whatever. Accordingly, one might say that the function of this button is to activate the exploding mechanism and be making a design-function statement. On the other hand, someone impressed by the fact that the button because of its extreme danger never is or will be used might claim that this button has no function at all; it is "dysfunctional." Such a person would be making, or rather denying, a use-function statement.

Wright's watch example is subject to a similar treatment. To be sure, the function the second hand was *designed* to serve is to indicate the seconds and not to sweep the dust from the watch. So one might say, following Wright, that sweeping dust away is not its function. On the other hand, since dust is in fact swept from the watch by means of the second hand and this is beneficial to its owner, we can say that one of the things the second hand does, the performance of which is a function it serves, is to sweep the dust from the watch, and hence that this is one of its functions, even though this is not a function it was designed to serve.

Wright insists that if x does y by coincidence then y cannot be the or a function of x. True, if x is designed to do y then generally if x does y it is not by coincidence.[29] But in the case of service-function sentences, x may come to perform a beneficial service which it was not designed to perform and in this way is coincidental; yet the performance of this service can be a function that x acquires. Wright also emphasizes the distinction between x's function and what x functions as. Although the watch's second hand functions as a dustsweeper, its function is not to sweep dust away, he claims. I agree that we might say that the second hand functions as a dustsweeper; but I suggest we would speak this way not because this isn't a function of the second hand but because it isn't a function the second hand was *designed* to serve. The "functioning as" locution seems particularly appropriate when what x is doing contrasts with what it was designed to do or when x was not designed at all. Yet there is no general incompatibility between x's functioning as a y and y's being one of x's functions. The heart functions as a blood pump, but pumping blood is one of its functions.

Turning to Hempel's heartsounds example, the claim is made by certain opponents of the good consequence doctrine that it is not a function of the heart to produce heartsounds, even though this confers a good upon the heart's owner since it aids doctors in diagnosing and treating heart disease. But if it does then I suggest that the critics are mistaken. One of the things which the heart does, the performance of which is a function it serves, is to produce a beating sound, since the production of such a sound is beneficial. Perhaps what confuses these critics is a failure to distinguish what x does which is of great benefit from what it does which is beneficial but less so (more important vs. less important functions); or what x does which is directly beneficial and what it does which is less directly so. The pumping of the blood is of greater and more direct benefit to the body than the producing of heartsounds. Still in doing both the heart is performing two functions. What may also be moving the critics is the fact that although the heart's pumping the blood is always beneficial to the body in all mammals, the heart's producing heartsounds is of diag-

---

29. But this need not be so in all cases. I may design a divining rod to enable me to find water underground. Now I use the rod, and it does jerk downwards on several occasions when water is present underground, but only because the rod is magnetic and there happens to be metals also beneath the surface. Accordingly, the rod enabled me to find water underground, but only by coincidence. Still that is the function it was designed to serve.

nostic value mainly in humans, and only in recent times for those rich or wise enough to have medical checkups. Accordingly, one can say that what the heart does, the performance of which is and always has been a function that it serves in all mammals, is to pump the blood; one cannot make a claim this strong about the production of heartsounds. Because of these differences the former activity of the heart may be classifiable as more centrally its biological function; and this consideration may be weighing heavily in the minds of the critics as well.

The baseball counterexample to Wright's analysis is appropriate if we are making a design-function statement. Jones was not *put on the team* to serve the function of continuing to bat over .300. Can a service-function statement be made here, e.g., that one of the things Jones does which is a function he performs on the team is to continue to bat over .300? This is inappropriate, I suggest, because continuing to bat over .300 is not something done *by means of* Jones, but simply something which he does (more of this in the next section). On the other hand, we might say that Jones's continuing to bat over .300, or his batting skill, has served a function on the team, viz. that of enabling the team to win consistently. This would be appropriate to say because the team has been enabled to win consistently by means of Jones's continuing to bat over .300, or his batting skill, and winning consistently is of benefit to the team. But note that here the function is attributed not to Jones, as it could be on Wright's analysis, but to his continuing to bat over .300 or to his batting skill; and the function itself is not continuing to bat over .300 but enabling the team to win consistently.

Another counterexample used against Wright involved the divining rod whose function is to detect the presence of water and the paint on the faces of the savages whose function is to ward off the evil spirits, although neither function is or can be performed by these items. Wright's response is that these are non-standard uses of function sentences. My rejoinder is that one who makes these claims would generally be making not a service-function statement but a perfectly ordinary design- or use-function statement. He might not be claiming that what the divining rod does, the performance of which is a function it serves, is to detect water, but that this is the function it was designed or is used to serve.

6. FUNCTIONS, ENDS, AND MEANS

Functions are intimately related to ends of certain kinds, and the items to which functions are attributed are means to those ends. The latter are doing-ends rather than thing-ends. Although money or fame might be spoken of as ends they cannot be functions. Rather, catching mice, pumping blood, enabling one to climb mountains are functions. Indeed, Geach has argued that no ends can be things but that all ends have a propositional structure.[30] If I say that my end is money what I want is to make and retain money, and this should be understood as saying that my end is *that I make and retain money*. I shall not here try to say whether sentences reporting ends are referring to objects, activities, propositions, or nothing at all. I am in agreement with Geach, however, that end-sentences can be transformed into ones in which the end is given in a propositional form, and that such sentences will often clarify the end in question.

Thus earlier we noted that my stopwatch was designed by its designers to resist water, but that resisting water is not one of its functions. The end which the designers want to achieve here is *that this stopwatch resists water*. And this end is not something designed to be achieved *by means of* this stopwatch, but rather by mean of certain water-resistant sealing material. (Thus we *could* claim that the function of this material in my stopwatch is to enable the watch to resist water.) Similarly, in the baseball example the end which Jones or his manager wants to achieve is *that Jones continues to bat over .300*. And this end is not something designed to be achieved *by means of* Jones, but, say, by means of constant practice on his part. (Thus we could claim that the function of constant practice on the part of Jones is to enable him to continue to bat over .300.)

Although functions are intimately related to ends the two should not be identified. Functions as well as ends can be given by infinitive and verb + ing nominals; yet ends but not functions can be given propositionally. My end might be that I make money, but my function (as chief fundraiser, say) is not that I make money or that money be made by me but simply to make money or making money. Nevertheless, for any function there is an associated end which can be formulated in a propositional way by a that-clause nominal. And the item with the function is a means to this end. For this to be the case the associated end will need to be a "generalized" one in this sense: its propositional formulation will include no reference to the item x to which the function is attributed. If my function in this organization is to make money for the organization then the associated end is *that money be made for the organization*, not that I make money for the organization or that money be made for the organization by me. And if this is my function then I am the (or a) means by which this generalized end is to be achieved. The association between functions and ends, then, comes to this. If x's function is to do y then that y is done is a

30. Peter Geach, "Teleological Explanation," in S. Körner, ed., *Explanation* (Oxford, 1975), pp. 76–95.

generalized end (given in propositional form) for which x is a means.

The ends with which functions are associated can be *pure*, i.e., ends which no one or thing actually desires or intends.[31] The function of the button on the sewing machine is to activate the exploding mechanism even if no one desires or intends *that the exploding mechanism be activated* (the emphasized words give the associated end). What, then, makes something an end with which a function can be associated?

Our discussion in Section 4 provides an answer. Doing y—or that y is done—is an end with which x's function can be associated if (but not only if)[32] one or more of the following conditions is satisfied:

(1) x was designed (produced, etc.) to be or to serve as a means of doing y;

(2) x is used as a means of doing y;

(3) y is in fact done by means of x and either (1) or (2) or y's being done confers a good.[33]

If that button is designed to serve as a means of activating the exploding mechanism then activating the exploding mechanism (that the mechanism be activated) is an end for which the button is designed, even if no one desires or intends that the mechanism be activated. If bodily wastes are removed by means of the kidneys and this being done confers a good upon their owner, then removal of such wastes is an end served by the kidneys, even if their owner neither desires nor intends this end.

Can the concept of *means* be eliminated from this account in favor of something else? For example, is x a means of doing y when and only when x (or x's presence or occurrence) is *necessary* or *sufficient* for y to be done? Obviously not. It is not necessary since the same end can often be achieved by a variety of means. Nor is it sufficient, since x can be a means of doing y without y's being done. (The electric chair is a means of executing prisoners even if prisoners are not executed.) Nor for the same reason is x a means of doing y when and

---

31. Geach, *ibid.*, p. 80, speaks of purely teleological propositions as "propositions affirming that something happened to a certain end with no reference to some desire or intention, nor yet to any subject who desires or intends."

32. These conditions do not exhaust the possibilities, since function sentences can also be understood prescriptively. But the prescriptive uses are subject to analogous conditions.

33. Woodfield, *Teleology*, attempts to provide a completely unified analysis of ends by in effect identifying all ends as *goods*. To do this, however, he must assume (a) that if someone designed something to do y then he wants y to be done; and (b) that if someone wants y to be done then he believes that y is a good thing to do (p. 203). Woodfield thinks that there are senses of "want" and "good" which make (a) and (b) true. I disagree. A reluctant designer may have designed this particular machinery to execute the prisoner even though he neither wants the prisoner to be executed nor believes this is a good thing to do, in any sense of "want" or "good."

only when x (or x's presence) *causes* y. Should we then weaken the claim to say that x is a means of doing y when and only when x *can* cause y. This is also too strong, since although ropes are a means by which people climb mountains I doubt that we would say that ropes can cause people to climb mountains. And if we weaken the claim still further by saying that x is a means of doing y when and only when x can be *causally relevant* for y (without necessarily being something that can cause y) we allow too broad a class of means. The strap on the electric chair can be causally relevant for the execution of prisoners in so far as it keeps prisoners in the chair. But it is not a means by which prisoners are executed. Some might want to appeal to the concept of *making possible* (or enabling). The heart makes it possible for the blood to be circulated; ropes make it possible for people to climb mountains. And x can make possible (or enable) the doing of y without being necessary or sufficient for, or causing, y to be done. However, this concept is also too broad for the job. The presence of these straps on the electric chair may make it possible for the prisoner to be executed without being a means by which he is executed.

These brief observations are not meant to demonstrate the impossibility of eliminating talk of means in favor of one or more of these other concepts. If such a reduction can be accomplished, however, I doubt that it will be a simple or straightforward one. Moreover, if the concept of means is not so reducible, I am not suggesting that this is so because it is irreducibly teleological, i.e., associated, e.g., with the concept of an end. Although means are typically associated with ends they are not always or necessarily so. Smith may have been killed by means of his own carelessness, even if Smith's being killed is not an end (e.g., it is not something which anyone intended or desired, it was not something which anything was designed to be, or was used as, a means of achieving, and it did not confer a good). In a similar way one can speak of a certain reaction as occurring by means of a catalyst, or of chemical bonds forming by means of electrons, without assuming that the reaction's occurrence or the formation of the bonds is an end. Although functional talk involves both the idea of a means and an end, what makes such talk teleological is the commitment not to a means but to an end. In this respect reference to means is on a par with other possible appeals to necessary and sufficient conditions, causation, and making possible.

The fact that functional talk involves not only the idea of a means but that of an end in the way I have suggested explains why we resist such talk in certain instances. There are cases in which y is done by means of x and yet doing y is not x's function since doing y (or that y is done) is not an end. Although the formation of a chemical bond between a sodium atom and a chlorine atom occurs by means of the transfer of an electron from the former atom to the latter, it seems objectionable to say that the function of a such a transfer is to form a chemical bond between these atoms. This is because normally we do not regard the formation of such a bond as an end.

In the light of the discussion in this and the previous sections we ought to recognize certain elements of truth in the good consequence, goal, and explanation doctrines. Conferring a good can be an important consideration in assigning a function to something, though it is neither necessary nor sufficient in general. But it is necessary in cases in which talk of design or use is inappropriate. And if y is done by means of x, then the fact that y's being done confers a good is sufficient for characterizing the doing of y as a function that x serves. The goal doctrine is important because it stresses the idea of a goal and thus brings out the teleological character of function sentences. It goes awry, however, in requiring the doing of y to be or to contribute to some goal which x or its user, owner, or designer has. The goals—or more broadly the ends—associated with function sentences, even with design- or use-function sentences, can be pure ones. Finally, as will be seen in what follows, the explanation doctrine is correct in its claim that function sentences can be invoked in providing explanations, even of x's existence or presence, although contrary to this doctrine these explanations are not in general causal.

7. FUNCTIONAL EXPLANATIONS

Do function sentences explain anything, and if so what? Or, to put this in the framework of the illocutionary theory, if one knows or assumes the truth of some function sentence what, if anything, can one explain? What kinds of explanations are possible?

There is a lever on the dashboard of my car whose function puzzles people since when it is pulled nothing seems to happen. I explain its function by saying that it opens the reserve fuel tank. My explanation is

(The function of that lever is to open the reserve fuel tank; explaining what the function of that lever is).

Employing the ordered pair theory, this is an explanation in virtue of the fact that its second member is a type of explaining act whose constituent question is a content-question, and the first member is a complete content-giving proposition with respect to that question.

More generally, a function sentence of the form "The function of x is to do y," can be used to construct an explanation of the form

(1) (The function of x is to do y; explaining what the function of x is).

Are other explanations possible on the basis of function sentences? There is a type of explanation which, traditionally, has been of most concern to philosophers and scientists alike. It involves explaining the existence of the item with the function. Returning to our present example, the audience might not understand why that lever exists. If I know its function then I can explain why it exists by saying: to open the reserve fuel tank. My explanation is

(The reason that lever exists is to open the reserve fuel tank; explaining why that lever exists).

What kind of an explanation is this, and how is one able to construct it knowing the function of the lever?

This is what in Chapter 7 I called a reason-giving explanation (except now, the reason need not be given by a that-clause nominal, but, e.g., by one that is an infinitive phrase). In the present case the blank in the complete answer form

The reason that lever exists is ———

is filled by an expression giving an *end* for which the lever exists. This is one type of reason that, or for which, something can exist. How does knowing the function of that lever enable me to construct this explanation? As emphasized in Section 6, for any function there is an associated end. If the function of that lever is to open the reserve fuel tank then the associated end is that the reserve fuel tank be opened. Knowing what this end is—and assuming that the reason that lever exists is to serve it (i.e., to open the tank)—we can readily construct the explanation above.

More generally, a function sentence of the form "The function of x is to do y" can be used, in certain cases, as a basis for constructing an explanation of the form

(2) (The reason that x exists is to do y; explaining why x exists).

It can be used because if the function of x is to do y, then the doing of y is an end associated with x; and given this information, and the assumption that the reason x exists is to serve its associated end, it follows that the reason x exists is to do y. A function sentence of the form "The function of x is to do y" can be used to construct explanations of type (2) in those cases in which it can be assumed that the reason x exists is to serve its associated end. When can this assumption be made?

One of the main themes of earlier sections is that a sentence of the form "The function of x is to do y" can be used to make a number of different functional claims. We distinguished three broad types: design, use, and service. Recall the example of the regal chair which was designed to serve as a means of seating the king, which is used (unsuccessfully) to block a doorway, and which, in fact, serves as a means of attracting visitors to the palace. Now consider the question

> Why does the regal chair exist? (or: For what reason does that regal chair exist?)

Information about its design and its use seems particularly relevant here. In responding to this question, I suggest that our procedure would be to transform the question into these

> For what reason was that regal chair designed?
>
> For what reason is that regal chair used?

Now, if the function that regal chair was designed to serve is to seat the king, then the reason for which the chair was designed is to seat the king. And if the function that regal chair is used to serve is to block a doorway, then the reason for which the chair is used is to block a doorway.

More generally, if we are justified in asserting a function sentence of the form "The function of x is to do y," and this is elliptical for a design-function sentence, or for a use-function sentence, then we can assume that the reason x exists—i.e., the reason for which it was designed (produced, placed where it is, etc.), or the reason for which it is used—is to serve its associated end. Under these conditions, given the sort of function sentence in question, we can construct an explanation of form

> (2) (The reason that x exists is to do y; explaining why x exists).

However, this is not true universally for service-function sentences. Suppose we are justified in asserting "The function of x is to do y" where this is elliptical for "What x in fact does, the performance of which is a function that it serves, is y." In cases in which the service function is based on the conferring of a good, we cannot necessarily assume that the reason x exists is to do y. Where biological functions of bodily organs are involved such an assumption is usually made. (For example, we are willing to say that the reason the heart exists is to pump the blood.) In the case of service functions involving artifacts—where such functions are not based on design or use—we are unwilling to make such an assumption. Thus, despite the fact that one of the things the regal chair does, the performance of which is a function it serves, is to draw crowds to the palace, we are not willing to say that the reason for which the chair exists is to do this.

What the essential difference is between such service functions I shall not speculate here. Of concern to me is the nature of existence explanations of form (2) in which "to do y" gives an end for which x exists. Are they causal? Not in the sense defined in Chapter 7.

Consider

(3) (The reason that the regal chair exists is to seat the king; explaining why the regal chair exists).

Its constituent question is not a what-causal question. (It is not expressible by an interrogative of the form "What (———) caused ———?) There is, however, a what-causal transform of this question:

What caused that regal chair to exist?

Suppose, then, that we construct the following explanation:

(4) (The seating of the king is what caused the regal chair to exist; explaining what caused the regal chair to exist).

Now by the definition of Chapter 7, (3) will be a causal explanation if (4) is a what-causal equivalent of (3). And this will be so only if the constituent proposition of (4) is true if and only if that of (3) is. But this equivalence does not obtain. The constituent propositions in (3) and (4), respectively, are

(3)' The reason that the regal chair exists is to seat the king;

(4)' The seating of the king is what caused the regal chair to exist.

If "the seating of the king" in (4)' is to be understood as elliptical for "the fact that the king is seated (in that chair)," then (4)' is false, since (in the example) there is no such fact. (The king is never seated in that chair.) And even if there were such a fact, it is absurd to suppose that the fact that the king is seated in that chair is what caused it to exist. Since we are assuming (3)' to be true, under this interpretation, (3)' and (4)' cannot be equivalent.

Are there other interpretations of (4)'? Perhaps (4)' is to be understood as elliptical for

The fact that the king is to be seated (in some appropriate regal chair) is what caused that regal chair to exist.

But in our example we need not suppose even that this is true. We may assume that the king is never to be (and, by custom or law, is never supposed to be) seated, or at least seated in any regal chair. Even in these circumstances (3)' can be true, if the designer of the chair (unaware of this custom or law) designed it to be or to serve as a means of seating the king.

No doubt (4)' can be transformed into a true causal statement, e.g.,

(5) Certain actions of the designer of the chair (or such and such actions) caused the regal chair to exist.

But (5) is not equivalent to (3)', since (5) could be true, even if (3)' were false. (The designer of the chair might have designed it to seat the queen, not the king.)

I conclude that (4) is not a what-causal equivalent of (3), and more generally, that (3) does not have a what-causal equivalent. Therefore, (3) is not a what-causal explanation. It can also be shown that (3) has no counterpart what-causal explanation; but I will omit the argument. (For the definition of counterpart what-causal explanation, see Chapter 7, Section 4.) If this is right, then (3) is not a causal explanation in the sense defined in Chapter 7. (3) is a non-causal reason-giving explanation. It gives the reason for which the regal chair exists by giving the end for which it exists.

Some writers who say that function sentences can be used to construct existence explanations of form (2) believe that such explanations are causal. (See, e.g., Wright.) On this view, the fact that x does y, or the doing of y, causes x to exist. Against this, Robert Cummins correctly argues that function sentences cannot always be used to provide causal explanations of x's existence.[34] My heart was not caused to exist by the fact that it pumps my blood. Cummins, however, mistakenly concludes from this that, at least in the case of non-artifacts, function sentences cannot be used to explain x's existence or presence. What he (as well as the causal interpreters of (2)) overlooks is that one can explain x's existence by answering the question "Why does x exist?" in a way that is not causal. The child who asks why hearts exist, or why they are present in humans, may simply want to know an end for which they exist; in this case, what benefit to the body is accomplished by means of the heart for which it exists. He is not, or need not be, asking a causal or etiological question about how a heart came to be present in humans, what causes it to remain in the human species, or to be where it is in the body. Similarly, if the earth doctor dissects a Martian's body and discovers an organ different from any he knows, he might ask: "Why does that organ exist?" In answering this question—in giving a reason for its existence—he may not want to know what causal, evolutionary processes resulted in its being present in Martians, but simply for what end it exists.

To this it might be replied that the doing of y is an end for which x exists only if it is a cause of x's existence. But this cannot be so. As noted above, it is not the *seating of the king* which caused the regal chair to exist, but rather the actions of certain agents (who may have designed or built the chair to serve that end). Accordingly, a second reply might be that the doing of y is an end for which x exists only if there is some act or event, in which the doing of y is an end, which

34. Robert Cummins, "Functional Analysis," *Journal of Philosophy* 72 (1975), pp. 741–65.

caused x to exist. The seating of the king is an end for which that regal chair exists only if agents who designed or built the chair to serve that end acted in certain ways. Pumping the blood is an end for which the heart exists, it might be said, only if nature acts in such a way as to select in favor of those in whom the benefit of pumping the blood is achieved by means of the heart. Furthermore, to establish that doing y is an end for which x exists we must discover what caused x to exist, or at least make certain causal assumptions.

The validity of this second reply need not be discussed, since even if it is accepted the point I am trying to make is not vitiated. The questions

(6) For what end does x exist?
(7) What caused x to exist?

are different questions. An answer to (6) is not, and need not contain, an answer to (7). An answer to (6) may contain the information that an end for which x exists is the doing of y, without identifying any act or event that caused x to exist; and this is so even if in order to establish the correctness of this answer one must be able to identify or make some assumption about such an act or event. To be sure, having received an answer to a question of form (6) one is often in a position to construct at least some answer to (7). If I am informed that the end for which this chair exists is to seat the king, then knowing that it is an artifact I know that it was some action of an agent that caused it to exist. But if I get much more specific I risk the danger of falsehood. For example, if I assume that it was designed by its designer to serve that end, I may be making a mistake (as in the car bolt example of Section 4). Moreover, some answers to a question of form (6) may not put one in a position to answer every or any of form (7). One who is ignorant of the doctrines of natural selection and heredity may be informed about the end for which that strange organ in the Martian's body exists and yet be unable to causally explain how the organ came into existence or to say what, if anything, causes it to remain or to exist where it does.

Contemporary philosophers of science tend to adopt the view that to explain x's existence or presence is to cite a cause or causes. Within this tradition opinion then divides over functional explanations. On the one hand there are those who say that function sentences can be used to explain x's existence because they cite causes; on the other hand there are those who deny that function sentences provide causes for, and hence can be used to explain, x's existence. By contrast, I have been defending a position within a more ancient tradition, the Aristotelian, according to which one can explain x's existence—one can explain why x exists, why x is present, why S's have x—not only by answering a question of form (7) but by answering one of form (6), even if an answer to the latter does not contain an answer to the former.

CHAPTER 9
The Limits of Explanation

Occasionally the claim is made that something is unexplainable. This may simply be an expression of frustration over the inability to discover an explanation of a desired sort. Or it may reflect a belief that there exists no explanation of that sort to be found. It might also stem from the belief that no explanation is required, that explanation has come to an end, that what has been invoked is ultimate.

Stephen Toulmin believes that each science or theory postulates or presupposes some "ideal of natural order" which it regards as unexplainable—indeed, as requiring no explanation.[1] In Newton's mechanics, the law of inertia is not subject to explanation, he thinks, because it describes the natural motion of bodies. It is accelerations, i.e., deviations from the natural motion, that are explainable. In his preface to the second edition of Newton's *Principia*, Roger Cotes says roughly the same thing about gravitation. He declares that no explanation for this is possible or indeed needed, since gravitation, like extension, mobility, and impenetrability, is an ultimate and simple disposition of bodies.

It has been urged by Robert Causey, among others, that scientific identities are unexplainable.[2] On this view it is not possible, or required, to explain why water is (i.e., is identical with) H_2O, or why the temperature of a gas is identical with its mean molecular kinetic energy. Identities are explanatorily ultimate. (Causey makes important use of this claim in developing his theory of scientific micro-reduction.)

It is a consequence of Hempel's view of explanation that low probability random events are unexplainable. For example, in accordance

---
1. Stephen Toulmin, *Foresight and Understanding* (London, 1961).
2. Robert Causey, *Unity of Science*.

with quantum theory, if an electron in a diffraction experiment hits the screen in a region of low probability, then (assuming there is no deeper deterministic theory) why the electron hit the screen where it did is explanatorily ultimate.

I have cited these examples to illustrate various claims regarding the limits of explanation. In this chapter I shall try to clarify the possible import of such claims. Several strands might be separated: the idea of unexplainability; that of something being ultimate or ultimately unexplainable; that of a fact, or law, or theory being fundamental; and the idea of natural behavior or states. What sense, if any, can be made of these ideas? How, if at all, can various disputes about whether something is unexplainable be settled? Is it true, as Toulmin urges, that something unexplainable is presupposed by every science?

1. EXPLAINABILITY AND UNEXPLAINABILITY

As I have been doing throughout, I shall assume that what is, and is not, explainable can be formulated as an indirect question q, whose direct form is Q. And I shall begin with explainability relative to instructions I.

Using a product-view of explanation, an explainability claim might be understood as follows:

(I) q is explainable in a way that satisfies I if and only if $(\exists x)(x$ is an explanation of q that satisfies I).

On the ordered pair theory,

x is an explanation of q that satisfies I if and only if x is an ordered pair whose first member is a complete content-giving proposition with respect to Q that satisfies I and whose second member is the act-type explaining q.[3]

In the weak sense of explainability expressed by (I), how species were created is explainable in a way that satisfies instructions calling for an answer found in the Bible. This is so in virtue of the existence of the explanation

(1) (Species were created by God's willing them into existence; explaining how species were created),

whose constituent proposition satisfies instructions calling for an answer found in the Bible.

---

3. Using the no-product view of Chapter 3, Section 10, instead of (I) we could say that q is explainable in a way that satisfies I if and only if Possible $(\exists a)(\exists S)(\exists u)(a$ is an act in which S explains q by uttering u, and u expresses a proposition that satisfies I). In what follows, however, I shall employ the ordered pair theory.

However, when we say that q is explainable in a way that satisfies I, we may mean something stronger, viz. that it is correctly explainable:

(II) q is (correctly) explainable in a way that satisfies I if and only if $(\exists x)(x$ is an explanation of q that satisfies I, and x is correct).

In Chapter 4 we noted that if (p; explaining q) is an explanation of q, then it is correct if and only if p is true. Although how species were created is explainable in a way that satisfies instructions calling for an answer found in the Bible, it is not correctly explainable in this way. Explanation (1) we believe to be incorrect, since its constituent proposition is false.

Finally, an explainability claim might be construed to mean not only correctly explainable but appropriately explainable as well, i.e., explainable in such a way that appropriate instructions are satisfied.[4] So we have

(III) q is (correctly and appropriately) explainable in a way that satisfies I if and only if $(\exists x)(x$ is an explanation of q that satisfies I, and x is correct, and I is a set of appropriate instructions to follow for explaining q).

This is the strongest of the three conditions for explainability, since it entails, but is not entailed by, (I) and (II). In what follows, unless otherwise specified, I shall use it, and will understand "*un*explainability in a way that satisfies I" as its denial.

Something might be explainable in one way but unexplainable in another. Descartes believed that why a projectile continues to move after it has left the hand is *un*explainable in terms of physical causation, since, on his view, there are no physical causes for this motion. He also held that it *is* explainable in a way that calls for a more general law of which this regularity is a special case.[5] It is explainable in a way that satisfies instructions calling for a "special-case-of-a-law" explanation (which, as we argued in Chapter 7, is not a causal explanation). Finally, it is quite possible that Descartes believed that a projectile's motion is explainable in a way satisfying instructions calling for a cause, though not a physical cause (by appeal to God, who conserves motion).

There are different reasons that q might not be explainable in a way that satisfies I. One possibility is that I is not satisfiable by anything. Second, even if instructions I can be satisfied, any answer which satisfies I may be incorrect; in which case q would not be explainable in senses (II) and (III). How God created the first woman would be

---

4. See Chapter 4, Section 4.
5. See Chapter 7, Section 7.

declared unexplainable in senses (II) and (III) by an atheist, even if the instructions call simply for recounting the Biblical story. For an atheist this question, with its false presupposition, has no correct answer.

Third, even if Q can be answered correctly in a way that satisfies I, the latter may not be a set of appropriate instructions to follow for explaining q. If so, q is unexplainable in sense (III)—even if it is explainable in senses (I) and (II). Recall from Chapter 4 that I is a set of appropriate instructions to follow for explaining q to an audience if

(a) The audience does not understand q_I;
(b) There is a correct answer to Q, that satisfies I, the citing of which will enable the audience to understand q_I by producing the knowledge, of that answer, that it is correct;
(c) The audience is interested in understanding q in a way that satisfies I (or would be under certain conditions);
(d) Understanding q in a way that satisfies I, if it could be achieved, would be valuable for this audience.

These conditions were proposed as sufficient but not necessary, since there is also an epistemic criterion of appropriateness, according to which I is a set of appropriate instructions to follow for explaining q to an audience if it is reasonable for the explainer to believe that conditions (a) through (d) above are satisfied. The epistemic criterion requires relativization to an explainer as well as to an audience. However, I will assume that whether q is explain*able* to an audience—i.e., capable of being explained to it correctly and appropriately—is not to depend on, and vary with, who is doing the explaining. (III), then, will be taken to involve a non-epistemic criterion of appropriateness. (No doubt it is possible to consider epistemic evaluations of the form "q is capable of being explained to A *by S;* but for the purposes of (III), instructions will be considered appropriate if and only if conditions (a) through (d) above are satisfied.)

The non-satisfaction of the first part of condition (b)—that there is a correct answer to Q that satisfies I—has already been mentioned. The remaining conditions all pertain to the audience, and so their satisfaction may vary from one context to another. If the audience already understands q_I, then (barring certain pedogogical situations in which it is reasonable to treat the audience as if it does not understand q_I) we might say that q is not explainable, in a way satisfying I, to that audience. Similarly, if the audience is incapable of understanding q_I, then q is not explainable to it in a way satisfying I. Finally, if the audience would have no interest in understanding q_I, or if understanding q_I would be of no value to it, then q_I is not (correctly and appropriately) explainable to it in a way satisfying I. (To be sure, there are also senses in which q_I may be explainable, viz. (I) and (II).)

An explainability claim that is not relativized to instructions can be understood existentially, as follows:

(2) q is explainable if and only if $(\exists I)$(q is explainable in a way that satisfies I).

Since there are different senses of "q is explainable in a way that satisfies I," there are different interpretations of (2). In what follows, unless otherwise indicated, the strongest sense, viz. that afforded by (III), will be understood. That is,

(3) q is (correctly and appropriately) explainable if and only if $(\exists I)$(q is correctly and appropriately explainable in a way that satisfies I).

Let us return to Cotes to see how all this works. Cotes claimed that why bodies gravitate (q) is unexplainable. In saying this he was not asserting that there are no instructions with respect to which q is explainable in sense (I). (He would have agreed, e.g., that q is explainable in a sense that does not require a correct explanation.) Cotes was also, I think, not asserting that there are no instructions with respect to which q is explainable in sense (II). He might have supposed that there is a correct explanation of q satisfying certain theological instructions, e.g., (Bodies gravitate because God ordained it; explaining why bodies gravitate), or (Bodies gravitate because God ordained it for the reason ———; explaining why bodies gravitate). But if so, he might well have thought that the theological instructions satisfied by such explanations are not appropriate. A simple appeal to the Deity is not appropriate, he might have said, because conditions (c) and (d) for appropriate instructions are not satisfied. The answer "Bodies gravitate because God ordained it" would be of no interest or value to an audience that takes this for granted anyway. And we may assume that, since God's reasons are unintelligible to human audiences, an understanding of q that satisfies instructions calling for the specification of God's reasons will be impossible; hence condition (b) will not be satisfied with respect to such instructions. I take Cotes then to be making the claim that there are no instructions which are such that why bodies gravitate is *correctly* and *appropriately* explainable in a way that satisfies those instructions.

Next, consider Hempel's doctrine, a consequence of which is that an event such as an electron's hitting the screen where it did in a diffraction experiment is unexplainable (assuming that the event is a low probability one). Hempel is not saying that such an event is not explainable in the sense (I). (There are indeed explanations of q that satisfy some instructions or other.) Nor, I suggest, is Hempel saying that this event is not explainable in sense (II). The explanation (The electron landed where it did because it was diffracted by a slit; explaining why the electron landed where it did) is correct (in the sense of Chapter 4, Section 1); and it satisfies some instructions or other for explaining q (e.g., "Say why it landed in a position not on its original straight path"). Rather, I interpret Hempel to be claiming that why the electron landed where it did is not explainable in sense (III): there is no correct explanation of q that satisfies appropriate instructions for explaining q. Such instructions, on Hempel's view, would require a set of laws that determine the electron's final position with high probability. And there are no such laws.

2. EXPLANATORILY ULTIMATE QUESTIONS

The examples with which we began the chapter involve questions that are not only said to be unexplainable, but ultimate as well. I shall now define a concept of an explanatorily ultimate question which can be used to characterize these and other examples. Since our interest is in questions that can appear in the context "q is (not) explainable," I will restrict the definition to content-questions (see Chapter 2, Section 9).

(1) Where Q is a content-question, Q is *explanatorily ultimate* if and only if
  (a) Q is an ultimate question of its kind;
  (b) All propositions in the body of Q are true;
  (c) q is not (correctly and appropriately) explainable.

To understand (1), we will have to define concepts in (a) and (b); the concept of explainability in (c) has already been introduced.

Consider a syntactically well-formed interrogative sentence that contains an interrogative pronoun (e.g., "what," "why") or prepositional phrase (e.g., "in what way"), and by means of parentheses isolate the interrogative pronoun (or prepositional phrase) and whatever noun or noun phrase the interrogative pronoun modifies. For example,

(What) caused the acceleration?

The acceleration was caused (by what)?

(What contact force) caused the acceleration?

(For what reason) did the body accelerate?

(How) do bodies gravitate?
(Why) do bodies gravitate?

The expressions in parentheses will be called interrogative expressions.[6]

Let Q_1 and Q_2 be meaningful interrogative sentences each of which expresses a different content-question. Interrogative sentence Q_1 will be said to be of *the same kind as interrogative sentence Q_2 but more general* if Q_1 is obtainable from Q_2 by deleting one or more words from the interrogative expression of Q_2. Thus, the interrogative sentence

Q_1: What force caused the acceleration?

is of the same kind as, but more general than, the interrogative

Q_2: What contact force caused the acceleration?

Q_1 and Q_2 express different content-questions, and Q_1 is obtainable from Q_2 by deleting the word "contact" from the interrogative expression "what contact force" in Q_2.

Where Q is an interrogative sentence that expresses a content-question, Q will be said to be an *ultimate interrogative of its kind* if and only if there is no interrogative sentence of the same kind as Q but more general. Q_2 is not an ultimate interrogative of its kind, since there is an interrogative of the same kind as Q_2 but more general, e.g., Q_1. Nor is Q_1 an ultimate interrogative of its kind, since there is an interrogative of the same kind as Q_1 but more general, viz.

Q_3: What caused the acceleration?

There is, however, no interrogative of the same kind as Q_3 but more general. Q_3 is an ultimate interrogative of its kind.

Similarly,

Q_4: What pumping function does the heart have?

is not an ultimate interrogative of its kind, but

Q_5: What function does the heart have?

is. (If "function" is deleted from Q_5 the resulting sentence does not express a content-question.)

Now, as before, I am treating questions as abstract entities that are expressed by interrogative sentences. Q will be said to be an *ultimate question of its kind* if and only if it is expressible only by an ultimate interrogative sentence of its kind. Thus the interrogative Q_3 expresses

[6]. For simplicity I am considering interrogative sentences with only one interrogative expression. An interrogative such as "How and why did the acceleration occur?" can be treated as a conjunction of interrogatives each conjunct of which contains only one interrogative expression.

an ultimate question of its kind, but Q_1 and Q_2 do not.[7] Furthermore, although

What function does the heart have that is a pumping one?

is an ultimate interrogative of its kind, it does not express an ultimate question of its kind, since the question it expresses is also expressible by interrogative Q_4 which is not an ultimate interrogative of its kind. Similarly, although

What is the pumping function of the heart?

is an ultimate interrogative of its kind, it does not express an ultimate question of its kind; the question it expresses is also expressible by Q_4 which is not an ultimate interrogative of its kind. But Q_5 expresses an ultimate question of its kind since the question it expresses is expressible only by ultimate interrogatives of their kind, e.g.,

What is the function of the heart?

What function does the heart have?

The function of the heart is what?[8]

In order to characterize the notion of the *body* of a question used in definition (1), a preliminary definition will be needed. An *interrogative transform* of an interrogative sentence Q will be any meaningful interrogative sentence Q' obtained from Q by replacing the interrogative expression in Q by any other interrogative expression. Thus

What force caused the acceleration?

and

---

7. It might be objected that Q_3 does not express an ultimate question of its kind since the question it expresses is also expressible by

Q_3': What cause caused the acceleration?

And Q_3', it might be said, is not an ultimate interrogative of its kind, since there is an interrogative of the same kind as Q_3' but more general, viz. Q_3. The latter is not the case, since for Q_3 to be of the same kind as Q_3' but more general Q_3 and Q_3' must express different questions.

8. It might be said that Q_5 is not ultimate since it is equivalent in meaning to

Q_5': What activity does the heart engage in which is its function?

which is not an ultimate interrogative of its kind, since "activity" can be deleted to obtain

Q_5'': What does the heart engage in which is its function?

However, even if Q_5 and Q_5' are equivalent in meaning (which I think is dubious), for this to be a genuine counterexample Q_5'' must be of the same kind as Q_5' but more general. And the latter obtains only if Q_5' and Q_5'' express different propositions, which is not the case here. An analogous objection and response to Q_3 was considered in the previous footnote.

Who caused the acceleration?

are interrogative transforms of

What contact force caused the acceleration?

And

How do bodies gravitate?

is an interrogative transform of

Why do bodies gravitate?[9]

Now certain presuppositions of an interrogative sentence may also be presuppositions of *all* of its interrogative transforms. For example, "There was an acceleration" and "The acceleration was caused" are presuppositions of all of the interrogative transforms of "What contact force caused the acceleration?" By the *body* of an interrogative sentence I shall mean the set of sentences each member of which is presupposed by all of its interrogative transforms. Thus, the body of the interrogative sentence "What contact force caused the acceleration?" contains, among other sentences, "There was an acceleration," and "The acceleration was caused." It does not contain "Some contact force caused the acceleration," since the latter is not presupposed by all the interrogative transforms of the original interrogative sentence.

By the body of the *question* Q, I shall mean the set consisting of all those propositions expressed by sentences that are in the body of each interrogative sentence expressing Q. The body of the question expressed by the interrogative sentence "What contact force caused the acceleration?" contains, among other propositions, those expressed by the sentences "There was an acceleration" and "The acceleration was caused."

Condition (b) in definition (1) of an explanatorily ultimate question Q requires that all propositions in the body of Q are true. The question

Why do bodies gravitate?

satisfies this condition, since its body contains the proposition

Bodies gravitate,

and its logical consequences, all of which are true. The question

Why do bodies gravitate in accordance with an inverse cube law?

---

9. Note that to obtain an interrogative transform of Q the interrogative expression in Q must be replaced by another interrogative expression. Thus, "Why is it true that nothing caused the acceleration?" is not an interrogative transform of "What contact force caused the acceleration?"

does not satisfy this condition, since its body contains the false proposition

Bodies gravitate in accordance with an inverse cube law.

All of the concepts invoked in the definition of an explanatorily ultimate question have now been defined. In order to better understand why each of the three conditions of this definition has been imposed, let us turn to more examples. Consider first the question

Q_6: What contact force causes the motion of the planets?

Is Q_6 an explanatorily ultimate question? Why or why not? Q_6 satisfies condition (b) of definition (1)—that all propositions in the body of Q are true. The body of Q_6 contains the proposition "The motion of the planets is caused" (and its entailments), which is true. Q_6 also satisfies condition (c)—that q is not correctly and appropriately explainable—since no contact force causes the motion of the planets. Despite this, we are tempted to say, we have not here reached the limits of explanation. This is because condition (a) is not satisfied. Q_6 is not an ultimate question of its kind. There is a question of the same kind as Q_6 but more general, viz.

Q_7: What causes the motion of the planets?

(One *question* is of the same kind as another but more general if and only if some interrogative expressing the former is of the same kind as, but more general than, some interrogative expressing the latter.) Moreover, q_7 is correctly and appropriately explainable (by invoking the gravitational force of the sun on the planets). Even though q_6 is not explainable, there is a more general question of the same kind which is, viz. q_7. This is why although Q_6 is not correctly and appropriately explainable, we have not yet reached the limits of explanation. Q_6 is not explanatorily ultimate.

Consider, next, this question:

Q_8: Why does a contact force cause the motion of the planets?

Q_8 is an ultimate question of its kind; so condition (a) of definition (1) is satisfied. It is also not explainable (since a contact force does not cause the motion of the planets); so (c) is satisfied. Yet (b) is not satisfied since the body of Q_8 contains the proposition "A contact force causes the motion of the planets," which is false. So Q_8 is not explanatorily ultimate. As with Q_6 something can be done that is not possible with explanatorily ultimate questions. In the case of Q_8 we can correct a false proposition in the body of the question.

Finally, consider once again

Q_7: What causes the motion of the planets?

It is an ultimate question of its kind, satisfying condition (a). All propositions in its body are true; so condition (b) is satisfied. Yet it fails to satisfy (c), since q_7 is (correctly and appropriately) explainable by appeal to the sun's gravitational force.

Questions Q_6, Q_7, and Q_8 all fail to be explanatorily ultimate, since each violates one of the conditions of definition (1). With each of these questions something can be done that cannot be done with explanatorily ultimate questions. The question may itself be explainable, as with Q_7. Or there may be a more general question of that kind which is explainable, as with Q_6. Or it can be pointed out that the body of the question contains a false proposition, as with Q_8; and a quite similar question may be formulatable which is correctly and appropriately explainable and whose body contains only true propositions (e.g., "Why does a force cause the motion of the planets?").[10]

Compare these questions with

Q_9: Why do bodies gravitate (i.e., why do bodies exert gravitational forces on other bodies)?

This is an ultimate question of its kind. Propositions in the body of Q_9 are true. And—according to Cotes—q_9 is not (correctly and appropriately) explainable. In short, for Cotes, it is explanatorily ultimate. According to him, we cannot answer the question correctly and appropriately. Nor can we do so when it is transformed into a more general question of the same kind, since it is already an ultimate question of its kind. Nor can we do so when we correct some false proposition in its body, since there is none. To be sure, Cotes might admit, there is a correction that is possible, since Q_9 falsely presupposes that

(2) Bodies gravitate for some reason (i.e., for some reason that is correct and appropriate).

But (2) is a complete presupposition of Q_9, and any question that is explanatorily ultimate will have a complete presupposition that is false.[11] This is so because any such question will have no (appropriate) answer that is correct.

Can we alter Q_9 so that it becomes correctly and appropriately ex-

---

10. There are various questions that might be expressed by this interrogative. One of them, with emphasis on "force," is correctly and appropriately answered (according to Newtonian physics) by saying that the motion of the planets is accelerated and all accelerations are produced by forces.

11. This explains why we cannot substitute for condition (b)—that all propositions in the body of Q be true—the condition that every complete presupposition of Q be true. If we want a definition that makes Q_9 explanatorily ultimate (according to Cotes), we cannot make this demand, since (2), which is a complete presupposition of Q_9, is false, according to Cotes.

plainable? Since its failure to be so now does not depend on false propositions in its body, there is no need to change the body of Q_9. Nor is it possible to make it a more general question of its kind, since it is already an ultimate question of its kind. What we can do to obtain a question that is not explanatorily ultimate is to change the interrogative expression in Q_9, e.g., by writing

How (i.e., in what manner) do bodies gravitate?

This, Cotes would agree, is correctly and appropriately explainable—by appeal to a law of gravitation which gives the manner of the gravitation. But now we have fundamentally changed the original question. It is not a more general question of the same kind as Q_9; nor is it a question obtained by making a correction in the body of Q_9.

More generally, if a question Q is explanatorily ultimate, then there is no correct and appropriate explanation of q; nor can such an explanation be obtained by transforming Q into a more general question of its kind or by correcting some mistake in its body. To be sure, if a question is explanatorily ultimate a "correction" is possible. The falsity of a complete presupposition of the question can be noted. But if this is "corrected" a very different question emerges.

3. APPLICATIONS

At the beginning of the chapter claims regarding the limits of explanation were noted by various writers, including Toulmin, Cotes, and Hempel. I suggest that the concept of a question's being explanatorily ultimate provides a means to understand such claims. I have already mentioned the question "Why do bodies gravitate?" in the case of Cotes. Toulmin suggests that in Newtonian physics the principle of inertia is an ideal of natural order that requires no explanation. Using my terminology, the question "Why do bodies exhibit inertia?" is explanatorily ultimate, according to (Toulmin's conception of) Newtonian physics. It is not explainable and cannot be made so by transforming it into a more general question of its kind, or by correcting some false proposition in its body (since there is none).

Or, to revert to a position to which Hempel is committed, consider an electron diffraction experiment in which the electron hits the screen at a point of low probability, according to quantum mechanics. The question

Q_1: Why did the electron hit the screen at the point it did?

is an ultimate question of its kind, and all the propositions in its body are true. Furthermore, if we follow Hempel, Q_1 is not explainable, since there are no laws that determine the electron's final position with high probability. Therefore, for Hempel, q_1 is explanatorily ultimate. All we can do, according to Hempel, is to say that, given the truth of quantum mechanics, there is no correct and appropriate answer to Q_1. Every proposition in the body of Q_1 is true (e.g., there was an electron; it did hit the screen; and it did so at a certain point.) What is false is only the complete presupposition of Q_1, viz. that there is a reason that the electron hit the screen where it did.

Of course, whether a question is explanatorily ultimate is not simply a matter of what a scientist claims to be so. For Cotes the question

Q_2: Why do bodies gravitate?

is explanatorily ultimate. But he may have been mistaken. Physicists today are working on a theory of "supergravity" which explains the four known forces in nature, including gravity, by appeal to a more basic symmetry principle. If such a theory turns out to be correct, then q_2 is correctly and (presumably) appropriately explainable, and hence not explanatorily ultimate.

A question may be posable relative to a given theory for which neither the theory nor its theorist has an answer. It does not follow that such a question is explanatorily ultimate according to that theory or theorist. Why bodies gravitate is not explained within Newton's mechanics. But Newton did not regard this as explanatorily ultimate. Why not? What is the difference, e.g., between gravitation and inertia? Consider

Q_3: Why do bodies subject to no net force exhibit inertia (i.e., the tendency to remain at rest or to continue in motion with uniform speed in a straight line)?

Newtonian mechanics does not explain either q_2 or q_3. Yet, I shall assume, Newton regarded q_3 but not q_2 as explanatorily ultimate. The difference for Newton rests on condition (c) for explanatory ultimateness. Newton believed that q_2 is correctly and appropriately explainable by appeal to contact forces, although he himself could not produce the explanation. On the other hand, I shall suppose, following Toulmin, that Newton regarded q_3 as not being correctly and appropriately explainable.

What kind of claim is one that says that q is (or is not) explanatorily ultimate? Can it be decided a priori? With some questions, e.g., "Why does the law of identity hold?" this may be possible. In general, however, whether a question is explanatorily ultimate is not (completely) settleable a priori. Whether

Q_2: Why do bodies gravitate?

is explanatorily ultimate depends, for one thing, on whether bodies do gravitate—a fact not determinable a priori. It also depends on whether q_2 is explainable—again not something completely determinable a priori.[12]

4. BEHAVIOR NOT REQUIRING EXPLANATION; FUNDAMENTAL FACTS AND DISPOSITIONS

It may be said that I have still not captured certain central ideas associated with the earlier examples. Cotes is not claiming simply that why bodies gravitate is explanatorily ultimate. He is also saying that it *does not require explanation,* that the fact that bodies exert gravitational forces is a *fundamental fact* about, or *fundamental disposition* of, bodies. The emphasized words here express concepts which, it will be objected, cannot be adequately understood in terms of ones already introduced. Yet they are concepts important for understanding the claims being made.

I want to argue that the concept of a question's being explanatorily ultimate contains the idea of something's not requiring explanation, and that it can be used to understand fundamental facts and dispositions.

If q is explanatorily ultimate, then q is not correctly and appropriately explainable. This means that there are no instructions with respect to which q is correctly and appropriately explainable (see (3), Section 1); hence there is no correct answer to Q which satisfies any instructions that are appropriate to follow for explaining q. Now according to the account given in Chapter 2, A does *not* understand q_I only if Q_I is sound, i.e., only if there is a correct answer to Q that satisfies I. Employing the "appropriate instructions" use of "understand," as I have been doing, this means that A does not understand q (without relativization) only if there is a correct answer to Q which satisfies appropriate instructions. But if Q is explanatorily ultimate there is no such answer to Q, and so it cannot be true that A does not understand q. If Q is explanatorily ultimate, then it is not possible for someone to be in the position of not understanding q. Now whatever else we might say about *requiring* explanation, I suggest that we can say at least this:

(1) q requires explanation only if it is possible for someone to be in the position of not understanding q.

Equivalently,

---

12. With some questions, e.g., "Why is there something rather than nothing?" it may not be clear whether their explanatory ultimateness (or lack of it) is determinable a priori. Is the proposition "There is something rather than nothing" which is in the body of this question knowable a priori? If there are appropriate instructions for this question do they make empirical assumptions? If there are no appropriate instructions is this determinable a priori?

(2) q does not require explanation if it is impossible for someone to be in the position of not understanding q.

But, we have just seen, if Q is explanatorily ultimate, then it is impossible for someone to be in the position of not understanding q. Therefore, *if Q is explanatorily ultimate, then q does not require explanation.* Given (1) and (2), and the earlier assumptions about understanding, the concept of being explanatorily ultimate does yield the idea of requiring no explanation. New concepts are not needed for this purpose.

The concept of being explanatorily ultimate can also be used to define the concept of a fundamental fact or disposition. The sense of fundamentality here is explanatory:

The fact that p is a fundamental fact if and only if the question "Why is it the case that p?" is explanatorily ultimate.

d is a disposition that is ultimate if and only if the question "Why do the items which have d have it?" is explanatorily ultimate.[13]

If the fact that p is a fundamental fact then, according to the above definition, there is a certain question about that fact, viz. "Why is it the case that p?" which is not correctly and appropriately explainable. (This, as we have just seen, entails that it does not require explanation.) But, furthermore, its incapability of being correctly and appropriately explained is not due to any false proposition in its body; for if it is explanatorily ultimate, then all propositions in its body are true. Nor is it due to some false assumption not in its body (but in its interrogative expression) which can be removed by generalizing the question in a certain way. (For example, "What contact force causes the acceleration of the planets?" has the false presupposition that a contact force causes this acceleration.) Such false assumptions will be eliminated by the requirement that Q be an ultimate question of its kind. The only false assumption associated with an explanatorily ultimate question is its complete presupposition. So if the question "Why is it the case that p?" is explanatorily ultimate, then p is true (it is a fact that p), since p is contained in the body of the question, and so must be true; and the only false assumption associated with this question is that there is a reason that p is true.

The definitions above impute a certain relativity to fundamental facts and dispositions. Whether p is a fundamental fact depends on whether "why p" is correctly and appropriately explainable. Whether it is appropriately explainable is, in part, contextual. It depends upon whether there is a correct answer to "why p" that satisfies instructions the satisfaction of which will enable the audience to understand why p in a

---

13. These definitions can readily be extended to other items such as laws and theories which are claimed to be ultimate.

way that is of interest and value to it (see Section 1). For some audiences there may be such instructions, for others not; which means that for some audiences, but not others, the fact that p may be fundamental. Cotes may have believed that there is a correct answer to "Why do bodies gravitate?" that appeals to God's reasons, but that ordinary mortals would not be able to understand, in that way, why bodies gravitate. For beings more intelligent than we are, the fact that bodies gravitate may not be fundamental. For us it is.

It is also possible to introduce a non-relativized notion of fundamental fact and ultimate disposition by employing the criterion for correct explainability ((II) of Section 1) rather than that for correct and appropriate explainability (III). In the definition of an explanatorily ultimate question, simply change the third condition to: Q is not (correctly) explainable.

5. IDENTITIES

The explanatorily ultimate questions produced up to now have been ones that would have been claimed to be so by one thinker, but not necessarily by another. Are there questions whose explanatory ultimateness would be agreed to by everyone? In this section and the next, two such putative classes will be considered.

Scientists frequently make claims of the following sort:

Water is H_2O;

The temperature of a gas is its mean molecular kinetic energy;

An alpha particle is a doubly positively ionized helium atom;

The element of atomic number 29 is copper.

Many philosophers interpret these as expressing contingent identities, and they hold that such identities will need to be invoked when one theory or science is reduced to another. Causey, e.g., argues that the reduction of classical thermodynamics to statistical mechanics involves, among other things, the derivation of the laws of the former from the laws of the latter together with suitable identities (such as that "temperature of a gas = mean kinetic energy of gas molecules"). Furthermore, the claim is made that this is a genuine reduction because the only propositions subject to further explanation are the laws of the more basic ("reducing") theory. Although identities are needed to effect the reduction, these, it is claimed, are not subject to explanation.[14] I believe that Causey and others would say that questions such as

---

14. Causey, *Unity of Science*. Another philosopher who makes this claim is Clark Glymour, "Explanation, Tests, Unity, and Necessity," *Noûs* 14 (1980), pp. 31–50; see p. 34.

Why is water = H_2O?

Why is the temperature of a gas = its mean molecular kinetic energy?

Why is an alpha particle = a doubly positively ionized helium atom?

Why is the element of atomic number 29 = copper?

and more generally

Why is x = y? (where x = y)

are explanatorily ultimate.

Such questions satisfy the first two criteria of explanatory ultimateness. They are ultimate questions of their kind, and the propositions in their bodies are true. The only issue that remains is whether they are not (correctly and appropriately) explainable. Causey says they are not, and his claim seems to be a priori, rather than empirical. Genuine identities, he urges, are not explainable, no matter what particular assertions such identities make. To be sure, the claim that these questions are explanatorily ultimate is not (entirely) a priori, since part of such a claim is that the bodies of these questions contain only true propositions (which, in the present examples, is an empirical matter). But that part of the claim according to which these questions are not *explainable* is, I believe, intended by Causey to be a priori, not empirical.

Far from being a priori true, the latter claim may well be empirically false. Here is one way to explain why alpha particles = doubly positively ionized helium atoms:

(1) Neutral helium atoms have 2 electrons, so that if such atoms are doubly positively ionized, i.e., lose both electrons, what remains are helium nuclei; but alpha particles are simply helium nuclei. Therefore, alpha particles are (the same things as) doubly positively ionized helium atoms.

The following instructions are satisfied by this explanation:

I_1 (a) Say what scientific category ("natural kind") alpha particles fall under. (The explanation informs us that alpha particles are helium nuclei.)

(b) Say what positive double ionization of an atom is. (According to the explanation, it is the removal of two electrons from the atom.)

(c) Indicate what remains when a helium atom is doubly positively ionized ("a helium nucleus").

The explanation also satisfies these more general instructions:

I_2 Indicate features of X's and Y's in virtue of which they are identical.

Indeed, I_2 can be satisfied in explaining many identities, e.g., why the element of atomic number 29 = copper.

(2) The identity holds because copper is the element each of whose atoms contains 29 protons. But any element whose atomic number is n is such that its atoms contain n protons. Therefore, the element whose atomic number is 29 is copper, the element whose atoms contains 29 protons.

In this case, we have explained why the element of atomic number 29 = copper by indicating a feature of both—being an element whose atoms contain 29 protons—in virtue of which the identity holds.

Causey responds to such putative explanations in various ways, of which two will be noted. First, he considers these *justifications* of the claim that $x = y$, not *explanations* of why such a claim is true. I would agree that justifying the claim that $x = y$ and explaining why $x = y$ are different illocutionary acts with different products. Employing the ordered pair view, the product of the latter is an ordered pair whose first member is a complete content-giving proposition with respect to "Why is $x = y$?" and whose second member is the act-type "explaining why $x = y$." The product of the illocutionary act of *justifying* the claim that $x = y$ is an ordered pair whose first member is some proposition and whose second member is the act-type "justifying the claim that $x = y$." So the products of these acts are indeed different. Still the first members of these products—the propositions—can be identical. If someone doubts that alpha particles are doubly positively ionized helium atoms and asks me to justify my claim that they are, I might utter (1) with an intention appropriate for justification, not explanation. But if someone does not understand why alpha particles are the same things as doubly positively ionized helium atoms—if, e.g., he is not in a knowledge-state with respect to this question and instructions I_1 or I_2—I might also utter (1), with an intention appropriate for explaining.

Causey's second response is to note that (1) and (2) do not provide *causal* explanations, and to restrict his thesis to these. His claim, then, is that why x is identical with y cannot be causally explained. Unfortunately, Causey does not provide a general criterion for what counts as a causal explanation. Nevertheless, there does seem to be some intuitive sense in which (1) and (2) fail to provide causal explanations. In Chapter 7 various types of *non*-causal explanations were characterized, including *classification* explanations. One form the latter can take is this:

(The reason that x is a P_1 is that x is a P_2, and according to such and such a definition, or criterion, or system of classification, something is a P_1 if and only if it is a P_2; explaining why x is a P_1).

Now I suggest that (1) and (2) are best construed as classification explanations. Thus, (1) is telling us, in effect, that the reason that alpha particles are identical with doubly positively ionized helium atoms is that alpha particles are the same things as helium nuclei, and (according to a definition or system of classification in physics and chemistry) something is a doubly positively ionized helium atom if and only if it is a helium atom that has lost both of its electrons and hence is a helium nucleus. An analogous claim can be made in the case of (2). In Chapter 7 reasons were provided for concluding that classification explanations are non-causal.

More importantly, even if we grant that why some (or indeed any) identity holds is not *causally* explainable, it does not follow that it is not explainable or that the question "Why is x = y?" is explanatorily ultimate. What Causey is doing here might be described in one of two ways. First, he might be restricting explanatory questions about identities to ones of the form

Q_1: What caused it to be the case that x = y?

If nothing caused x to be identical with y, then, of course, this question is not correctly explainable. But it does not follow that Q_1 is explanatorily ultimate, since there is a proposition in its body that is not true, viz. "Something caused it to be the case that x = y." By contrast, the latter proposition is not in the body of

Q_2: Why is it the case that x = y?

And, as has just been argued, Causey has not succeeded in showing that all questions of form Q_2 are explanatorily ultimate. Even though a question of form Q_2 is an ultimate question of its kind, and even if all propositions in its body are true, Causey has not shown that such a question is not explanatorily ultimate, since he has not demonstrated that it is not explainable. Indeed, if I am right, he could not demonstrate this.

An alternative interpretation of Causey's procedure is that he is claiming that, at least in science, only causal explanations are appropriate. Since he assumes that causal explanations of identities are not possible, he concludes that identities are not scientifically explainable.

But again, I urge, this has not been demonstrated, since it has not been demonstrated that all appropriate scientific explanations must be causal. In Chapter 7, the latter general claim was examined and rejected. And in the present section the specific claim was made that (1)

and (2) do provide explanations of identities, and that these, being classification explanations, are not causal.

6. NATURAL BEHAVIOR

A second broad class of questions that might be deemed explanatorily ultimate concerns types of behavior or states that are *natural*. Toulmin holds that every science invokes some ideal of natural order—some progression of natural behavior—which is regarded as unexplainable and requiring no explanation. In Newtonian physics, inertial motion is natural, so that the question of why bodies exhibit that motion is explanatorily ultimate; and it is so, *because* inertial motion is natural.

More generally, there are two related theses that bear examination:

(i) If N is a type of natural behavior or state, then why items exhibit behavior of type N (or persist in that type of state) is explanatorily ultimate;[15]

(ii) It is explanatorily ultimate because N is a type of natural behavior or state (i.e., an appeal to naturalness helps to explain why certain questions are explanatorily ultimate).

What are the criteria for natural behavior or states? In what follows some possibilities will be considered.

a. Some behavior or state is natural if and only if it is uncaused. But unless some restrictions are placed on what counts as a cause, this proposal will not yield the behavior and states its proponents generally regard as natural. Descartes, e.g., had a concept of inertial motion of a body which was not caused by the body's impact with other bodies, but, at least indirectly, by God, the ultimate cause and conserver of all motion in the universe. With no restriction on types of causes the present proposal will yield the conclusion that for Descartes inertial motion is not natural. Even in Newton's case it is not impossible to say that inertial motion has a cause, viz. the absence of (unbalanced) forces acting on the body. Causes can be absences as well as presences.

Robert Cummins, who defends a version of the present criterion, admits that causal explanations are possible in connection with natural states.[16] But such explanations, he claims, indicate causes not of the

---

15. This does not preclude explaining some particular instance of natural behavior. Thus one might explain why that particular body continued to move in the absence of forces by saying that all bodies do this in the absence of forces. But—assuming inertial motion is natural behavior—the question of why all bodies do this is explanatorily ultimate.

16. Robert Cummins, "States, Causes, and the Law of Inertia," *Philosophical Studies* 29 (1976), pp. 21–36.

state itself but of its onset. He also admits the possibility of causally explaining a natural state by citing a cause that neutralizes certain disturbing factors. Suppose that we push a block of wood that is resting on a table, and we continue to do so in such a way that the block moves in a straight line with uniform speed. The uniform motion of the block is an instance of inertial motion. Here, according to Cummins, a causal explanation of the uniform motion is possible—an explanation that appeals to the pushing force—since the latter force neutralizes the force of friction acting on the block in the opposite direction. But, says Cummins, the uniform motion of the block is not an *effect* of the pushing force. Cummins's thesis, then, is that natural states are not the effects of any causes.

This thesis places great pressure on the notion of an effect of a cause—pressure I doubt it will bear. Cummins admits that the uniform motion of the block can be causally explained by my pushing the block. For this to be so the force I exert by pushing it must cause the uniform motion of the block, or at least it must be part of the cause of this motion. Yet, Cummins must say, the uniform motion is not an effect of the pushing force. Why not? Perhaps he is thinking that what is an effect of the pushing force is the neutralization of the force of friction acting in the opposite direction. I fail to see why this precludes the uniform motion of the block from also being an effect of the pushing force (perhaps in conjunction with the force of friction). But for the sake of argument, let us suppose it is not. Still there is, as Cummins admits, a neutralization of the force of friction. And there seems to be no reason for denying that the uniform motion of the block is an effect of this neutralization, which is its cause. If so, we have what (in Newtonian physics) Cummins takes to be a natural state—inertial motion—which is the effect of a cause. More generally, even in cases where forces are not being neutralized, I see no reason for supposing that inertial motion cannot be described as an effect of the absence of a net force acting on the body.

To generate the sorts of examples defenders of the present criterion regard as natural some restriction on causes will have to be made which precludes the absence of "disturbing" factors from counting as a cause. Although this restriction is obviously a vague one, I shall not try to clarify it, since, in any case, the present criterion will not do the job for which I invoked it in the first place.

The present criterion of natural behavior and states does not support claim (i)—nor, therefore, (ii), which depends on (i)—given at the beginning of this section. If N is a type of natural, i.e., uncaused, behavior, it does not necessarily follow that why items exhibit behavior of type N is explanatorily ultimate. (To permit behavior generally regarded as natural to be classified as such, I shall preclude the absence of "disturbing" elements from being counted as a cause of natural behavior.)

In geometric optics the rectilinear propagation of light can be said to be natural behavior, in the present sense of "uncaused behavior." Light's traveling in straight lines is not caused by anything (barring the absence of disturbing factors as a cause). Yet within geometric optics the question

Why does light travel in straight lines?

is not treated as explanatorily ultimate, since it (together with the laws of reflection and refraction) is explainable by appeal to what is often regarded as a more fundamental principle, viz. Fermat's principle of least time—that light takes the path which requires the shortest time. (The explanation would be what in Chapter 7 I call a complex derivation-explanation.) Again, within Newtonian mechanics, that a system of bodies conserves momentum might be taken to be natural behavior of the system, in the present sense of "uncaused behavior." (Or its having the total momentum it does might be considered to be a natural state it is in.) Yet according to Newtonian mechanics, the question

Why do mechanical systems conserve momentum?

is not explanatorily ultimate, since it is explainable by appeal to Newton's second and third laws of motion.

In general, the fact that a law describes uncaused behavior does not preclude its explanation by appeal to more fundamental laws.

b. *Some behavior or state is natural if and only if it will persist forever if no external cause interferes.* This is not the same as (a). Some behavior or state might be uncaused without persisting forever; and some behavior or state might persist forever, in the absence of external causal interference, without being uncaused. (God, e.g., may have caused the behavior which will persist forever in the absence of causal interference.) Although the present notion emphasizes the persistence of the behavior rather than its lack of causation, it, like its predecessor, will not support claims (i) and (ii). If N is a type of behavior that will persist forever unless some external cause interferes, it does not necessarily follow that why items exhibit behavior of type N is explanatorily ultimate. The examples given above can be used here as well. The rectilinear propagation of light, or the total momentum of a system of particles, will persist forever (unless some external cause interferes). But this does not mean that why light is propagated in straight lines, or why systems conserve momentum, is explanatorily ultimate.[17]

17. Accordingly, such examples can be used against a criterion which combines (a) and (b).

c. *Some behavior or state is natural if and only if it is caused by the item whose behavior or state it is.* Variations of this are possible, but all contain the theme that the behavior is due to the nature of the substance behaving rather than to factors external to that substance.[18] The general view—historically perhaps the best known—derives from Aristotle. Nature, for Aristotle, is a source or cause of motion and rest (*Physics*, Book 2). Each thing has a nature, which is the source of its behavior (to use a broader term for motion and rest, which is probably more in accordance with Aristotle's intentions). Natural behavior, then, is behavior whose cause lies in the nature of the item behaving. Aristotle says that this nature can be associated either with the matter of the substance or with its form (though he regards the latter association as preferable). A substance is behaving naturally, then, when its behavior is in accordance with its nature, i.e., its form (and matter). Keeping in mind Aristotle's doctrine of the four causes, which includes the formal and the material, we can say that a substance is behaving naturally when it is caused to behave in that manner by its form (and/or possibly its matter). It is of the nature of any substance composed of the element "earth" to be in the center of the universe; so if external causes are not preventing it from doing so, such a substance not at the center will move toward the center. This is its natural motion. By contrast, it is not part of the nature of a cart on a road to move in a direction parallel to the road; any movement in such a case is caused by an external mover, and is thus not "natural."

Aristotle's views about natural behavior in Book 2 of the *Physics* are complicated by his discussion in Book 8. In the latter he admits that the motion of a body can be natural even if it has an external cause. Perhaps his doctrine is this. Bodies are exhibiting natural behavior if they are behaving in accordance with their form and matter—if they are actualizing their potential. This potential, in certain cases, can be actualized by an external cause. This is so in the case of falling bodies. An external cause—the removal of a hindrance from falling—actualizes the potentiality of a stone to fall toward the earth. The potentiality—as determined by its form and matter—is internal to the stone. If the stone is falling it is exhibiting natural behavior since it is behaving in accordance with its form and matter—even if there is also an external cause which activates this potential.

There is no need here to further explicate Aristotle's doctrine, or to offer criticisms, since one thing seems quite clear. The doctrine, like its predecessors, will not serve the purpose for which we invoked it, since it will not support (i), nor therefore (ii). According to (i), if N is a type of natural behavior, then why items exhibit behavior of type

---

18. One variation of the present idea is that some behavior or state is natural if and only if it is the manifestation of some disposition of the substance whose behavior or state it is—where that manifestation is not caused by factors external to the substance.

N is explanatorily ultimate. For Aristotle, the falling of a stone when released from the hand is a type of natural behavior. Yet for Aristotle the question

Why does a stone fall when released from the hand?

is not explanatorily ultimate. It is explainable (according to Aristotle) by saying (something like) this: a stone is composed (primarily) of the element earth, and it is the nature (i.e., in accordance with the form) of the element earth to move to the center of the universe, unless external causes intervene; if a stone is released from the hand, then the releasing is an external cause that actualizes the stone's potential to fall.

7. ELLIS'S THEORY

A final view of natural behavior is this:

d. Some behavior or state is natural if its persistence does not require causal explanation. This view, invoking as it does the idea of causal explanation, is perhaps a variation of (a) in Section 6. However, I want to consider a specific version due to Brian Ellis. In it a particular theory of causal explanation is adopted according to which such an explanation involves subsuming what is explained under a so-called law of succession. Ellis writes:

> . . . a system is considered to be in an unnatural state if and only if we consider that its continuance in that state requires what has been called "causal explanation". . . . The behavior of a given system is considered to require causal explanation if and only if we feel that this behavior is not sufficiently explained by its subsumption under a law of succession. A law of succession is any law that enables us to predict the future states of any system (or given class of systems) simply from a knowledge of its present state, assuming that the conditions under which it exists do not change.[19]

The law of inertia is a law of succession. Inertial motion is natural, according to Ellis, because we regard a body's inertial motion as sufficiently explained by subsumption under the general law of inertia. Ellis's notion of causal explanation is quite different from the one in Chapter 7. According to the latter, an explanation of some particular body's inertial motion by subsumption under the law of inertia is not a causal explanation (but a "special-case-of-a-law" explanation, which, it was argued, is not causal). However, to avoid controversy regarding what should and should not be counted as a causal explanation, we

---

19. Brian Ellis, "The Origin and Nature of Newton's Laws of Motion," in R. Colodny, ed., *Beyond the Edge of Certainty* (Englewood Cliffs, N.J., 1965), p. 45.

may simply formulate Ellis's position to be that *some particular behavior is natural if it is sufficiently explained by subsumption under a law of succession*.

Like the law of inertia, Galileo's law of falling bodies (that the distance an unsupported body falls toward the earth is proportional to the square of the time) is also a law of succession. However, according to Ellis, the gravitational acceleration of a falling body is not natural because we do not regard a body's accelerative motion as sufficiently explained by subsumption under Galileo's law.[20] Ellis goes on to suggest that "there appears to be no objective criterion for distinguishing between natural and unnatural successions of states."[21] And he seems to think of this as conventional. He asserts that "we *choose to regard* [his italics] some succession of states as an unnatural succession."[22] His position seems to be that when behavior is subsumable under a law of succession we decide by convention whether that behavior is *sufficiently* explained by such subsumption. If it is, the behavior is natural; if not it is unnatural. But this is a conventional decision (which presumably can be changed).

Ellis leaves the notion of *sufficient* explanation undefined, but possibly what he means by saying that behavior is "sufficiently explained" by subsumption under a law of succession can be expressed using concepts previously introduced, as follows: the behavior is (correctly and appropriately) explainable by appeal to a law of succession (see Section 1), and why that law of succession holds is explanatorily ultimate (see Section 2). If so, Ellis's position might be formulated like this:

(1) b is an instance of natural behavior if and only if why b occurred is, by convention, taken to be (correctly and appropriately) explainable under a law of succession, and why that law of succession holds is, by convention, taken to be explanatorily ultimate.

However, this formulation will not support thesis (i) nor therefore (ii), given at the beginning of Section 6. What (i) requires is that if b is an instance of natural behavior then why items exhibit that type of behavior—i.e., why this law holds—is explanatorily ultimate. But the fact that a law is taken by convention to be explanatorily ultimate does not suffice to make it so. As noted earlier, whether a question is explanatorily ultimate is, in general, an empirical issue which cannot be decided by appeal simply to conventions, whether conventions of language or any other.

20. See *ibid.*, pp. 45–46.
21. *Ibid.*, p. 46.
22. *Ibid.*

On the other hand, perhaps Ellis means to emphasize not convention but context. Indeed, one passage suggests that this is what he has in mind:

> That subsumption under such a law [a law of succession] is considered to be a final explanation is shown by the fact that we should reject any request for an explanation of why systems behave in that way or remain in that state as inappropriate *in the context of the given inquiry*.[23]

Three ideas seem to be present here: that a given piece of behavior is natural only if it is explainable under a law of succession; only if why that law holds is not itself explainable; and only if what is and is not explainable depends, in part, on features of the explanatory context. Perhaps these ideas can be captured by means of the following definition:

(2) b is an instance of natural behavior if and only if why b occurred is (correctly and appropriately) explainable by appeal to a law of succession, and why that law of succession holds is explanatorily ultimate.

For example, the uniform motion of a particular body is natural if why this is occurring is correctly and appropriately explainable by appeal to the law of inertia, and why that law holds is explanatorily ultimate. That (2) captures the first two ideas noted above should be obvious. It captures the third, because whether why b is occurring is appropriately explainable depends on standards appropriate for the context of explanation; and whether some q is explanatorily ultimate depends on whether it is appropriately explainable, which again is affected by context.

Both (1) and (2) employ the idea of a law of succession. Unfortunately, this idea is not a particularly clear one as used by Ellis. It presupposes the notion of a state of a system (which he does not clarify), and the assumption that "conditions under which it [the state] exists do not change." Consider the law of free fall, which Ellis claims to be a law of succession. Let us write the law in the form

$s = \frac{1}{2} 32 t^2$

where s is the distance through which the body freely falls and t is the time during which it falls. To what does the concept of the "state" of the body here refer? Perhaps the state of a freely falling body at a given time is given by the distance through which it has fallen and the time taken to fall that distance, i.e., s and t. Then the law of free fall enables us to predict a future state of the body from its present state. Suppose it is presently in a state $t = 1$ second and $s = 16$ feet, i.e., at

23. *Ibid.*, my emphasis.

the present moment it has fallen freely for 1 second and it has fallen 16 feet. And suppose we want to determine its state two seconds from now. Then given a knowledge of its present state (t = 1, s = 16), and the fact that in 2 seconds from now t = 3, from the above law we can determine its state in 2 seconds (t = 3, s = 128).

However, a freely falling body has many other properties besides the distance through which it has fallen and the time of the fall. The body may be positively, negatively, or neutrally charged; it may be rotating through various possible axes; and so forth. If these properties are included in its "state," then from a knowledge of its present "state," Galileo's law will not enable us to predict its future "state." If we want Galileo's law to be a law of succession, in Ellis's sense, it seems that the notion of state will have to be understood by reference to the law itself; the state of a body will be determined by the variables related by the law, in this case by s and t.

Yet if this is what is to be done, then, given what else Ellis says, any law relating variables will be a law of succession. For example, the law of the simple pendulum

$$T = 2\pi \sqrt{L/g}$$

relates the state variables T (the period of the pendulum) and L (the length of the pendulum). Given the values of T and L (or just one of them) at any time, their values are determined for all times, under the assumption that Ellis makes, viz. "that the conditions under which it [the state] exists do not change." (Presumably, the latter means or entails that nothing external interferes with the system which is in that state other than what is represented in the law itself; for example, no contact force is exerted on the bob of the pendulum.) If, on the other hand, the "state" of the pendulum at a given time does not consist of its length and (or) period, then Ellis owes us some instructions for defining this state.

An obvious remedy is simply to drop Ellis's notion of a law of succession and its attendant concept of a state, and generalize (2) to cover any law whatever. We would then obtain

(3) b is an instance of natural behavior if and only if why b occurred is (correctly and appropriately) explainable by appeal to a law, and why that law holds is explanatorily ultimate.

I don't know whether (3) adequately captures the Ellisian spirit. But this much is certain. If this is to be the definition of natural behavior, then of the two claims (i) and (ii) at the beginning of Section 6, (i) holds but (ii) is indefensible. This is so since (3) invokes the notion of an explanatorily ultimate question in defining natural behavior. Claim (i), we recall, is that if N is a type of natural behavior, then why items exhibit behavior of that type is explanatorily ultimate. If (3) is the definition of natural behavior, then why some given piece of natural behavior b occurred is explainable by appeal to a law, but why that law holds—and hence, why behavior of that type generally occurs—is explanatorily ultimate. So (i) obtains. Claim (ii), we recall, is that if N is a type of natural behavior, then why items exhibit behavior of that type is explanatorily ultimate *because* N is a type of natural behavior. But using (3) as our definition of natural behavior, the latter explanatory claim becomes circular. It amounts to the claim that if the question of why items exhibit behavior of type N is explanatorily ultimate, then it is so because this question is explanatorily ultimate. The present definition of "natural behavior" renders this notion powerless to explain why certain questions are explanatorily ultimate.

Perhaps there are other criteria for natural behavior and states. But the ones given in this and the previous section—which I think are quite typical of those to be found in the literature—do not support claims (i) and (ii) which relate explanatorily ultimate questions to natural behavior and states. Natural behavior (or, more specifically, why items exhibit such behavior) will be explanatorily ultimate, I suggest, only if—as with (3) above—we build the notion of explanatory ultimateness into the definition of natural behavior. The usual criteria do not do this, and as a result do not yield the conclusion that why items exhibit natural behavior is explanatorily ultimate. But if explanatory ultimateness is built into the definition of natural behavior, then we forgo an appeal to natural behavior to explain why certain questions are explanatorily ultimate.

8. DOES EXPLAINING PRESUPPOSE A LIMIT?

I turn finally to an issue that emerges from Toulmin's claim that each science or theory postulates or presupposes some "ideal of natural order" which it regards as both unexplainable and requiring no explanation. Let me say that a question of the form "why p" is *related to* a science, theory, or explanation, if the latter asserts or presupposes p. Using this notion, as well as that of an explanatorily ultimate question, there are two theses that are suggested to me by Toulmin's claim:

(1) Every science, or theory, or scientific explanation assumes of some particular related why-question that it is a related why-question that is explanatorily ultimate. (For example, Newtonian mechanics assumes that why the law of inertia holds is explanatorily ultimate.)

(2) Every science, or theory, or scientific explanation assumes that there is some related why-question that is explanatorily ulti-

mate—though no particular related why-question need be assumed by it to be such. (For example, Newtonian mechanics assumes that there is at least one related why-question that is explanatorily ultimate, but it does not necessarily assume that "Why does the law of inertia hold?" is such a question.)

I will deal with the second thesis only, for, if this is not true, neither is the first, since (1) entails (2). What argument might be given in defense of (2)?

One invokes an idea of completeness. A science, or theory, or scientific explanation is complete, it might be urged, only if every proposition p which it invokes or presupposes has a related why-question (why p) which is either explained by something else it invokes or is explanatorily ultimate. Now if we suppose, what seems reasonable, that not every related why-question is explained by something else invoked in the science, theory, or explanation (if we assume that the science has some "axioms"), then if the science is complete, it must have some related why-questions that are explanatorily ultimate. We get thesis (2) immediately if we suppose that every science, theory, or explanation assumes that it satisfies the above condition for completeness. But clearly this is an unwarranted supposition. It is no assumption of Newtonian mechanics that every proposition which it invokes has a related why-question which is either explained by something else it invokes or is explanatorily ultimate. Newtonian mechanics invokes the proposition that bodies gravitate toward one another, but there is no assumption (at least in Newton's own version) that why bodies gravitate is explained by something else the system invokes or is explanatorily ultimate. Newton makes no assumption that his theory is complete in the above sense.

It might be supposed that (2) ought to be accepted because it is so weak. It says only that a science, or theory, or scientific explanation assumes that at least one related why-question is explanatorily ultimate. And surely that is a reasonable assumption for any science to make. For example, Newton's mechanics not only asserts the laws of motion and gravitation but makes simple assumptions such as that there are bodies, that bodies exist in space and time, and that bodies exert forces on one another. It seems reasonable to suppose that at least some related why-questions, e.g.,

Why are there bodies?

Why do bodies exist in space and time?

Why do bodies exert forces on one another?

are explanatorily ultimate. Perhaps some or all of these questions are explanatorily ultimate. It does not follow that Newtonian mechanics must assume that they are. (And this is what must be shown if (2) is to be established.) Even if every science in fact has some related why-question that is explanatorily ultimate, why does that science have to assume or presuppose that it has such related questions? Why can't Newtonian physics simply leave it an open question whether or not there are correct and appropriate explanations for all of its related why-questions?

Defenders of (2) may reply that it is not an open question because the very activity of explaining requires the existence of explanatorily ultimate questions. That is,

(3) Explaining is possible only if some questions are explanatorily ultimate.

Since any science, or theory, or explanation obviously presupposes that explaining is possible, it also presupposes that some questions are explanatorily ultimate. Newtonian physics being a science or theory thus is committed to the existence of at least some explanatorily ultimate questions (even if it is not committed to the existence of any particular ones).

Whether (3) is true I shall consider in a moment. Even if it is, (2) is not established. It may well be that every science assumes that explaining is possible, and hence, by (3), that some questions are explanatorily ultimate. But this is not enough to establish (2), since it is not enough to ensure that these explanatorily ultimate questions will be why-questions *related to that science*. It does not guarantee that the explanatorily ultimate questions will be "why-p" questions where p is some proposition presupposed by that particular science.

In the absence of further arguments I see no reason to believe that (2) is true. If it is not, this does not mean that there are theories that assume that none of their related questions are explanatorily ultimate. For (2) to be false it suffices that there be theories that make no assumptions one way or the other about whether or not there are related why-questions that are explanatorily ultimate.

Is the weaker thesis (3) acceptable? I shall construe it as a thesis about correct and appropriate explaining, and reformulate it like this:

(3)' (q)(q is (correctly and appropriately) explainable ⊃ (∃q')(q' is explanatorily ultimate)).

Thesis (3)' seems very weak indeed, since all it says is that if any question is correctly and appropriately explainable then some question or other is explanatorily ultimate. Could something be correctly and appropriately explainable even if nothing is explanatorily ultimate? My answer is yes. It follows from conditions (III) and (3) formulated in Section 1 that q is correctly and appropriately explainable if and only if there is a correct explanation of q that satisfies some set of appropriate instructions to follow for explaining q. But whether an explanation (p; explaining q) is correct depends just on the truth of p, and not on whether there are explanatorily ultimate questions. And, as can be ascertained by examining conditions (a) through (d) in Section 1, whether instructions I are appropriate for explaining q also does not depend upon whether there are any explanatorily ultimate questions.

If (3)' is false, this does not mean that there are no explanatorily ultimate questions. All it means is that whether something is explainable does not depend on the existence of questions that are explanatorily ultimate. If explanatory "limits" are understood as explanatorily ultimate questions, then the activity of explaining (correctly and appropriately) does not require such limits. Whether explaining has limits is, in general, an empirical issue. Explaining, as an activity, does not require them.

9. CONCLUSIONS

In the present chapter I have introduced various notions of explainability, and in terms of the strongest of these (Q's being correctly and appropriately explainable), a concept of a question's being explanatorily ultimate. This concept is helpful, I suggest, in understanding what various scientists and philosophers mean or might mean when they claim that explanation has reached a limit—e.g., when they say that a law, such as Newton's law of inertia, is fundamental and requires, and can receive, no explanation. This analysis puts us in a better position to assess the truth of "explanation-limit" claims about particular laws, as well as ones about large classes of statements, such as those affirming empirical identities or those describing natural behavior and states. It also enables us to evaluate the thesis that every science, and indeed the act of explaining itself, presupposes explanatory limits.

CHAPTER 10
Evidence and Explanation

Evidence is a concept for which many have sought a definition. In the present chapter it will be argued that definitions which have become standard in the literature are faulty, and that a more adequate account can be offered by appeal to a concept of explanation which we have defined earlier.

1. THREE CONCEPTS OF EVIDENCE

Alan's skin has yellowed, so on Monday he sees the doctor, who examines him and declares that he has jaundice, i.e., the visible expression of an increased concentration of bilirubin in the blood (which I shall abbreviate as an i.c.b.). Some tests are made as result of which on Friday, although Alan's yellowness remains, the doctor declares that Alan does not have an i.c.b. but that his yellow skin was produced by a dye with which he was working. On Friday which of the following propositions, if any, should the doctor affirm?

(i) Alan's yellow skin was evidence that he has an i.c.b. and still is.

(ii) Alan's yellow skin was but no longer is evidence that he has an i.c.b.

(iii) Alan's yellow skin is not and never was evidence that he has an i.c.b.

The doctor might be tempted to assert (i) on the ground that Alan's yellow skin is typically the kind of skin associated with an i.c.b. On the other hand, (ii) might be tempting to say since the doctor now has additional information which makes the original evidence efficacious no longer. Finally, he might be tempted to assert (iii) on the ground that false or misleading evidence is no evidence at all. He might say that Alan's yellow skin is not and never was (real or genuine) evidence of an i.c.b., though on Monday he mistakenly thought it was.

I begin with a notion which I shall call *potential* evidence. The presence of Alan's yellow skin is potential evidence that he has an i.c.b. since yellow skin of that sort is generally associated with an i.c.b. That 35 percent of those sampled in this district said they would vote for the Democratic candidate is potential evidence that roughly 35 percent of all those voting in the district will vote for him, since samples of that size are usually accurate. Without here trying to define this concept let me indicate several of its features.

First, e can be potential evidence that h even if h is false.[1] Second, potential evidence is objective in the sense that whether e is potential evidence that h does not depend upon anyone's beliefs about e or h or their relationship. That Alan has yellow skin is potential evidence that he has an i.c.b. even if no one believes that it is or knows or believes that he has yellow skin or an i.c.b. In these two respects potential evidence is akin to the concept Hempel seeks to define in "Studies in the Logic of Confirmation"[2] and to one Carnap calls the classificatory concept of confirmation which he defines using his theory of probability.[3]

Although Hempel and Carnap in addition allow e as well as h to be false, I am inclined to think that if there is a concept of potential evidence in use it is one that requires e to be true. (That Alan has yellow skin is potential evidence that he has an i.c.b. only if he does in fact have yellow skin.) This, then, is the third feature I attribute to this concept. The concept that Hempel and Carnap seek to analyze which has no such requirement could be described as "doubly potential" ("e would be potential evidence that h if e were true"). Finally, although both of these authors allow e to entail h (as a "limiting" case), I doubt that there is such a concept of evidence in use. The fact that Alan has yellow skin is not evidence that he has skin; it is too good to be evidence.

Can a concept of evidence with these characteristics be defined, and

---

1. In what follows the evidence sentences that will be considered are, or are transformable into, ones of the form

  (E) The fact that e (or that e) is evidence that h

in which e and h are sentences. However, when speaking schematically about such evidence sentences I shall follow the usual custom and simply write "e is evidence that h." This does not mean that I am subscribing to the view that evidence sentences of form (E) should be construed as relating sentences, or indeed anything at all. This chapter will not be concerned with the "logical form" of E-sentences, although the reader is referred to Chapters 3 and 6 where such questions about explanation and causation sentences are discussed the answers to which have a bearing on E-sentences.
2. Reprinted in Carl G. Hempel, *Aspects of Scientific Explanation*.
3. Rudolf Carnap, *Logical Foundations of Probability* (2nd ed., Chicago, 1962), p. xvi.

if so will it support proposition (i)? Various definitions of potential evidence will be examined in later sections, after which this question will be addressed.

I turn now to a second concept, *veridical* evidence, that sanctions proposition (iii) above. e is veridical evidence that h only if e is potential evidence that h and h is true. However, this is not yet sufficient. Suppose that Alan's having yellow skin is potential evidence of an i.c.b. and that Alan in fact has an i.c.b. But suppose that Alan's yellow skin did not result from an i.c.b. but from the chemical dye with which he was working. We would then conclude that his having the yellow skin he does is not (veridical) evidence of an i.c.b. Veridical evidence requires not just that h and e both be true but that e's truth be related in an appropriate manner to h's. Alan does not have the yellow skin he does *because* he has an i.c.b. but because he has been working with a yellow dye. More generally, I shall speak of an *explanatory connection* between e's being true and h's being true and say that

(1) e is veridical evidence that h if and only if e is potential evidence that h, h is true, and there is an explanatory connection between e's being true and h's being true.

Before proposing a definition of "explanatory connection" some initial clarification is in order. The concept of evidence characterized in (1) does not require that h's being true correctly explain e's being true; the converse is also possible. That Jones has a severe chest wound can be veridical evidence that he will die, even though the hypothesis that he will die does not explain the fact that he has a severe chest wound. Rather the reverse explanation is correct: he will die because he has a severe chest wound. Alternatively, there may be some hypothesis which correctly explains why e is true as well as why h is true. The fact that the gases hydrogen and oxygen combine in a simple ratio by volume may be evidence that the gases nitrogen and oxygen do too. (Gay-Lussac indeed took it to be so.) In this case h does not explain e, nor conversely. Still both h and e are explained by appeal to the fact that (it is a law that) all gases combine in simple ratios by volume. (This is an example of what, in Chapter 7, Section 7, we called special-case-of-a-law explanations.)

The present ideas can be used, together with the ordered pair theory of explanation, to formulate the following definition of "explanatory connection":

(2) There is an explanatory connection between (the truth of) sentences h and e if and only if either
   (i) (The reason that h is that e; explaining why h) is a correct explanation, *or*

(ii) (The reason that e is that h; explaining why e) is a correct explanation, *or*
(iii) There is a sentence h' such that both (The reason that h is that h'; explaining why h) and (The reason that e is that h'; explaining why e) are correct explanations.

On the ordered pair view,

E is an explanation of q if and only if Q is a content-question and E is an ordered pair whose first member is a complete content-giving proposition with respect to Q and whose second member is the act-type *explaining q*.

On this definition, ordered pairs of the sorts appearing in (i) through (iii) of (2) above are explanations.

Now, recall from Chapter 4, Section 1, that

(3) If (p; explaining q) is an explanation, then it is correct if and only if p is true.

This notion of a correct explanation will suffice for definition (2). Using (2) we can say, e.g., that there is an explanatory connection between "Jones will die" and "Jones has a severe chest wound," if

(The reason that Jones will die is that Jones has a severe chest wound; explaining why Jones will die)

is a correct explanation. And the latter, which is an explanation, is correct if and only if its constituent proposition is true.

On definitions (2) and (3), whether there is an explanatory connection between h and e does not depend on what anyone believes. Whether (p; explaining q) is an *explanation* depends solely on whether Q is a content-question and p is a complete content-giving proposition with respect to Q. Whether the explanation is *correct* depends solely on whether p is true. Neither of these depends upon what beliefs anyone may have regarding p or Q or the explanation (p; explaining q). (We are not here concerned with illocutionary evaluations of explanations, where such beliefs can be relevant.) Accordingly, veridical evidence as defined by (1) through (3), like potential evidence, is an objective concept. Moreover, it is a concept in accordance with which proposition (iii) should be asserted. If Alan does not have an i.c.b. (i.e., if h is false) then by (1) the fact that he has yellow skin is not and never was (veridical) evidence that he has an i.c.b.

Turning to a third concept of evidence, we speak not only of something's being evidence that h but also of something's being *so-and-so's* evidence that h. On Monday the doctor's evidence that Alan has an i.c.b. was that Alan has yellow skin. I take this to involve at least the claim that on Monday the doctor believed that Alan's yellow skin is potential evidence of an i.c.b. However, this is not sufficient if on Friday Alan's yellow skin is potential evidence of an i.c.b.; for the fact that Alan has yellow skin is not on Friday the doctor's evidence that Alan has an i.c.b., even if on Friday the doctor believes that it is potential evidence. Accordingly, one might be tempted to say that the fact that Alan has yellow skin is the doctor's evidence that Alan has an i.c.b. only if the doctor believes that this fact is *veridical* evidence of an i.c.b. More generally,

(4) e is X's evidence that h only if X believes that e is veridical evidence that h; i.e., X believes that e is potential evidence that h, that h is true, and that there is an explanatory connection between the truth of h and e.[4]

However, (4) may be too strong in requiring that X believe that h is true and that there is an explanatory connection between h and e. Suppose that on Monday the doctor is unsure about whether Alan has an i.c.b. He thinks it probable but he does not know whether to believe it, so he orders tests. Later when the tests reveal no i.c.b. and Alan indignantly asks the doctor "What was your evidence that I have an i.c.b.?" the doctor might reply: "The fact that you have yellow skin." Even if on Monday the doctor was not sure whether to believe that Alan has an i.c.b., at least he believed that this is probable and that it is probable that this explains his yellow skin. Accordingly, (4) might be weakened as follows:

(4′) e is X's evidence that h only if X believes that e is potential evidence that h, that it is probable that h is true, and that it is probable that there is an explanatory connection between the truth of h and e.

Neither (4) nor (4′), however, supplies a sufficient condition. For e to be X's evidence that h it is necessary in addition that

(5) X believes that h is true or probable (and does so) *for the reason that e.*

The fact that Alan is receiving a certain medical treatment T may be (veridical) evidence that he has an i.c.b. (since treatment T is given only to such people). Even if Alan's doctor knows and therefore believes that the fact that Alan is receiving treatment T is (veridical) evidence that he has an i.c.b., this fact is not the *doctor's* evidence that Alan has an i.c.b. His reason for believing this is not that Alan is re-

---

4. "X's evidence that h" is ambiguous. It can mean (roughly) "what X takes to be evidence that h"—as in (4) above—or "what X takes to be evidence that h *and* is evidence that h," which is a combination of (4) with either "e is potential evidence that h" or "e is veridical evidence that h."

ceiving treatment T. Accordingly, I would add condition (5) to (4) and (4′) to obtain sufficient conditions. (4) and (5) can be said to characterize a strong sense of "X's evidence," (4′) and (5) a weak one.

Both (4) and (4′) (with (5) added) sanction proposition (ii). When we say that Alan's yellow skin was but no longer is evidence that he has an i.c.b. we may be understood to be referring to *someone's* evidence, in this case the doctor's. We may mean that on Monday the fact that Alan has yellow skin was the doctor's evidence that Alan has an i.c.b., but on Friday it is no longer so. On Friday due to other facts he has learned the doctor no longer believes (it probable) that Alan has an i.c.b. This concept of evidence is thoroughly subjective. Whether e is X's evidence that h depends entirely on what X believes about e, h, and their relationship, and not on whether in fact e is potential or veridical evidence that h.

This subjectivity means that one cannot draw an inference from the fact that e is X's evidence that h to the claim that e is at least some good reason to believe h, or even for X to believe h. It is commonly supposed that evidence bears some relationship to what it is reasonable to believe. Although this may be expressed in a variety of ways perhaps the following simple formulation will suffice for our purposes:

A *Principle of Reasonable Belief.* If, in the light of background information b,[5] e is evidence that h, then, given b, e is at least some good reason for believing h.[6]

This principle is satisfied by the two objective concepts of evidence. If, in the light of the background information (b) that yellow skin of that type is generally associated with an i.c.b., the fact that Alan has yellow skin is potential (or veridical) evidence that he has an i.c.b., then, given b, the latter fact is at least some good reason for believing that Alan has an i.c.b. The subjective concept, on the other hand, does not satisfy this principle. The fact that Max has lost ten fights in a row may be *his* evidence that his luck will change and he will win the eleventh. But this fact is not a good reason at all, even for Max, to believe this hypothesis.

To summarize, then, the three concepts of evidence here characterized provide a way of answering the question of whether the fact that Alan has yellow skin is evidence that he has an i.c.b. It is potential evidence since that kind of skin is typically associated with an i.c.b. It is not veridical evidence since the hypothesis is false and his yellow

---

5. The role of background information here and its relationship to evidence statements will be discussed in Section 5.
6. Note that this is not a principle relating evidence to what anyone is justified in believing. e can be a good reason for believing h even though Smith, say, is not justified in believing h for the reason e (since, e.g., he is not justified in believing e).

skin is correctly explained not by his having an i.c.b. but by the fact that he was working with a dye. On Monday but not on Friday it was the doctor's evidence that Alan has an i.c.b., since on Monday but not on Friday the doctor believed that Alan has an i.c.b. for the reason that he has yellow skin which he believed was veridical evidence of an i.c.b.

If potential evidence can be defined, then so can the other two concepts via (1), (4), and (5). Of various definitions of potential evidence that appear in the literature two general types will be discussed here. Each by itself is not sufficient but if appropriately altered and combined the result may be. The first and most popular type defines evidence in terms of probability, the second in terms of explanation.

2. THE PROBABILITY DEFINITIONS

According to one probability definition e is potential evidence that h if and only if the probability of h given e is greater than the prior probability of h:

(1a) e is potential evidence that h if and only if $p(h,e) > p(h)$.

Or, if b is background information,

(1b) e is potential evidence that h if and only if $p(h, e \& b) > p(h,b)$.

A definition of this sort is offered by many writers.[7] However, despite its widespread acceptance it cannot possibly be correct if "evidence" and "probability" are being used as they are in ordinary language or science. For one thing, neither (1a) nor (1b) requires that e be true; and this, as noted earlier, seems to be necessary for evidence. That Alan has yellow skin is not evidence that he has an i.c.b., if he does not have yellow skin. However, even with the addition of a truth-requirement the resulting definition is unsatisfactory. I shall concentrate on (1b), since this is the most prevalent form of the definition, and note three types of counterexamples. The first shows that an increase in probability is not sufficient for evidence, the second and third that it is not necessary.

The first lottery case. Let b be the background information that on Monday 1000 lottery tickets were sold and that John bought 100 and Bill bought 1. Let e be the information that on Tuesday all the lottery tickets except those of John and Bill have been destroyed but that one ticket will still be drawn at random. Let h be the hypothesis that Bill

---

7. E.g., Carnap, *Logical Foundations of Probability*, p. 463; Mary Hesse, *The Structure of Scientific Inference* (Berkeley, 1974), p. 134; Richard Swinburne, *An Introduction to Confirmation Theory* (London, 1973), p. 3.

will win. The probability that Bill will win has been increased approximately tenfold over its prior probability. But surely e is not evidence that Bill will win. If anything it is evidence that John will win.

Reverting to the principle of the previous section which relates potential (as well as veridical) evidence to a reason for belief, assume for the sake of argument that in the light of b, e is potential evidence that (h) Bill will win. Then according to the principle of reasonable belief, given b, e is at least some good reason for believing h. But surely it is not. In the light of the background information that on Monday John bought 100 and Bill bought 1 of the 1000 lottery tickets sold, the fact that on Tuesday all of the tickets except those of John and Bill have been destroyed but one ticket will still be drawn at random is not a good reason at all for believing that Bill will win. Someone who believes that Bill will win for such a reason is believing something irrationally.

Events often occur which increase the probability or risk of certain consequences. But the fact that such events occur is not necessarily evidence that these consequences will ensue; it may be no good reason at all for expecting such consequences. When I walk across the street I increase the probability that I will be hit by a 1970 Cadillac; but the fact that I am walking across the street is not evidence that I will be hit by a 1970 Cadillac. When Mark Spitz goes swimming he increases the probability that he will drown; but the fact that he is swimming is not evidence that he will drown.

What these examples show is that for e to be evidence that h it is not sufficient that e increase h's (prior) probability. The next two examples show that it is not even necessary.

The paradox of ideal evidence.[8] Let b be the background information that in the first 5000 spins of this roulette wheel the ball landed on numbers other than 3 approximately $35/36$ths of the time. Let e be the information that in the second 5000 spins the ball landed on numbers other than 3 approximately $35/36$ths of the time. Let h be the hypothesis that on the 10,001st spin the ball will land on a number other than 3. The following claim seems reasonable:

$p(h, e \& b) = p(h,b) = 35/36$.

That is, the probability that the ball will land on a number other than 3 on the 10,001st spin is unchanged by e, which means, according to (1b), that e is not evidence that h. But it seems unreasonable to claim that the fact that the ball landed on numbers other than 3 approximately $35/36$ths of the time during the second 5000 spins is not evi-

8. The expression is Karl Popper's, *The Logic of Scientific Discovery* (London, 1959), p. 407, but I am changing his example to suit my purposes here.

dence that it will land on a number other than 3 on the 10,001st spin, even though there is another fact which is also evidence for this. More generally, e can be evidence that h even if there is other equally good evidence that h. To be sure, if we have already obtained the first batch of evidence there may be no need to obtain the second. But this does not mean that the second batch is not evidence that h.

The second lottery case. Let b be the background information that there is a lottery consisting of 1001 tickets, one of which will be drawn at random, and by Tuesday 1000 tickets have been sold, of which Alice owns 999. Let e be the information that by Wednesday 1001 lottery tickets have been sold, of which Alice owns 999, and no more tickets will be sold. Let h be the hypothesis that Alice will win. Now, I suggest, e is evidence that h in this case. The information that Alice owns 999 of the 1001 tickets sold and that no more tickets will be sold is evidence that Alice will win. (The principle of reasonable belief, e.g., is clearly satisfied, since, given b, e is a good reason for believing h.) However, notice that in this case $p(h, e\&b) < p(h, b)$. ($p(h, e\&b) = {}^{999}/_{1001}$, $p(h,b) > {}^{999}/_{1001}$).[9] In the case of the "paradox of ideal evidence," we have a situation in which information is evidence for a hypothesis even though it does not increase the probability of the hypothesis. In the second lottery case, we have a situation in which information is evidence for a hypothesis even though *it lowers the probability of the hypothesis from what it was before.* (In this case, on Tuesday there is evidence that Alice will win, and on Wednesday there is also evidence that Alice will win, which is slightly weaker than Tuesday's evidence.)

It may be replied that in the last two examples e is not evidence that h but only evidence that h is probable. Even if some others might wish to pursue this (dubious) line, it is doubtful that defenders of the increase-in-probability definition can. Let us change the second lottery case a bit by introducing e′ = the information that by Wednesday 1001 tickets have been sold, of which Alice owns 1000, and no more tickets will be sold. Now, since $p(h, e'\&b) > p(h, b)$, using definition (1b), increase-in-probability theorists will conclude that e′ is evidence that h. They will not insist here that e′ is evidence only for the hypothesis that h is probable. How could they then insist that e in the second lottery case is evidence only for the hypothesis that h is probable? (The probabilities conferred on h by e and e′ are only marginally different, and both, though less than 1, are extremely high.)

In the light of these three examples perhaps it will be agreed that

---

9. The latter inequality holds assuming that p(Alice will buy the remaining ticket on Wednesday) > 0. $p(h,b) = {}^{999}/_{1001} + {}^{m}/_{n}({}^{1}/_{1001})$, where ${}^{m}/_{n}$ = the probability that Alice will buy the remaining ticket on Wednesday.

"e is evidence that h" cannot be defined simply as "e increases h's probability." But it may be contended that a related concept can be so defined, viz. "e increases the evidence that h." Thus,

e increases the evidence that h if and only if $p(h, e\&b) > p(h,b)$.

However, increasing the evidence that h is not the same as increasing the probability of h. To increase the evidence that h is to start with information which is evidence that h and add to it something which is also evidence that h or at least is so when conjoined with previous information. But to do this it is neither sufficient nor necessary to increase h's probability. The first lottery example shows that it is not sufficient, while the paradox of ideal evidence shows that it is not necessary. In the first lottery example there is no increase in evidence that Bill will win, since in the first place there is no evidence that he will win, and the combined new and old information is not evidence that he will win, even though the probability that he will win has increased. In the paradox of ideal evidence there is an increase in evidence that the ball will land on a number other than 3 on the 10,001st spin, but there is no increase in the probability of this hypothesis.

At this point a second definition of evidence in terms of probability might be offered, viz.

(2) e is potential evidence that h if and only if $p(h,e) > k$ (where k is some number, say $1/2$).

Some writers, indeed, claim that the concept of evidence (or confirmation) is ambiguous and that it can mean either (1) or (2).[10] One of these meanings is simply that given e, h has a certain (high) probability.

This proposal has the advantage of being able to handle both the first and second lottery cases and the paradox of ideal evidence. In the first lottery case, although the probability that Bill will win is increased by e, the probability that Bill will win, given e and b, is not high. (It is $1/101$.) Therefore, by (2), e&b is not evidence that Bill will win. On the other hand it is evidence that John will win, since p(John will win, e&b) = $100/101$. And this is as it should be. In the second lottery case we judge that the information given is evidence that Alice will win (despite the fact that the probability that Alice will win is lower than before). Such a judgment is justified on the basis of (2), since p(Alice will win, e) = $999/1001$.

The paradox of ideal evidence is also avoided by (2) since the fact

---

10. See Carnap, *Logical Foundations of Probability*, pp. xv–xx; Wesley C. Salmon, "Confirmation and Relevance," in G. Maxwell and R. Anderson, Jr., eds., *Minnesota Studies in the Philosophy of Science* 6 (Minneapolis, 1975), p. 5; Hesse, *The Structure of Scientific Inference*, pp. 133–34.

that (e) the ball landed on numbers other than 3 approximately $^{35}/_{36}$ths of the time during the second 5000 spins makes the probability very high that it will land on a number other than 3 on the 10,001st spin. In this case p(h,e) > k, and therefore, by (2), e is evidence that h, even though p(h, e & b) = p(h,b).

However, (2) is beset by a major problem of its own.

The Wheaties case (or the problem of irrelevant information).[11] Let e be the information that this man eats the breakfast cereal Wheaties. Let h be the hypothesis that this man will not become pregnant. The probability of h given e is extremely high (since the probability of h is extremely high and is not diminished by the assumption of e). But e is not evidence that h. To claim that the fact that this man eats Wheaties is evidence that he will not become pregnant is to make a bad joke at best.

Such examples can easily be multiplied. The fact that Jones is drinking whisky (praying to God, taking vitamin C, etc.) to get rid of his cold is not evidence that he will recover within a week, despite the fact that people who have done these things do generally recover within a week (i.e., despite the fact that the probability of recovering in this time, given these remedies, is very high). It may well be for this reason that some writers prefer definition (1) over (2). On (1) e in the present examples would not be evidence that h because p(h,e) = p(h), i.e., because e is probabilistically irrelevant for h. I would agree that the reason that e is not evidence that h is that e is irrelevant for h, but this is not mere probabilistic irrelevance (as will be argued later).

A defender of (2) might reply that in the Wheaties example we can say that the probability of h given e is high only because we are assuming as background information the fact that no man has ever become pregnant; and he may insist that this background information be incorporated into the probability statement itself by writing "p(h, e & b) > k." In Section 5 contrasting views about the role of background information with respect to probability (and evidence) statements will be noted, only one of which insists that such information always be incorporated into the probability statement itself. However, even if the latter viewpoint is espoused the Wheaties example presents a problem for (2) if we agree that information that is irrelevant for h can be added to information that is evidence that h without the result being evidence that h.

Suppose that b is evidence that h and that p(h,b) > k. There will be some irrelevant e such that p(h, e & b) = p(h,b), yet e & b is not evi-

---

11. Salmon, *Statistical Explanation and Statistical Relevance*, uses a similar example against the D-N model of explanation, but not as an argument against (2).

dence that h. Thus let h be the hypothesis that this man will not become pregnant. Let b be the information that no man has ever become pregnant, and let e be the information that this man eats Wheaties. We may conclude that p(h, e & b) > k, which, as demanded above, incorporates b into the probability statement. But although b is evidence that h we would be most reluctant to say that e & b is too.[12]

Wesley Salmon[13] has claimed that the notion of evidence that confirms a hypothesis can be understood in terms of Bayes's theorem of probabilities, a simple form of which is

$$p(h,e) = \frac{p(h) \times p(e,h)}{p(e)}.$$

According to this theorem, to determine the probability of h on e we must determine three quantities: the initital probability of h (p(h)), the "likelihood" of h on e (p(e,h)), and the initial probability of e (p(e)). Salmon criticizes a view of evidence which says that if a hypothesis entails an observational conclusion e which is true, then e is evidence that h. This view, he points out, considers only one of the probabilities above, viz. p(e,h) = 1. To determine whether e is evidence that h one must also consider the initial probabilities of h and e.

This may be a valid criticism but it does not avoid the previous problems. Suppose that Bayes's theorem is used to determine the "posterior" probability of h, i.e., p(h,e), by reference to the initial probabilities of h and of e and the likelihood of h on e. We must still determine what, if anything, this has to do with whether e is evidence that h. If definitions (1) or (2) are used to determine this we confront all of the previous difficulties even though we have used Bayes's theorem in calculating p(h,e). Thus let e be the information that this man eats Wheaties, and h be the hypothesis that this man will not become pregnant. Assume the following probabilities, which do not seem unreasonable: p(h) = 1, p(e,h) = $^1/_{10}$, p(e) = $^1/_{10}$. Then by Bayes's theorem, p(h,e) = 1. Using definition (2) we must conclude that e is evidence that h.

Can the problems with (1) and (2) be avoided by combining these definitions? Using a background-information formulation, it might be tempting to say that

---

12. An analogous problem arises for those who defend the D-N model of explanation. Kinetic theory entails the ideal gas law, but so does kinetic theory conjoined with laws from economic theory. Not wishing to say that this enlarged set explains the ideal gas law, the D-N theorist requires an elimination of the irrelevant laws in the D-N explanans.
13. "Bayes's Theorem and the History of Science," in R. Stuewer, ed., *Minnesota Studies in the Philosophy of Science* 5.

(3) e is evidence that h if and only if *both* p(h, e&b) > p(h, b) *and* p(h, e&b) > k.

Unfortunately, this does not provide a necessary condition for evidence. As demonstrated by the paradox of ideal evidence and the second lottery case, e can be evidence that h even though e does not raise h's probability (indeed, even if e lowers it). Does (3) provide a sufficient condition for evidence? Let us consider one more lottery, as follows.

The third lottery case. Let b be the background information that there is a lottery consisting of 1 million tickets, one of which will be drawn at random, and by Tuesday all but one of the tickets have been sold, of which Eugene owns 1. Let e be the information that on Wednesday Jane, who owns no other tickets, bought the remaining ticket. Let h be the hypothesis that Eugene will not win. Since p(h, e&b) = $^{999,999}/_{1,000,000}$ > p(h, b), (3) permits us to conclude that the fact that Jane, who owns no other tickets, bought the one remaining ticket is evidence that Eugene will not win. But that seems much too strong a claim to make. The fact that Jane bought the one remaining ticket is not a good reason at all for believing that Eugene will lose.

It is not my claim that probability is irrelevant for evidence, but only that the particular probability definitions (1), (2), and (3)—the standard definitions and their conjunction—are inadequate.

3. AN EXPLANATION DEFINITION

I turn then to a very different proposal which appeals to the concept of explanation:

(1) e is potential evidence that h if and only if e is true and h would correctly explain e if h were true.

Using the ordered pair theory of explanation, this idea can be formulated as follows:

e is potential evidence that h if and only if e is true, and if h is true then (The reason that e is that h; explaining why e) is a correct explanation.

Definition (1) can be closely associated with at least two views. One is Hanson's account of retroductive reasoning, which takes the form

Some surprising phenomenon P is observed;
P would be explicable as a matter of course if h were true;
Hence, there is reason to think that h is true.[14]

14. N. R. Hanson, *Patterns of Discovery* (Cambridge, 1958), p. 72.

The fact that phenomenon P has been observed is then potential evidence that h; it is so because h would correctly explain P if it were true. (1) is also closely associated with the hypothetico-deductive account of theories, according to which if hypothesis h is a potential explanans of e (which, on this view, means roughly that h contains a lawlike sentence and entails e), then if e turns out to be true, e is confirming evidence for h. Sympathy with (1) might then lead to the following simple definition of veridical evidence:

(2) e is veridical evidence that h if and only if h correctly explains e (i.e., e is potential evidence that h and h is true).[15]

Despite the emphasis in recent years on the role of explanation in inference from evidence, neither (1) nor (2) provides a necessary or a sufficient condition for potential or veridical evidence. Neither provides a necessary condition since, as noted earlier, e may be evidence that h even if h does not, and would not if true, correctly explain e. The fact that Jones has the chest wound he does may be potential or veridical evidence that he will die, even though the hypothesis that he will die does not, and would not if true, correctly explain why he has that chest wound. The explanation condition if necessary at all should be changed to require only some explanatory connection between h and e.

Nor do these definitions provide sufficient conditions. Suppose my car won't start this morning. The hypothesis

h: At precisely 2:07 last night 5 boys and 2 girls removed the 18.9 gallons of gas remaining in my tank and substituted water

would if true correctly explain why my car won't start this morning; indeed, suppose that h is true and that it does correctly explain this. In either case the fact that (e) my car won't start this morning is not evidence that h is true. There is too much of a gulf between this e and h for e to be evidence that h, even if h does or would if true correctly explain e. What this gulf amounts to I shall try to say later.

It is worth noting here that the earlier principle of reasonable belief is violated. According to this principle, if the fact that my car won't start this morning were evidence that h, then this fact would be at least some good reason for believing h. But (given the "normal" background information one might imagine for such a case) the fact that my car won't start this morning is far too meager a reason to believe

---

15. (2) might be associated with a view, in addition to the above, that in any inductive inference one infers that a hypothesis correctly explains the evidence. This is Gilbert Harman's view in "The Inference to the Best Explanation," *Philosophical Review* 74 (1965), pp. 88–95. Later he revised it by requiring only that h correctly explains something, and also by stressing the global nature of inference, viz. that one infers to the best overall system.

the very specific hypothesis h. Indeed, innumerably many hypotheses in addition to h can be invented which if true would correctly explain why my car won't start. The hypothesis that

h′: At precisely 3:05 last night 2 monkeys removed the remaining 3.7 gallons of gas in my tank and substituted crushed bananas

if true would explain why my car won't start. Is the fact that my car won't start evidence that h′ is true? Does this fact provide any reason to believe such a hypothesis?

4. A NEW PROPOSAL

Although neither the probability nor explanation definitions are adequate, if these are combined in a certain way the outcome may be more successful. Here are my proposals:

(1) e is potential evidence that h if and only if (a) e is true, (b) e does not entail h, (c) $p(h,e) > k$, (d) p(there is an explanatory connection between h and e, $h \& e) > k$.

(2) e is veridical evidence that h if and only if e is potential evidence that h, h is true, and there is an explanatory connection between (the truth of) h and e. (This is simply (1) of Section 1.)

For e to be potential evidence that h we require, in addition to e's being true and not entailing h, the satisfaction of two probability conditions. One is that the probability of h given e be high. The other is that the probability that there is an explanatory connection between h and e, given that h and e are both true, be high; i.e., that it be probable, given h and e, that h is true because e is, or conversely, or that some one hypothesis correctly explains why each is true. (Recall the definition of "explanatory connection" given in (2) of Section 1.) Veridical evidence requires, in addition to this, that h be true and that there be an explanatory connection between h and e.

In Section 1 the following features of potential evidence were cited: (i) e can be potential evidence that h even if h is false; (ii) potential evidence is objective, i.e., whether e is potential evidence that h does not depend on whether anyone believes e or h or anything about their relationship; (iii) e is potential evidence that h only if e is true; (iv) e is potential evidence that h only if e does not entail h. Features (i), (iii), and (iv) are obviously satisfied by (1) above. So is feature (ii), provided that the concept of probability used in conditions (1c) and (1d) is construed as an objective one. (The concept of an explanatory connection, as defined earlier, is objective.)

To assess these definitions let us reconsider the counterexamples to the previous probability and explanation definitions.

Evidence and Explanation   337

The first lottery case. The fact that all the lottery tickets except those of John and Bill have been destroyed, that of the original 1000 tickets John has 100 and Bill has 1, and that one ticket will be drawn at random, does not make it probable that Bill will win. Hence by condition (1c) this fact is not potential (and therefore not veridical) evidence that Bill will win. On the contrary, as previously indicated, it ought to be potential evidence that John will win. And indeed it is on definition (1), since given the fact in question it is probable that John will win (1c); and given the same fact and the fact that John will win it is probable that a correct explanation of why John will win is that all the 1000 lottery tickets except those of John and Bill have been destroyed, that of the 101 remaining tickets John has 100 and Bill 1, and that one ticket will be chosen at random (1d). Definition (1) above gives us a reasonable analysis of this case in a way that the probability definition (1) of Section 2 does not.

The paradox of ideal evidence. The fact that (e) the roulette ball has landed on numbers other than 3 approximately $^{35}/_{36}$ths of the time during the second 5000 spins ought to be potential evidence that (h) it will land on a number other than 3 on the 10,001st spin, even if the probability of this hypothesis has not increased over its prior probability. Definition (1) gives us what we want, since the probability of h on e is high. And given that h and e are both true it is probable that both h and e are correctly explained by the hypothesis that on the roulette wheel there are 36 places of equal size for the ball to land, 35 of which show numbers other than 3.

The second lottery case. The fact that (e) by Wednesday 1001 lottery tickets have been sold, of which Alice owns 999, and no more tickets will be sold, ought to be potential evidence that (h) Alice will win—despite the fact that Alice's chances have diminished slightly from what they were on Tuesday. Again, definition (1) yields the desired result, since the probability of h on e is high. And given that h and e are both true, it is probable that the reason that h is that e: it is probable that the reason that Alice will win is that she owns 999 of the 1001 lottery tickets that have been sold and that no more will be sold.

The Wheaties case. The fact that (e) this man eats Wheaties should not be (potential or veridical) evidence that (h) this man will not become pregnant, even though the probability of h given e is high. And it is not evidence on definition (1) since condition (d) is violated. Given that h and e are both true, it is not probable that there is an explanatory connection between h and e: it is not probable that this man will not become pregnant because he eats Wheaties, or that he eats Wheaties because he will not become pregnant, or that there is some hypothesis that correctly explains both why he eats Wheaties and why he does not become pregnant. His eating Wheaties and his not becoming pregnant are not only probabilistically independent, they are (probably) explanatorily independent.

However, we might alter the example as follows. Let e' be the information that this man who wants to become pregnant believes that he never will and as a consequence becomes anxious and eats Wheaties to reduce his anxiety. In this case, it might be urged, given both h and e' it is probable that h does correctly explain e': it is probable that this man believes what he does, becomes anxious and eats Wheaties because in fact he will not become pregnant. Then by definition (1), e' is potential evidence that h. Assuming that h is true and that there is such an explanatory connection between e' and h, then e' is also veridical evidence that h, by definition (2). Is this reasonable?

A claim that it is not might be made on the ground that we do, after all, have extremely good evidence that h, viz. that this man is indeed a *man,* not a woman. And this is both potential and veridical evidence that there will be no pregnancy. However, the fact that there is other evidence whose support for h is stronger than that given by e' does not by itself mean that e' is not evidence that h. Here it is important to recall the distinction between something's being evidence that h and its being *someone's* evidence that h. In our earlier example at the end of Section 1, the fact that Alan is receiving treatment T might be (potential or veridical) evidence that he has an i.c.b. without its being the *doctor's* evidence. (The fact that Alan is receiving this treatment may not be the doctor's reason for believing that he has an i.c.b.) Similarly, the fact that this man, who wants to become pregnant, is anxious because he believes he never will and eats Wheaties to reduce his anxiety can be evidence that he will not become pregnant without its being *anyone's* evidence for this. No one who believes that this man will not become pregnant may do so for the reason just given.

The third lottery case. The fact that (e) on Wednesday Jane bought the remaining ticket in the lottery should not be potential evidence that (h) Eugene will lose. This is so even if we assume as background information (b) that there is a lottery consisting of one million tickets and by Tuesday all but 1 of the tickets have been sold, of which Eugene owns one. By definition (1), e is not evidence that h, since condition (1d) is violated. Given that Eugene will lose and that Jane bought the remaining ticket, it is not probable that the reason Eugene will lose is that Jane bought the remaining ticket, or that the reason that Jane bought the remaining ticket is that Eugene will lose, or that some hypothesis correctly explains both why Eugene will lose and why Jane bought the remaining ticket.

The case of the stalled car. The fact that (e) my car won't start this morning is not potential evidence that (h) at precisely 2:07 last night 5 boys and 2 girls removed the 18.9 gallons of gas remaining in my tank and substituted water. This is so because condition (1c) is not satisfied. The gulf mentioned earlier which prevents e from being evidence that h is a probabilistic one: it is not the case that h is probable given e.

The proposed definition of potential evidence thus avoids the counterexamples to previous definitions, and yields those evidence claims in the examples that we are willing to assert. Will it sanction other evidence statements that we are usually prepared to make? Is the fact that all the observed crows have been black potential evidence that all crows are black? Is the fact that the gases hydrogen and oxygen combine in a simple ratio by volume potential evidence that the gases nitrogen and oxygen do too (as Gay-Lussac thought)? Is the fact that electrons produce tracks in cloud chambers potential evidence that they carry a charge? Is the fact that Alan has yellow skin potential evidence that he has an i.c.b.? Whether these claims can be made depends on whether certain probability statements can be asserted. The fact that Alan has yellow skin is potential evidence that he has an i.c.b. only if it is probable that he has an i.c.b., given that he has yellow skin, and it is probable that there is an explanatory connection between his having yellow skin and his having an i.c.b., given that he has both. And whether these probability claims can be made depends on what background information is being assumed, and on the general relationship between probability statements and background information. What view we take of this relationship will determine what evidence statements we can assert, as will be shown next.

5. BACKGROUND INFORMATION

Two views about the role of background information are possible. One is that probability statements should be relativized to the background information to which any appeal is made. The other is that no such relativization is necessary. According to the former view the background information must be incorporated into the probability statement itself. If appeal is made to b in determining that the probability of h on e is r then we should write "$p(h, e \& b) = r$." In the case of evidence, we could say that the conjunction of e and b is evidence that h, provided that the definitions of Section 4 are satisfied. Or we can continue to say that e is evidence that h, provided that we relativize the evidence statement to the background information by writing "e is evidence that h, given b," and reformulate the definitions of Section 4 as follows:

(1) e is potential evidence that h, given b, if and only if (a) e and b are true, (b) e does not entail h, (c) p(h, e&b)>k, (d) p(there is an explanatory connection between h and e, h&e&b)>k;

(2) e is veridical evidence that h, given b, if and only if e is potential evidence that h, given b; h is true; and there is an explanatory connection between the truth of h and e.

In what follows I will discuss that version of the relativization view given by (1) and (2). According to it the fact that Alan has yellow skin can be shown to be potential evidence that he has an i.c.b., given the doctor's background information on Monday (which includes the fact that people with that kind of skin usually have an i.c.b. and usually have that kind of skin because they have an i.c.b.). However, the fact that Alan has yellow skin is not potential evidence that he has an i.c.b., given the doctor's background information on Friday (which includes the results of tests).

By contrast to the relativization view, one might claim that background information need not be construed as a part of the probability statement itself, but only as information to which one appeals in *defending* or *justifying* that statement. Thus

(3) The probability that Alan has an i.c.b., given that he has yellow skin, is high

might be defended by appeal to the empirical fact that

(4) In most cases people with (that kind of) yellow skin have an i.c.b.

But on this view the fact that (3) is defensible by appeal to the empirical fact that (4) is true shows that (3) itself is an empirical statement. It does not show that (3) is an incomplete version of the (perhaps) a priori statement that

The probability that Alan has an i.c.b., given that he has yellow skin and that in most cases people with (that kind of) yellow skin have an i.c.b., is high.

On Monday the doctor defends (3) by appeal to (4). On Friday he has accumulated new information, which includes the results of tests, and he then defends the negation of (3) by appeal to this new information. By contrast, the relativist must say that on Monday the doctor is asserting a probability statement of the form "p(h, e&b₁)>k," while on Friday he is asserting one of the form "p(h, e&b₂)<k." And these are not incompatible statements.

Returning to evidence, a non-relativist with regard to background information can accept the definitions of evidence as these are given in Section 4, and need not relativize them to the background information by writing either "e&b is evidence that h" or "e is evidence that h, given b." He can consider statements of the form

(5) The fact that Alan has yellow skin is potential evidence that he has an i.c.b.

to be complete, even though appeals to background information will be made in defending (5) or its denial. On Monday, on the basis of the information available to him, the doctor affirms (5); on Friday, on the basis of the new information, he denies (5). The relativist, on the other hand, regards (5) as incomplete. If (5) is relativized to the background information on Monday it is true, and if relativized to the background information on Friday it is false.

I shall not here try to arbitrate between these views. Perhaps each reflects different tendencies in the way we speak about probability and evidence. Perhaps one is more dominant than the other in linguistic practice or is more advantageous for other reasons. But which of these views about background information we employ will affect what claims about potential and veridical evidence we are prepared to make.

Thus, in Section 1 it was asked whether there is a concept of potential evidence according to which

(6) Alan's yellow skin was evidence that he has an i.c.b. (on Monday) and still is (on Friday)

is true, despite the fact that on Friday the doctor's tests prove negative. Using (1), a background-relative concept of potential evidence, (6) might be understood by reference to

(7) The fact that Alan has yellow skin is potential evidence that he has an i.c.b., given the background information of the doctor on Monday.

And (7) is as true on Friday as it is on Monday, since it is true timelessly. Of course if (6) is relativized to the information of the doctor on Friday then it is false timelessly. But the relativist who wants to explain the sense in which (6) is true can use definition (1) and relativize his evidence statement to the doctor's information on Monday.

The non-relativist will not be able to regard (6) as true, if construed as it has been so far. He will say that whereas on Monday he believed

(5) to be true, on Friday he realizes that it is false; but he cannot assert that (5) was true on Monday and remains true on Friday, as (6) suggests.

Nevertheless, the non-relativist can resurrect (6) by claiming that it is true if it is construed as making a *general* claim rather than a particular one, viz.

(8) Having the kind of yellow skin which Alan has is (timelessly) evidence of (having) an i.c.b.

And he can provide the following set of necessary and sufficient conditions for statements of this type, using as a guide the previous definition of potential evidence:

(9) Having F is potential evidence of having G (for A's) if and only if (a) "X is F" does not entail "X is G"; (b) the probability of something's (an A's) having G, given that it has F, is high; (c) the probability that there is an explanatory connection between something's (an A's) having F and its having G, given that it has both, is high.

Using (9) the non-relativist can argue that (8) is true and therefore that a sense of potential evidence can be provided which sanctions (6). There is then a certain analogy between the relativist's and the non-relativist's response to (6), since according to both (6) has different interpretations. The relativist argues that (6) is true if construed as (7) but false if construed as

The fact that Alan has yellow skin is potential evidence that he has an i.c.b., given the background information of the doctor on Friday.

The non-relativist argues that (6) is false if construed as (5) but true if construed as (8).

6. THE CASE OF THE LOCH NESS MONSTER

Several objections to the definitions of Section 4 will be considered in this section and the two that follow.

There is a well-known photograph taken by a London surgeon in 1934 which purports to depict the Loch Ness monster.[16] Even if the existence of the monster is very improbable despite the photograph, isn't the existence of the photograph evidence that the monster exists? If so there is a violation of the probability condition for potential evidence that p(h,e) be high. Those who favor the "increase in probability" condition for evidence may cite examples such as this in defense

---

16. This photograph as well as three others taken between 1934 and 1960 are reproduced in David James, *Loch Ness Investigation* (London, n.d.).

of their position and in criticism of mine. Let us examine this criticism.

We are being asked to consider the claim that

(1) The existence of this photograph which purports to depict the Loch Ness monster is evidence that the monster exists.

Someone who believes that (1) is true might defend it by providing information about the surgeon who took the picture, from what position, at what time of day, etc. He might also point out how the camera works and that photographs are generally reliable, i.e., usually when a photograph depicts what seems clearly to be an X and the photographer has not made efforts to be deceptive, then an X exists.[17] But if this is how he defends (1) then, I think, he should not accept the claim that the existence of the monster is very improbable despite the photograph. On the contrary, he should believe that it is probable. (Indeed, this is what Sir Peter Scott, a leading British naturalist, claimed about new photographs taken in 1972.) New information can cancel the negative effects of background information. Even if the existence of the Loch Ness monster is improbable on the background information, a suitable photograph can make its existence very probable.

However, two other situations are possible. First, we may not know the reliability of the photographer, the conditions under which the photograph was taken, or indeed whether it is a genuine photograph (rather than a clever drawing). If so, and if there are independent reasons for doubting the existence of X, then we may be very unsure of the truth of the claim (1). We should then assert not that the existence of this photograph *is* evidence that a monster exists, but that it *may be* evidence (we don't yet know), or that it is evidence that there *may be* a monster (a claim to which I will return in a moment). Second, we may know that the photographer was unreliable or that the conditions under which he took the photograph were. Given this and other background information, the existence of the Loch Ness monster is very improbable, let us suppose. But given the unreliability of the photographer and photographic conditions would we make claim (1)? I seriously doubt it.

The importance of the subjective concept of evidence should not be minimized here. A question such as "Is there any evidence that there is a Loch Ness monster?" might be understood as "Is there anything that people take to be evidence that the monster exists?" To which the answer is: emphatically yes! It is not that those who believe that the

---

17. A non-relativistic notion of evidence is here being assumed in accordance with which this background information is being used to defend (1). However, the same argument could proceed with a relativized notion by relativizing (1) to this background information.

monster exists are unable to appeal to any facts as their reason for so believing. Quite the reverse. The existence of this photograph, among other things, is their evidence.

Finally, when the probability of h given e is low, although e is not evidence that h is true it can be evidence that h *may be* true (or that h is possible). One might claim that the existence of the surgeon's photograph is evidence that there may be a monster. The fact that Joe is one of 50 finalists in the state lottery is not evidence that he will win $1 million—that is too strong a claim to make. But it can be evidence that he may win $1 million or that his winning is a possibility. The fact that Bob is playing one round of Russian roulette is not evidence that he will die but that he may die, that his dying is a possibility. Such evidence claims, although different from the ones we have been considering, can be understood by altering the concept of potential evidence so as to require not high probabilities but non-negligible ones. That is,

> e is evidence that h may be true (or that h is possible) if and only if (a) e is true; (b) e does not entail h; (c) the probability of h given e is not negligible; (d) the probability that there is an explanatory connection between h and e, given h and e, is not negligible.

The fact that (e) Bob is playing one round of Russian roulette is evidence that h may be true, where h is the hypothesis that Bob will die. This is so since the probability of h, given e, is not negligible; and given that h and e are both true, the probability is not negligible that there is an explanatory connection between Bob's dying and his playing one round of Russian roulette.

7. THE ARTHRITIC BUS RIDER AND SEVERED HEAD CASES

Now let me consider two examples that might be used to argue against the "explanatory connection" condition (d).

Let e be the information that at time t Sam is riding on a certain bus. Let b be the background information that at time t there are 100 passengers on that bus of whom 90 have arthritis, although the fact that at t the bus contains a set of passengers 90 of whom have arthritis is just a coincidence. (The bus makes numerous stops, and at a later time t' the percentage of arthritics may be quite different.) We may be tempted to say that the fact that Sam is riding on that bus at t is evidence that (h) Sam has arthritis, given b. But if we say this, we have violated condition (d). Let us use (1) of Section 5—the background information form of the definition of potential evidence. Given that Sam is on that bus at t and that he has arthritis but that it is a mere coincidence that at t 90 percent of the passengers have arthritis, the probability is not high that there is an explanatory connection between Sam's having arthritis (h) and his being on that bus at time t (e).[18]

My reply is to agree that we have here a violation of condition (d), but to claim that it is a mistake to say that e is evidence that h in this case. If it really is a coincidence that, of the 100 passengers on the bus at time t, 90 have arthritis, then I suggest that the fact that Sam happens to be on the bus at t is no evidence whatever that he has arthritis. Suppose that Sam is informed that, by coincidence, at t, 90 of the 100 riders have arthritis (and suppose that Sam has no other reason to believe that he has arthritis and no reason to believe he does not). Should Sam now see his doctor? Would the ridership of the bus at time t provide Sam (or anyone else) with any reason to believe that he has arthritis? I suggest that the answer to these questions is No.

Consider the information

e': there is a class containing 100 people of whom 90 have arthritis, and Sam is a member of that class.

If we can assume that there are at least 10 non-arthritic people in the world, and at least 90 arthritic people, then we know that e' must be (trivially) true. (If Sam is non-arthritic, such a class exists in virtue of the existence of at least 9 other non-arthritics and at least 90 arthritics; if Sam is arthritic, it exists in virtue of the existence of at least 89 other arthritics and at least 10 non-arthritics.) The fact that there is a class of the sort described by e' is not evidence that Sam has arthritis. All that e&b in our previous example adds to e' above is that by coincidence one such class consists of riders on a certain bus at time t.

What may be tempting the objector to say in the original case that, given b, e is evidence that h is a confusion of this case with another. Let

e_1 = a person will be selected at random from among the passengers on that bus at time t

b = same as above

h_1 = the person randomly selected will have arthritis.

It does seem to me reasonable to say that e_1 is evidence that h_1, in the light of b. But this case satisfies the explanatory connection condition (d): given h_1 & e_1 & b, it is very probable that the person randomly selected will have arthritis because that person will be randomly selected from among the passengers on that bus at t (of whom, b tells us, 90 percent have arthritis). Indeed, the present example is like the second

18. An example of this general type was suggested to me by Adele Laslie.

lottery case, in which it is probable that one of Alice's tickets will be selected, because she owns 999 out of 1001.

The next example also questions the necessity of the explanatory connection condition, in the case of both veridical and potential evidence. Henry drops dead from a heart attack. Afterwards his head is severed by a fiendish decapitator.[19] Isn't the fact that his head is severed from his body veridical evidence that Henry is not living, even though there is no explanatory connection between his state of decapitation and his not being alive? If so the explanatory requirement is not necessary for veridical (or potential) evidence.

The latter conclusion is too hasty. There are various reasons one may not be alive, not all of which need be reasons for which one died. Anyone whose head is severed from his body is not alive for that very reason (since among other things, his brain is unable to receive oxygen from the rest of his body). Consider an analogous case. Tom's TV set is not working because one of the tubes burned out. Later Tom accidentally drops the set breaking the remaining tubes. Now among the reasons the set is not working is that all the other tubes are broken (although this is not among the reasons that it stopped working in the first place). There is an explanatory connection between the fact that these other tubes are broken and the fact that the set is not working; indeed, the former is veridical evidence that the latter is true.

To this one might respond that we should change the case a bit. Let e be the information that someone severed Henry's head. Instead of considering the hypothesis that Henry is not living, let h be that Henry died (i.e., that he went through some process that eventuated in death). Now e should be evidence that h despite the fact that there is no explanatory connection between e and h. I agree that we would speak of evidence in such a case, and yet there is something objectionable about it. The tension can be resolved, I suggest, by saying that the fact that someone severed Henry's head is potential, but not veridical, evidence that he died. The conditions for potential evidence are satisfied. For example, given background information which does not include the fact that Henry's death was due to a prior heart attack, and given that someone severed Henry's head and that Henry died, the probability is high that Henry died because someone severed his head. However, the fact that someone severed his head is not veridical evidence that he died, since there is no explanatory connection here. This is analogous to one of the examples in Section 1 involving Alan and the i.c.b., viz. that in which Alan has yellow skin and an i.c.b.

19. Brian Skyrms, "The Explication of 'X knows that p,'" *Journal of Philosophy* 64 (1967), pp. 373–89, uses this kind of case as a counterexample against certain causal analyses of "X knows that p."

although his yellow skin did not result from the i.c.b. but from the chemical dye with which he was working. Although his having the yellow skin he does is potential evidence that he has an i.c.b. it is not veridical evidence.

8. THE EVEN NUMBER CASE

The final example I shall mention questions the sufficiency of my conditions for potential evidence.[20] Let

e = 1766 is an even number
h = 12 is an even number.

The conditions for e to be potential evidence that h seem to be satisfied: (a) e is true; (b) e does not entail h (unless we assume, which I will not, that any a priori statement is entailed by every statement); (c) $p(h,e) > k$ (since the probability that 12 is an even number is high, and is not diminished by the assumption that 1766 is an even number); and (d) p(there is an explanatory connection between h and e, h&e) $> k$. Condition (d) is satisfied in virtue of the fact that, given h&e, the probability is high that some hypothesis (e.g., that even numbers are numbers divisible by 2) correctly explains why 12 is an even number and why 1766 is too. Nevertheless, the claim that e is evidence that h seems to be dubious.

My response is to agree that it is dubious, but to deny that condition (d) has been shown to be satisfied. Let us look more closely at the kind of explanatory connection between h and e suggested by the objection. Given that both 1766 and 12 are even numbers (and that we are dealing with arithmetic), it is, I shall assume, probable that the reason that 12 is an even number is that 12 has some property P which all even numbers have, and the reason that 1766 is even is that 1766 has that same property.

Let us make these explanations more specific by substituting "is divisible by 2" for P. What sorts of explanations are being envisaged here? I suggest these:

(1) (The reason that 1766 is an even number is that *1766 is divisible by 2;* explaining why 1766 is an even number);

(2) (The reason that 12 is an even number is that *12 is divisible by 2;* explaining why 12 is an even number);

where, by definition, any number divisible by 2 is an even number. Such explanations are ones which, in Chapter 7, Section 8, I call clas-

---

20. An example similar to the one that follows was suggested to me by William Friedman.

sification explanations. It is also possible to reformulate them by including the definitional fact in the constituent proposition, obtaining

(3) (The reason that 1766 is an even number is that *1766 is divisible by 2, and by definition any number divisible by 2 is even;* explaining why 1766 is an even number);

(4) (The reason that 12 is an even number is that *12 is divisible by 2, and by definition any number divisible by 2 is even;* explaining why 12 is an even number).

Now according to the relevant part of the definition of "explanatory connection" ((2iii) of Section 1), there is an explanatory connection between (the truth of) sentences h and e if

(iii) There is a sentence h' such that both (The reason that h is that h'; explaining why h) and (The reason that e is that h'; explaining why e) are correct explanations.

But neither explanations (1) and (2) nor (3) and (4) are of the right types to satisfy this condition. The hypothesis h' in (1), given by the emphasized words, is not the same as that given by the emphasized words in (2). Similarly the (emphasized) hypotheses in (3) and (4) are not the same, although they do contain a conjunct in common. In short, in these explanations, the hypotheses invoked to explain why 1766 is an even number and why 12 is an even number are not identical. They do not satisfy the appropriate condition for an explanatory connection.

It may be useful to compare this example with a similar one involving empirical rather than a priori claims. Let

e_1 = John has symptoms $S_1, \ldots, S_n$

h_1 = Bill has symptoms $S_1, \ldots, S_n$.

Let us suppose that, on the basis of background information we possess, the probability of h_1 is very high, and is not diminished by the assumption of e_1. But let us also suppose that the background information indicates nothing about any contact, direct or indirect, between John and Bill. I am assuming, then, that e_1 is *not* evidence that h_1. Yet it might be claimed that given my definition, e_1 must be evidence that h_1. Focusing on the explanatory connection condition (d), it might be thought that this condition is satisfied, in virtue of the fact that given e_1 and h_1, the probability is high that some hypothesis (e.g., that symptoms $S_1, \ldots, S_n$ are caused by disease D) correctly explains why John has symptoms $S_1, \ldots, S_n$ and also why Bill has them.

However, a more careful look at the type of explanation envisaged will lead to a retraction of this claim. The explanations are, I suggest, causal (in the sense expounded in Chapter 7):

(5) (The reason that John has symptoms $S_1, \ldots, S_n$ is that *John has disease D;* explaining why John has symptoms $S_1, \ldots, S_n$);

(6) (The reason that Bill has symptoms $S_1, \ldots, S_n$ is that *Bill has disease D;* explaining why Bill has symptoms $S_1, \ldots, S_n$).

And the emphasized hypotheses here are not the same. In Chapter 7, Section 13, I also noted that an explanation of a particular event which explicitly invokes as explanatory factors another particular event together with a causal law might be formulated as two explanations. Thus, in addition to (5) and (6) we might formulate

(7) (The reason that John's having disease D is what caused him to have symptoms $S_1, \ldots, S_n$ is that *it is a law that having disease D causes symptoms $S_1, \ldots, S_n$;* explaining why John's having disease D caused him to have symptoms $S_1, \ldots, S_n$);

(8) (The reason that Bill's having disease D is what caused him to have symptoms $S_1, \ldots, S_n$ is that *it is a law that having disease D causes symptoms $S_1, \ldots, S_n$;* explaining why Bill's having disease D caused him to have symptoms $S_1, \ldots, S_n$).

Note that in (7) and (8) the emphasized hypotheses are identical. What this allows us to do (under circumstances in which the background information indicates that both John and Bill have symptoms $S_1, \ldots, S_n$) is to say that

e_2: John's having disease D is what caused him to have symptoms $S_1, \ldots, S_n$

is evidence that

h_2: Bill's having disease D is what caused him to have symptoms $S_1, \ldots, S_n$.

This claim, unlike that involving e_1 and h_1, can be supported by appeal to the explanatory connection condition (d).

9. THE DESIRABILITY OF EVIDENCE

In concluding, I shall briefly turn to the question of why evidence is desirable. When one has a hypothesis h why should one seek evidence that h?

The answer, which is straightforward, is that one wants one's hypothesis to be true or at least probable, and one wants a reason for believing it. To have evidence is to satisfy both desires, at least on the theory of evidence in Section 4. If e is veridical evidence that h, then h is true, and if e is (only) potential evidence that h, then h is probable given e. Moreover, given background information b, if e is potential or veridical evidence that h, then, following the principle of reasonable belief, given b, e is at least some good reason for believing h. According to the theory of Section 4, evidence that h provides at least some good reason for believing h only if (a) h is probable given e, and (b) there is probably an explanatory connection between h and e, given h and e. The Wheaties example provides a case satisfying (a) but not (b), while the case of the stalled car satisfies (b) but not (a). And in neither case does e provide a good reason for believing h. The fact that this man eats Wheaties is not a good reason for believing that he will not become pregnant. And the fact that my car won't start this morning is not a good reason for believing the very specific hypothesis h of Section 3.

On the other hand, the alternative definitions of evidence that I rejected do not jointly satisfy the twin desires of truth/probability and reasons for believing. The "increase in probability" definition spawns the first lottery case in which the hypothesis that Bill will win is not probable (though its probability is increased over its prior probability); nor in this case does the information cited provide a reason for believing that Bill will win. The "high probability" definition, although it satisfies the desire for probability, generates the Wheaties case in which, as we have seen, the "evidence" provides no reason for believing the hypothesis. Finally, the explanation definition of Section 3 satisfies neither the truth/probability desire nor that for a reason for believing, as is shown by the case of the stalled car.

CHAPTER 11
Evidence: Additional Topics

There is a considerable philosophical literature on evidence (or "confirmation"), and in this chapter I shall relate the theory and examples developed in Chapter 10 to some of that discussion. Specifically, I will briefly examine (a) some probability definitions of evidence not treated in Chapter 10; (b) Glymour's bootstrap theory of evidence; (c) various "conditions of adequacy" frequently mentioned for a definition of evidence; (d) the notorious paradox of the ravens; (e) the "variety of instances" requirement; and (f) the question of how observations provide evidence for a scientific theory.

1. OTHER PROBABILITY DEFINITIONS

a. Mackie's relevance criterion.[1] Mackie writes:

> The relevance criterion of confirmation is the principle of which the basic form is that an hypothesis . . . is confirmed by an observation . . . in relation to a body of background knowledge or belief . . . if and only if what is observed would have been more likely to occur, given the hypothesis together with the backgound knowledge or belief, than it would have been given that background knowledge or belief alone.[2]

Mackie expresses this idea in terms of probabilities; it can be written as follows:

(1) e is evidence that h, given b, if and only if $p(e, h\&b) > p(e, b)$.

Whereas the increase-in-probability definition of Chapter 10 requires that e increase the probability of the hypothesis h, Mackie requires

---

1. J. L. Mackie, "The Relevance Criterion of Confirmation," *British Journal for the Philosophy of Science* 20 (1969), pp. 27–40.
2. *Ibid.*, p. 27.

that h increase e's probability. (Indeed, in an earlier article Mackie calls this the inverse principle.)[3]

I suggest that (1) provides neither a necessary nor a sufficient condition for evidence, as can be shown by invoking two examples from Chapter 10.

The case of the stalled car (to show that (1) is not sufficient for evidence). Let e = my car won't start this morning; b = "normal" background information; h = at precisely 3:05 last night 2 monkeys removed the remaining 3.7 gallons of gas in my tank and substituted crushed bananas. p(e, h & b) is approximately 1, whereas (let us suppose) p(e, b) is considerably less than 1. Thus Mackie's condition (1) is satisfied. Yet e is not evidence that h, given b.

The second lottery case (to show that (1) is not necessary for evidence). Let

- b = there is a lottery consisting of 1001 tickets, one of which will be drawn at random, and by Tuesday 1000 tickets have been sold of which Alice owns 999;
- e = by Wednesday 1001 lottery tickets have been sold, of which Alice owns 999, and no more tickets will be sold;
- h = Alice will win.

p(e, h & b) < p(e, b), since Alice's winning (h) makes it more likely (or at least not less likely) that she bought the remaining ticket, and hence that e is false. Yet e is evidence that h, given b; which violates (1).

The existence of such counterexamples should not be surprising, since Mackie's relevance criterion (1) can be shown to be equivalent to the increase-in-probability definition of Chapter 10 in cases in which $p(h, b) \neq 0$. By Bayes's theorem, assuming $p(h, b) \neq 0$,

$$p(e, h \& b) = \frac{p(h, e \& b) \times p(e, b)}{p(h, b)}.$$

It follows that

$$p(e, h \& b) > p(e, b) \text{ if and only if } \frac{p(h, e \& b)}{p(h, b)} > 1.$$

But $\frac{p(h, e \& b)}{p(h, b)} > 1$ if and only if $p(h, e \& b) > p(h, b)$.

So, assuming that $p(h, b) \neq 0$, we conclude that

$$p(e, h \& b) > p(e, b) \text{ if and only if } p(h, e \& b) > p(h, b),$$

---

3. "The Paradox of Confirmation," *British Journal for the Philosophy of Science* 13 (1963), pp. 265–77.

which shows that, where $p(h, b) \neq 0$, Mackie's criterion is equivalent to the increase-in-probability definition of Chapter 10.

b. Popper's probability definition.[4] Popper has suggested a definition which is an alternation of the increase-in-probability definition of Chapter 10 and Mackie's relevance criterion. It can be expressed as follows:

(2) e is evidence that h if and only if either $p(h, e) > p(h)$ or $p(e, h) > p(e)$.

The reason for this alternation, according to Popper, is that if h is a universal law then (he believes) $p(h) = 0 = p(h, e)$. In such a case e could still be evidence that h if the second disjunct in (2) obtains.

However, as we have just seen in the discussion of Mackie's criterion, where $p(h) \neq 0$, $p(e, h) > p(e)$ if and only if $p(h, e) > p(h)$. So in cases in which $p(h) \neq 0$, Popper's definition (2) reduces to the increase-in-probability definition of Chapter 10. It is thus subject to all of the counterexamples of the latter noted in that chapter. (In none of these counterexamples is h a universal law.)

c. Nozick's probability definition.[5] Very recently Robert Nozick has propounded the following definition of evidence:

(3) e is evidence that h if and only if (a) $p(e, h) = 1$, and (b) $p(e, \sim h) = 0$.

He first formulates the definition in non-probabilistic terms as follows:

(4) e is evidence that h if and only if (a) if h were true e would be true, and (b) if h were not true e would not be true.

His basic idea is that evidence for a hypothesis is something that would hold if the hypothesis were true and would not hold if the hypothesis were not true. (This is what he calls "strong evidence." e is weak evidence for h when (b) in (4) is changed to "if h were false e might be false.")[6] Using (4) as a basis Nozick goes on to formulate the definition in probability terms as (3), in which $p(e, h)$ is construed subjunctively as "the probability that e would be true, given that h is true."[7]

I suggest that neither (a) nor (b) in (3) (or (4)) is a necessary condition for evidence. To show that (b) is not necessary we may invoke

---

4. Karl R. Popper, *The Logic of Scientific Discovery* (New York, 1968), p. 389.
5. Robert Nozick, *Philosophical Explanations* (Cambridge, Mass., 1981), p. 252.
6. See *ibid.*, p. 250.
7. *Ibid.*, p. 252. Nozick states (fn. 79) that this is to be treated as a conditional probability, not as the probability of the subjunctive conditional "h→e."

the second lottery case (used above in the discussion of Mackie). Where e, b, and h are as given earlier,

p(e, ~h&b) ≠ 0.

(Indeed, p(e, ~h&b) is approximately equal to, if perhaps slightly more than, p(e, h&b); and the latter is not equal to, or even close to, 0.) If so, and if, as I suggest, e is evidence that h, given b, then Nozick's condition (3b) is not necessary. More precisely, a version of that condition relativized to background information (e is evidence that h, given b, only if p(e, ~h&b) = 0) is not necessary.

The following example shows that condition (3a) is not necessary. Let

e = there is a lottery consisting of 1000 tickets, of which Abel has 999, and one ticket will be drawn at random;

h = Abel will win;

I take e to be evidence that h, even though p(e, h) ≠ 1.

If p(h) ≠ 1, and p(e) ≠ 0, and the probability function Nozick is speaking of obeys the probability calculus, then his condition (3b)—that p(e, ~h) = 0—holds if and only if p(h, e) = 1. This is seen as follows. By Bayes's theorem, if p(~h) ≠ 0 (i.e., if p(h) ≠ 1), then

$$p(e, \sim h) = \frac{p(\sim h, e) \times p(e)}{p(\sim h)}.$$

So, p(e, ~h) = 0 if and only if $\frac{p(\sim h, e) \times p(e)}{p(\sim h)} = 0.$

But the latter is true if and only if

p(~h, e) × p(e) = 0.

And, where p(e) ≠ 0, this holds if and only if

p(~h, e) = 0, i.e., p(h, e) = 1.

So if p(h) ≠ 1 and p(e) ≠ 0, Nozick's condition (3b)—p(e, ~h) = 0—holds if and only if p(h, e) = 1. Accordingly, for those cases in which the prior probability of the hypothesis is not 1 and the prior probability of the evidence is not 0, Nozick's two conditions for evidence come to this:

p(e, h) = 1
p(h, e) = 1.

Let us now weaken these conditions to require only "high" probability (say greater than $^1/_2$), rather than a probability of 1. We thus have the following modified version of Nozick's definition:

(5) e is evidence that h if and only if (a) p(e, h) > k, and (b) p(h, e) > k.

Note that this version contains a high probability condition (b) of the sort advocated in my definition of (potential) evidence in Chapter 10. Although, as I have urged, (5b) is necessary for evidence (5a) is not. (In the lottery case above involving Abel, we need not suppose that p(e, h) > k.) Nor does (5) provide sufficient conditions for evidence. This can be shown by employing an example analogous to the Wheaties case of Chapter 10. Let e = that woman has no penis; h = this man will not become pregnant. p(e, h) > k, since p(e) > k and h is irrelevant to e. p(h, e) > k, since p(h) > k and e is irrelevant to h. So both conditions in (5) are satisfied. Yet e is not evidence that h![8]

2. GLYMOUR'S BOOTSTRAP THEORY

Clark Glymour offers an account of evidence which is quite different from any of those so far discussed.[9] Indeed, it includes no probability or explanation conditions whatever.[10] Glymour's claim is that something is evidence for a hypothesis only in relation to a theory consisting of a set of hypotheses. His basic idea (the "bootstrap condition") is that e is evidence that h with respect to theory T if and only if using T it is possible to derive from e an instance of h, and the derivation is such as not to guarantee an instance of h no matter what e is chosen. The specific conditions Glymour imposes are complex, and the account is most readily developed for hypotheses that are expressed quantitatively in the form of equations.

I shall use Glymour's example of a psychological theory formulated as a set of linear equations, as follows:

(1) $A_1 = E_1$
(2) $B_1 = G_1 + G_2 + E_2$
(3) $A_2 = E_1 + E_2$
(4) $B_2 = G_1 + G_2$

---

8. I have tried to choose a case in which the probability statements can be understood in the subjunctive manner that Nozick suggests. p(e, h) = the probability that "that woman has no penis" would be true given that "this man will not become pregnant" is true. Nozick offers little clarification of his use of subjunctives generally and of subjunctive probabilities in particular. He does indicate that a sentence of the form "if h were true e would be true" could be true in a case in which "e would hold anyway, even if h were not true" (p. 248). The above example illustrates the analogous possibility in the case of probabilities.
9. Clark Glymour, *Theory and Evidence* (Princeton, 1980), chapter 5.
10. For a later attempt to introduce probabilities see his "Bootstraps and Probabilities," *Journal of Philosophy* 77 (1980), pp. 691–99.

(5) $A_3 = G_1 + E_1$
(6) $B_3 = G_2 + E_2$.

The A's and B's are observable quantities whose values can be determined by experiment. The E's and G's are "theoretical" quantities whose values cannot be determined directly by experiment but only indirectly through the theory by determining values for A's and B's. (Glymour does not give the meanings of these quantities; he seems to think this is not necessary.)

How would one proceed to test—i.e., get evidence for—this theory? Let us concentrate on the first hypothesis of the theory

(1) $A_1 = E_1$.

As noted, by experiment we can directly obtain a specific value for the quantity A_1. By experiment we can also obtain specific values for the (other) A and B quantities. With these values—in particular, with values for B_1, B_3, and A_3 we can use the theory itself (specifically, hypotheses (2), (5), and (6)) to compute a value for E_1. Here is the computation

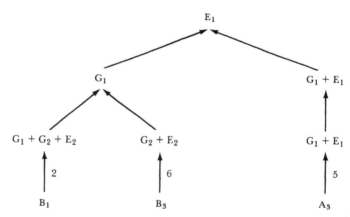

The numerals to the right of the arrows indicate which hypothesis in T is being used. Thus, using hypothesis (2), from a value for B_1 we can compute a value for $G_1 + G_2 + E_2$. Using hypothesis (6), from a value for B_3 we can compute a value for $G_2 + E_2$. The results of these computations can be combined by subtracting $G_2 + E_2$ from $G_1 + G_2 + E_2$ to obtain G_1. And this can be combined mathematically with a computation from A_3 to obtain a value for E_1. Accordingly, using hypotheses (2), (5), and (6) in the theory, from specific values for the "observables" B_1, B_3, and A_3 determined experimentally, we can compute a specific value for the "theoretical" quantity E_1. Now, if the specific value of the "observable" quantity A_1 (determined experimentally) is the same as the specific value of E_1 that has been computed, then we have obtained evidence for hypothesis (1) in the theory. We can say that the specific values of A_1, B_1, B_3, and A_3 obtained experimentally constitute evidence for hypothesis (1), relative to the theory consisting of (1) through (6).

By contrast, suppose that we obtain a specific value for A_1 but not for any of the other "observable" quantities. Using hypothesis (1) we may compute a specific value for the "theoretical" quantity E_1. We would then have values for A_1 and E_1 which are in accord with hypothesis (1). But these values would not suffice to confirm (1), since in this case a value for E_1 has been computed in such a way that no matter what the observed value of A_1 it could not fail to accord with (1). So Glymour imposes the following condition. For a set I of values of quantities to constitute evidence for hypothesis h with respect to theory T there must exist "computations (using hypotheses in T) from I of values for the quantities occurring in h, and there exists a set J of *possible* values for the same initial quantities such that the same computations from J result in a counterinstance to h—i.e., the values of the quantities occurring in h that are computed from J must contradict h."[11]

A hypothesis in a theory has what Glymour calls a representative equation (or just a *representative*) for a given computation. In the representative each quantity that appears is an "observable" one used as a basis for the computation. For example, the representative of hypothesis (1) for the computation which uses the "observables" B_1, B_3, and A_3 as a basis is

$$A_1 = A_3 + B_3 - B_1.$$

(In the computation in question we compute E_1 by computing $A_3 + B_3 - B_1$.) By contrast, if we compute E_1 from A_1 using hypothesis (1), then the representative of hypothesis (1) for this computation is

$$A_1 = A_1,$$

which is a mathematical identity. Now the requirement introduced one paragraph back can be formulated as follows: given the computations used for obtaining specific values for the "theoretical" quantities in h, the equation representing h for these computations must not be a mathematical identity.

Glymour's conditions for evidence for an equation can now be stated as follows. Where e describes a set of observed values of quantities, and h is an equation containing quantities, and T is a theory consisting of equations, e is evidence that h with respect to T if and only if

11. *Theory and Evidence*, p. 115.

1. e is consistent with T.
2. From the specific values in e a value for each quantity in h is computable.
3. The computation involves only equations in T and their consequences.
4. The equation representing h for the computations used is not a mathematical identity.
5. The values obtained from the computations used constitute a solution to the equation in h.[12]

These conditions invoke no probability requirements. In particular, for e to be evidence that h with respect to T there is no requirement that the probability of h be high, given e and T, or that this probability be greater than the probability of h given T. (Indeed, in the usual case of interest to us hypothesis h will be a part of T itself, so that p(h, e & T) = 1 and p(h, T) = 1. Thus h will have maximal probability, and its probability will not be increased by e.) Nor is there a requirement that the confirmed hypothesis h explain the observed values reported in e, or that given e & h & T the probability be high that there is an explanatory connection between h and e. (In a moment I shall consider an example violating such explanation requirements.) The basic idea is that observed values of quantities provide evidence for an equation in a theory if from these values together with equations in the theory computations can be made for all quantities in the equation which will yield (positive) instances of that equation.

Let me now suggest a simple example which will illustrate what I take to be a major shortcoming in the bootstrap approach. The theory that I will propose consists of a set of equations employing three quantities. The first two quantities are familiar ones which I shall take to be directly measurable:

A = the total force acting on a particle
B = the product of a particle's mass and acceleration.

The third quantity, the "theoretical" one, is not directly measurable:

C = the quantity of God's attention focused on a particle.

Theory T consists of two equations:

(i) $A = C$
(ii) $B = C$.

Experimentally we determine specific values for A (total force) and B (product of mass and acceleration). From a specific value for B, using

12. *Ibid.*, p. 117.

equation (ii) we compute a specific value for C (quantity of God's attention). The representative of (i) for this computation is A = B, which is not a mathematical identity. Therefore, if the value of C computed from B is the same as the experimentally determined value of A, we have confirmed the first equation A = C. In this case the statement describing the experimentally determined values of total force and mass × acceleration is evidence that the total force on a particle = the quantity of God's attention focused on the particle, with respect to theory T.

Similarly, a value for C can be computed from A using equation (i). If this computed value of C is the same as the experimentally determined value of B, then the second equation B = C has been confirmed. In this case the statement describing the experimentally determined values of total force and mass × acceleration is evidence that the product of a particle's mass × acceleration = the quantity of God's attention focused on the particle, with respect to theory T. Thus, the observed values of force and mass × acceleration provide evidence for each equation in this theory, with respect to that theory.

This result may strike one as absurd. How can information about experimentally determined values of the total force acting on a particle and its mass × acceleration provide evidence for how much of God's attention is focused on the particle? Given no other background information about God and his relationship to particles, and indeed given little if any idea of what it means to talk about "the quantity of God's attention focused on a particle," such an evidence claim seems unwarranted. It would be precluded by the account of evidence I propose in Chapter 10. In the absence of any other information, there is no reason to suppose that the probability that A = C is high, given that the specific values observed for force and mass × acceleration are the same. If not, the high probability condition is not satisfied. Neither is the explanation condition. There seems to be no reason to say that the probability is high of an explanatory connection between the fact that A = C and the fact that the observed values of A and B are equal, given these facts.

However, this objection may not be decisive. We must bear in mind that Glymour insists on relativizing evidence for a hypothesis to a theory. What he would (or could) say is that e (the sentence describing the observed values of A and B) is evidence for (i) *with respect to theory T* (consisting of equations (i) and (ii)); it need not be evidence for (i) with respect to some different theory T' (e.g., one that does not contain (ii)). Within T the observed values of force and mass × acceleration count as evidence for equations relating each of these quantities to the quantity of God's attention focused on a particle. Within some different T', say the theory consisting of Newton's first, second, and third laws of motion, the observed values of force and mass × acceleration will not count as evidence for equations relating each of these quantities to the quantity of God's attention focused on a particle. And—perhaps Glymour will say—it is because we are thinking of a theory of the latter sort, rather than of the former, that we find the present evidence claim so puzzling and objectionable.

This raises a fundamental question about the usefulness of the concept of evidence or confirmation that Glymour is defining. In Chapter 10, I took evidence to bear an important relationship to what it is reasonable to believe: if, in the light of the background information b, e is evidence that h, then, given b, e is at least some good reason for believing h. I also took a sentence of the form "e is evidence that h, given b" to require the truth of b as well as e (see Section 5). Accordingly, if e is evidence that h, given b, then we can say that in view of the truth of b, e is at least some good reason for believing h. Now a concept of evidence is a useful one only if it is possible for a person to come to know that e is true, that b is true, and that, given b, e is evidence that h—at least for a considerable range of cases in which, given b, e is evidence that h. Only then can persons have (legitimate) reasons for believing hypotheses. It would be useless for us to have a concept of evidence which is such that although e *is* evidence that h, given b, we would not be in a position to have e as our reason for believing h.

Returning to our "God's attention" example above, on Glymour's definition, e (the sentence describing the observed values of A and B) is evidence that A = C, with respect to theory T. It is not part of Glymour's view that the latter claim entails or presupposes the truth of theory T. (Glymour's "with respect to T" is not to be understood in the way I am construing "given b" in "e is evidence that h, given b.") So from the truth of the claim "e is evidence that A = C with respect to T" we cannot conclude that *in view of the truth of T* e is at least some good reason for believing that A = C. At most we can say this:

(7) If T is true, then e is at least some good reason for believing that A = C.

Now this conditional fact is a useful one in helping to form people's beliefs only if people can come to know the truth of T (or, at least, if they can come to have good reasons to believe that T is true). If they do come to know that T is true (or have good reasons to believe it is), and they do come to know that e is evidence that A = C with respect to T, then (I shall suppose) they have at least some good reason for believing that A = C. But how are people to come to know that T is true?

Presumably (since T is not a priori) by discovering that something is evidence that T; which means discovering that something is evidence for hypothesis (i) and that something is evidence for hypothesis (ii), or by discovering that something is evidence for the conjunction of (i) and (ii). Now we have shown that (on Glymour's definition) e, which reports the observed values of A and B, is evidence that (i) and also evidence that (ii). If this suffices to give us knowledge that T, then in view of (7) we get the absurd result that the observed values of force and mass × acceleration provide some good reason for believing that the total force acting on a particle = the quantity of God's attention focused on the particle. However, I doubt that Glymour will say that e's being evidence for (i) and (ii) suffices to give us knowledge that T, since e is evidence for (i) and for (ii) *with respect to T*. (It is not, e.g., evidence for (i) or (ii) with respect to T′, Newton's three laws of motion.) Although I know that (in Glymour's sense) e is evidence for (i) and for (ii) with respect to T, this fact does not enable me to know that T is true, Glymour will probably say. Nor, indeed, does this fact give me a reason to believe T. (It only gives me a reason to believe T with respect to T).

Glymour in his book does have a short section entitled "Comparing Theories" (pp. 152–55). He claims that "what makes one theory better than another is a diffuse matter that is not neatly or appropriately measured by any single scale" (p. 153). Glymour briefly mentions several criteria. For example, one theory may contain some hypotheses that are disconfirmed by the evidence whereas the other does not. (Presumably Glymour has in mind disconfirmation with respect to the theory in question.) One theory may contain more untested hypotheses than the other. The evidence may be more various for one theory than for another. And so forth. (My "God's attention" theory T contains only confirmed hypotheses with respect to T, and no untested ones, although the evidence for it is less various than that for Newtonian theory.) However, if one theory is better than another according to Glymour's criteria for comparing theories, this does not mean that the former is known to be true whereas the latter is not; nor does it mean that there is a good reason to believe the former to be true and the latter false. Indeed, so far as I can make out from his brief description of the criteria, we can conclude nothing about the (probable) truth or falsity of a theory from the fact that it is "better" or "worse" in Glymour's sense than another theory.

On Glymour's conception of evidence it is difficult to see how one can come to know that T is true, or to have good reasons to believe it is, and hence to establish the antecedent of the conditional in (7). If one cannot do this, then one cannot come to know whether e is at least some good reason for believing that A = C. If Glymour's theories were entirely observational—if the equations contained quantities all of which are directly observable—then this problem would not arise. We could come to know the truth of each equation by making direct measurements of each quantity. There would be no computations at all. Each equation could come to be known without using any other equation, without, indeed, using any theory. But this is not the sort of case that interests Glymour. In general, his theories will contain "theoretical" quantities whose values are determined not directly but only by computations from observable quantities using equations of that theory or some other. In such cases I fail to see how one comes to establish the truth of the theory, on Glymour's view of evidence. So even if the crazy theory T above is true, the fact that e is evidence that $A = C$, with respect to T, gives a person no reason whatever to believe that $A = C$ (or to act on this assumption). In the absence of a method of coming to know the truth of a theory, or at least for coming to have good reasons for believing that it is true, Glymour's concept of evidence, I suggest, is not of sufficient use to us.

3. CONDITIONS OF ADEQUACY

The idea that a definition of evidence should satisfy certain broad conditions of adequacy was propounded by Hempel in a classic article on confirmation.[13] Hempel discusses the following conditions, the first three of which he accepts:

1. *Consequence condition:* If e is evidence that each member of a set K of hypotheses is true, and K entails h, then e is evidence that h.

The consequence condition has two important corollaries, which are even more widely discussed than it:

   a. *Special consequence condition:* If e is evidence that h, and if h entails h', then e is evidence that h'.
   b. *Equivalence condition:* If e is evidence that h, and if h is logically equivalent to h', then e is evidence that h'.

2. *Entailment condition:* If e entails h, then e is evidence that h.
3. *Consistency condition:* e is logically compatible with the set of all hypotheses for which it is evidence.
4. *Converse consequence condition:* If e is evidence that h, and if h' entails h, then e is evidence that h'.
5. *Nicod's criterion:* If h is a universal hypothesis of the form $(x)(Fx \supset Gx)$, then e is evidence that h if and only if e has the form $Fx \& Gx$ (or if it is a conjunction of sentences of this form).

13. Carl G. Hempel, "Studies in the Logic of Confirmation," reprinted in *Aspects of Scientific Explanation*.

6. *Modified Nicod:* Change "if and only if" in (5) to "if." (That is, Nicod's criterion provides a sufficient but not a necessary condition for evidence.)

Hempel accepts the consequence, entailment, consistency, and modified Nicod conditions, and provides reasons for rejecting the converse consequence and (unmodified) Nicod conditions. I believe, contrary to Hempel, that all of the conditions—save for the consistency condition—should be rejected. Let us examine these in turn.

1. Consequence condition. I shall consider only the equivalence condition. If this is not acceptable then neither is the special consequence condition. The type of argument I propose to use against this condition employs emphasis—to which considerable attention has been given in Chapter 6. Let

b = there is a lottery consisting of 1000 tickets, one of which will be drawn at random on Tuesday;

e = Irving owns 999 of the lottery tickets;

h = *Irving* will win the lottery on Tuesday.

I assume that e is evidence that h, given b. Note, however, that h is logically equivalent to

h' = Irving will win the lottery *on Tuesday.*

And I question whether e is evidence that h', given b. That is, although given b, the fact that Irving owns 999 of the lottery tickets is evidence for who will win the lottery, it is not evidence for the day on which the winning will take place. "Evidence," like "explanation," is emphasis-sensitive. This should not be surprising, given the definition of (potential) evidence that I have proposed, since that definition invokes the concept of explanation in one of its conditions. Indeed, that condition in the definition precludes e from being evidence that h', given b. The explanation condition requires that

p(there is an explanatory connection between h' and e, h'&e&b) > k.

But given h'&e&b, the probability is not high that the reason that Irving will win the lottery *on Tuesday* is that Irving owns 999 of the lottery tickets, or that the reason that he owns 999 of the lottery tickets is that he will win the lottery *on Tuesday,* or that some hypothesis correctly explains both why he will win *on Tuesday* and why he owns 999 tickets.

What *would* be evidence that h'? Let

b' = there is a lottery consisting of 1000 tickets, one of which will be drawn at random;

e' = Irving owns 999 of the lottery tickets and the drawing will take place on Tuesday.

I suggest that e' is evidence that h', given b'. All of the conditions of my definition of evidence are satisfied, including the one that requires that

p(there is an explanatory connection between h' and e', h' & e' & b') > k.

Given h' & e' & b', the probability is high that the reason that Irving will win *on Tuesday* (h') is that Irving owns 999 of the lottery tickets and the drawing will take place on Tuesday (e').

2. Entailment condition. This, as I suggested in Chapter 10, is too strong. That Alan has an i.c.b. is not evidence that he has an i.c.b. From the fact that e is evidence that h we may conclude that there is evidence that h. But from the fact that "Alan has an i.c.b." entails itself we cannot conclude that there is evidence that Alan has an i.c.b. Nor can we draw this conclusion if in addition we assume the sentence to be true. The entailment condition is, of course, not satisfied by my definition of (potential) evidence. Indeed, the latter stipulates that if e entails h then e is not evidence that h.

3. Consistency condition. This condition is satisfied by my definition of evidence, if we assume that the high probability condition requires a probability greater than $\frac{1}{2}$. If e is evidence that h, then (by my definition, where $k = \frac{1}{2}$), $p(h, e) > \frac{1}{2}$. If h' is logically incompatible with h, then, in accordance with the probability calculus, $p(h', e) < \frac{1}{2}$. So e cannot be evidence that h' if it is evidence that h.

One philosopher who rejects the consistency condition is Carnap.[14] He presents the following example. Let

e = in a certain population of 10,000 individuals there are 8000 with a property P;

h = in a sample of 100 individuals drawn from that population 80 have P;

h' = in a sample of 100 individuals drawn from that population 79 have P.

Carnap claims both that e is evidence that h and that e is evidence that h'. Since h and h' are logically incompatible, the consistency condition is violated.

If in h and h' Carnap means that *exactly* 80 (79) individuals in the sample will have P, then h and h' make very strong claims, and I

14. Rudolf Carnap, *Logical Foundations of Probability,* pp. 476–77.

would deny that e is evidence that either of them is true. The probability that *exactly* 80 individuals in a 100 member sample will have P, given that 8000 of the 10,000 in the population have P, is low. (It is .0993, where the sampling is done with replacement.)[15] Accordingly, the high probability condition in my definition precludes e from being evidence that h (or that h'). If, on the other hand, we modify h and h' by inserting "approximately" in front of 80 and 79, then I would agree that e is evidence that h and that h'. With this modification the high probability condition is satisfied. However, with this modification h and h' are no longer logically incompatible.

The reason that Carnap gives for saying that e is evidence that both h and h' is that p(h, e) > p(h), and p(h', e) > p(h'). In short, Carnap is espousing the increase-in-probability definition of evidence, which, I argued in Chapter 10, is neither necessary nor sufficient for evidence.

4. Converse consequence condition. Hempel rejects this. If it were adopted, then, given the consequence condition (which Hempel accepts), any e would be evidence that any h. (Suppose e is evidence that h; then by the converse consequence condition, e is also evidence that h & h', where h' is any hypothesis whatever; by the consequence condition, it follows that e is evidence that h'.) That the present condition is much too strong can be seen by means of a modification on my stalled car example of Chapter 10. Let

e = my car won't start this morning although the battery is not dead;

h = my car is out of gas;

h' = at precisely 3:05 last night two monkeys removed the remaining 3.7 gallons of gas in my tank and substituted crushed bananas.

In the light of some reasonable background assumptions, e is evidence that h (let us suppose). Since h' entails h, the converse consequence condition requires that e is evidence that h', which seems absurd. The problem, as I suggested in Chapter 10, is that p(h', e) is low. More generally, the converse consequence condition is to be rejected since it violates the high probability condition. If p(h, e) = r, and if h' entails h, then p(h', e) ≤ r. So in general we cannot assume that if p(h, e) > k then p(h', e) > k.

5. Nicod's criterion. Hempel rejects the general criterion on the grounds that it fails to provide a set of necessary conditions for evidence. But he does accept "modified Nicod": if h is a universal hypothesis of the form $(x)(Fx \supset Gx)$, and if e has the form $Fx \& Gx$ (or if

---

15. With appreciation to Jonathan Achinstein for the calculation.

e is a conjunction of sentences of this form), then e is evidence that h.

Even "modified Nicod" is too strong to be acceptable. One way to show this is to appeal to Goodman's grue paradox. (Let hypothesis h be "all emeralds are grue"—where something is grue at a time if it is green and the time is before the year 2000 or it is blue and the time is after 2000; and let e be "a is an emerald and a is grue.") However, Hempel and others may seek to exclude "positional" predicates such as "grue" (which mention particular places, times, or things) from the sentences to which modified Nicod is applicable. (The problem of how to distinguish positional from non-positional predicates is notorious, and has spawned a considerable literature.) So let us consider instead certain hypotheses whose predicates are not positional ones like "grue" and indeed are the sorts that are common in science.

Suppose that a scientist makes many experiments to determine how one physical quantity, say the length of a solid rod, varies with another, say its temperature. He records his results as points on the

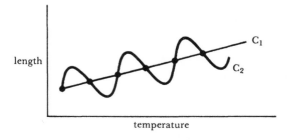

accompanying graph. Let h_1 be the hypothesis that (within a certain range of temperatures) all the points, not just the observed ones, lie on the curve C_1, and let h_2 be the hypothesis that all of the points lie on C_2. With a sufficient number of such points scientists will claim that the fact that all the observed points lie on the straight line C_1 (e) is evidence that the length of a solid rod increases linearly with its temperature (h_1). They will not claim that the fact that all the observed points lie on the sine curve C_2 is evidence that length varies with temperature in a way given by that curve (h_2). We thus have a violation of modified Nicod, assuming that the latter criterion is supposed to be applicable to hypotheses of this type. If it can be supposed that $p(h_1, e) > k$, and hence that $p(h_2, e) < k$, then my definition of (potential) evidence can be used to support the claim that e is evidence that h_1, and that e is not evidence that h_2.

Of the six conditions of adequacy mentioned in this section, only the consistency condition will obtain according to my definition of evi-

Evidence: Additional Topics 367

dence. The consequence condition will not, in general, be satisfied because of the explanation requirement that p(there is an explanatory connection between h and e, h&e)>k. As we have seen, explanatory contexts are emphasis-sensitive, so that although differently emphasized h and h' are logically equivalent (or although h' is a logical consequence of h), from the fact that there is an explanatory connection between h and e we cannot conclude that there is an explanatory connection between h' and e. Nor does the entailment condition hold under my definition, since according to the latter a necessary condition for e to be evidence that h is that e not entail h. The converse consequence condition, as we have seen, is precluded by the high probability requirement. Finally, modified Nicod is not satisfied by my definition if, as illustrated in the curve-fitting problem, we need not suppose that the probability of a universal hypothesis must be high given that all observed instances satisfy the hypothesis. (In the next section the latter claim will be further buttressed.)

4. THE PARADOX OF THE RAVENS

Hempel, in the article already mentioned, formulates the following intriguing paradox, based on several "conditions of adequacy" discussed in the previous section. Let

h = all non-black things are non-ravens;

e = all the observed non-black things are non-ravens;

h' = all ravens are black.

By modified Nicod, e is evidence that h. But h is logically equivalent to h'. Therefore, by the equivalence condition, e is evidence that h'. But this conclusion seems absurd, since it states that the fact that all the non-black items so far observed are non-ravens (say these are white shoes, green leaves, and red apples) is evidence that all ravens are black. Since Hempel accepts modified Nicod, the equivalence condition, and the fact that h and h' are logically equivalent, he accepts this paradoxical conclusion and attempts to show why it has only an air of paradox. In the previous section reasons were proposed for rejecting both modified Nicod and the equivalence condition. Hence, e cannot be shown to be evidence that h' using these conditions.

Can my definition of (potential) evidence be employed to show that e is not evidence that h'? By itself, no. (Nor can the standard probability and explanation definitions I reject in Sections 2 and 3 of Chapter 10.) Additional argumentation is necessary. Is the high probability condition in my definition satisfied? Consider

368 The Nature of Explanation

(1) p(all non-black things are non-ravens, all observed non-black things are non-ravens) < k

and

(2) p(all ravens are black, all observed ravens are black) > k.

If (1) and (2) are true, then my definition of evidence will yield the conclusion that while the fact that all observed ravens are black is evidence that all ravens are black, the fact that all observed non-black things are non-ravens is not evidence that all non-black things are non-ravens. But, of course, the definition of evidence I propose won't yield (1) or (2). Some independent argument is necessary.

In what follows I shall consider conditions that would make (1) and (2) true. The basic idea is that in choosing the non-black things to observe that we do we are employing a selection procedure that is strongly biased in favor of non-ravens. By contrast, in choosing the ravens to observe that we do we are employing a selection procedure that is not (or need not be) biased in favor of black. If this is so, then (1) and (2) may both be true. Let me explain this idea.

A selection procedure is a rule which the investigator is following in choosing items for observation. Suppose, e.g., that he is testing the hypothesis that all mice in North America are brown. Here are two different selection procedures he might use for choosing mice to observe:

S_1: Select mice from various sections of North America and at various times of the year;

S_2: Select North American mice from cages marked "brown mice."

If S_2 is the rule he is following we will say that his selection procedure for North American mice is biased in favor of brown. If S_1 is the procedure we are inclined to say that it is not biased in favor of brown. The following probability claims seem reasonable:

p(all observed North American mice are brown, North American mice are selected for observation in accordance with S_2) = 1;

p(all observed North American mice are brown, North American mice are selected for observation in accordance with S_1) = p(all observed North American mice are brown) << 1.

(I am supposing that whatever background information we possess does not indicate the color of North American mice. Also, I shall understand "all observed North American mice are brown" as "(x)(x is a North American mouse and x is observed ⊃ x is brown)" and "North American mice are selected for observation in accordance with S" as "(x)(x is a North American mouse and x is observed ⊃ x is selected for observation in accordance with S).")

Generalizing this, let us say that a selection procedure S for choosing F's to observe is (strongly) biased in favor of G if

p(all observed F's are G's, F's are selected for observation in accordance with S) ≈ 1 (is close to, or equal to, 1)

and

p(all observed F's are G's) ≉ 1.

If the background information is b, this can be (though need not be— see Chapter 10, Section 5) incorporated into the probability statements by writing

p(all observed F's are G's, b and F's are selected in accordance with S) ≈ 1:

p(all observed F's are G's, b) ≉ 1.

On some b's but not others S will be biased in favor of G.

A much weaker notion of bias is possible, according to which S is a selection procedure for choosing F's that is biased in favor of G if

p(all observed F's are G's, F's are selected for observation in accordance with S) > p(all observed F's are G's).

Like the previous criterion, this requires that the use of S increase the probability that all observed F's are G's. However, unlike the former it does not require that given that S is used the probability that all observed F's are G's is approximately 1.

In what follows I shall use the stronger notion. I believe that in the ravens paradox the selection procedures we have in mind for choosing non-black things to observe are biased in the stronger sense. And, most importantly, if a selection procedure for observed F's is biased in favor of G—in the stronger sense—then, given certain intuitive assumptions, we can demonstrate that p(all F's are G's, all observed F's are G's) < k.

Let

h = all F's are G's;

e = all observed F's are G's;

i = F's are selected for observation in accordance with selection procedure S.

By Bayes's theorem, assuming $p(e, i) \neq 0$,

$$p(h, e\&i) = \frac{p(h, i) \times p(e, h\&i)}{p(e, i)}.$$

Since h entails e, p(e, h&i) = 1. Furthermore, p(h, i) = p(h). (The probability that all F's are G's should not be affected by what procedures are used to select F's for observation—though this can affect the probability that all *observed* F's are G's.) So we have

$$p(h, e\&i) = \frac{p(h)}{p(e, i)}.$$

Now suppose that S is biased in favor of G (in the stronger sense, which is the one that I shall mean hereafter). Then p(e, i) ≈ 1. So

p(h, e&i) ≈ p(h).

Where h is a non-tautological universal statement of the form "All F's are G's" then (in many cases) we may take its prior probability p(h) to be low (much less than k ≥ ½). If so, then p(h, e&i) is also low (<k). Accordingly, in a case in which the prior probability of "All F's are G's" is low, and in which observed F's are selected by a procedure that is biased in favor of G, we may conclude that

p(all F's are G's, all observed F's are G's) < k.

In such a case, since the high probability condition for evidence is violated, we cannot conclude that the fact that all observed F's are G's is evidence that all F's are G's.

For example, selection procedure S_2 for selecting North American mice is biased in favor of brown since

p(all observed North American mice are brown, North American mice are selected for observation in accordance with S_2) = 1,

while

p(all observed North American mice are brown) ≠ 1.

If we can assume that the prior probability p(all North American mice are brown) is low, then, if all the observed mice are selected in accordance with S_2, we may conclude that

p(all North American mice are brown, all observed North American mice are brown) < k.

Hence, given that S_2 is used, and given the high probability condition for evidence, the fact that all observed North American mice are brown is not evidence that all North American mice are brown.

By contrast, if S is not biased in favor of G, then we cannot conclude that p(e, i) ≈ 1. What we have in such a case is

$$p(h, e\&i) = \frac{p(h)}{p(e, i)},$$

which can be a number greater than k, even if p(h) < k.

Turning to the ravens paradox, here is a selection procedure for non-black things that is biased in favor of non-ravens:

S_3: Select non-black things known in advance to be non-ravens.

p(all observed non-black things are non-ravens, non-black things are selected for observation in accordance with S_3) = 1. And p(all observed non-black things are non-ravens) < 1. Obviously there is bias if the criterion for selecting non-black things to observe is that in advance of the observation they be known to be non-ravens.

Here are other biased selection procedures:

S_4: Select non-black things from inside your house;
S_5: Select non-black things in your pocket;
S_6: Select non-black things that are inanimate.

Given background information which does not indicate whether all non-black things are non-ravens but does indicate that it is extremely likely that nothing in your house, or in your pocket, or inanimate, is a raven we can conclude that

p(all observed non-black things are non-ravens, non-black things are selected for observation in accordance with S_4 (S_5, S_6)) ≈ 1.

If we can assume that the prior probability p(all non-black things are non-ravens)—which equals p(all ravens are black)—is small, then, where the selection procedure is one of S_3 through S_6,

(1) p(all non-black things are non-ravens, all observed non-black things are non-ravens) < k.

Accordingly, with such selection procedures for non-black things, we cannot conclude that the fact that all observed non-black things are non-ravens is evidence that all non-black things are non-ravens.

More generally, I suggest that the reason we are unwilling to take the fact that all the observed non-black things are non-ravens to be evidence that all non-black things are non-ravens is that the selection procedure we are following in choosing non-black things is (strongly) biased in favor of non-ravens. It is a selection procedure that is, or is like, one of S_3 through S_6. We choose our non-black things to observe by selecting objects already known to be non-ravens, or by selecting in ways that are known or believed extremely likely to yield all non-ravens (e.g., by selecting objects inside the house).

Are there selection procedures for non-black things that are not biased in favor of non-ravens? Consider these:

S_7: Select non-black birds from different locales known to be frequented equally by ravens and birds other than ravens;

S_8: Select members from a large group that has been collected consisting of non-black birds of every species that has non-black birds;

S_9: Select members from a group of non-black birds that has been collected in the following way: if non-black ravens exist these comprise half the group; if they don't exist then the group consists of non-black birds other than ravens.

Of these three selection procedures, S_9 seems to be the one most clearly free of bias in favor of non-ravens. Using our (stronger) criterion for bias, it seems reasonable to say that

p(all observed non-black things are non-ravens, non-black things are selected for observation in accordance with S_9) $\neq$ 1.

If so, then S_9 does not satisfy our criterion for bias.

Possibly S_9 satisfies some other condition for bias in such a way as to enable us to conclude that

(1) p(all non-black things are non-ravens, all observed non-black things are non-ravens) < k.

(A selection procedure might be *weakly* biased in such a way that (1) is not inferable.) However, it must be emphasized, my main purpose is not to prove that S_9 (or S_7 or S_8) is not biased. If my opponent can show that S_9 is biased in such a way that (1) is inferable then so much the better for my purposes. What I want to argue is that even if S_9 is a selection procedure for non-black things that is not biased in favor of non-ravens, we are not in a position to use S_9. We do not know whether any group of non-black birds presented to us is of the sort described in S_9. If a group of non-black birds containing no ravens is presented, then, in the absence of information about the color of ravens, we do not know whether the group contains no ravens because no (non-black) ones exist, or because although some exist they have not yet been found. The same problem emerges with S_8. In the case of S_7, we must find locales frequented equally by ravens and non-ravens, which is not easy to do. (We cannot guarantee non-bias simply by stocking various locales equally with ravens and non-ravens; for our procedure for selecting ravens for stocking may well be biased in favor of black.)

When observing non-black things it is much easier to follow selection procedures such as S_3 through S_6; it is also quite easy to know *that* we are following S_3 through S_6—much easier than to know that we are following S_7 through S_9. It is easy to follow, and to know that we are following, a procedure that calls for selecting non-black objects we know in advance to be non-ravens, or for selecting non-black things from inside one's house or pocket. But such selection procedures for non-black things are biased in favor of non-ravens. If these are the selection procedures for the observed non-black things then we cannot infer that

(3) p(all non-black things are non-ravens, all observed non-black things are non-ravens) > k.

Hence with such selection procedures we cannot conclude that the fact that all observed non-black things are non-ravens is evidence that all non-black things are non-ravens.

Turning now to the hypothesis "All ravens are black," what about selection procedures for ravens? Compare these:

S_{10}: Select ravens from different locales and at different times of the year;

S_{11}: Select ravens from cages you know to contain only black birds.

S_{11} is clearly biased in favor of black. (p(all observed ravens are black, ravens are selected for observation in accordance with S_{11}) = 1 >> p(all observed ravens are black). Again, I assume that background information indicates nothing about the color of ravens.) Assuming that the prior probability p(all ravens are black) is low, then, if S_{11} is the selection procedure for the observed ravens, we may infer

(4) p(all ravens are black, all observed ravens are black) < k.

By contrast, S_{10} is not biased in favor of black. (We have no reason to suppose that p(all observed ravens are black, ravens are selected for observation in accordance with S_{10}) ≈ 1 >> p(all observed ravens are black).) So if S_{10} is the selection procedure for observed ravens, then (4) is not inferable; or at least it is not inferable on grounds of bias.

Furthermore, as was not the case with S_9 (which I have also supposed to be unbiased), we are in a position to use S_{10}. We can follow this procedure in selecting ravens for observation, and know that we are doing so. Since we have an unbiased selection procedure for ravens that we are in a position to use, and the selection procedures that we are in a position to use for choosing non-black things (e.g., S_3 through S_6) are biased, we prefer the former rather than the latter. We prefer to examine ravens rather than non-black things. This is reasonable to do if we want to draw conclusions of the form

p(all F's are G's, all observed F's are G's) > k,

which, I am claiming, are necessary if we are to say that the observed F's being G's is evidence that all F's are G's.

In his discussion of the ravens paradox Hempel makes a claim that bears some similarity to the ideas above. It can be put like this. When we observe a non-black object such as a white shoe we normally would be in a position to know that it is a shoe and hence not a raven. So the information that it is a non-raven adds nothing to what we already know. ("(It) becomes entirely irrelevant for the confirmation of the hypothesis ('all non-black things are non-ravens').")[16] More generally, when we observe a non-black object we usually know more than that it is non-black. We usually know that it is an item of a type T and we may know without making any further observations that T's (e.g., shoes) are not ravens. If so, observing such an object to be a non-raven provides no new information. Given background information which tells us that T's are not ravens,

p(all non-black things are non-ravens, x is non-black & x is a non-raven) = p(all non-black things are non-ravens, x is a T).

Since these probabilities are equal, Hempel seems to be saying, the information that x, which is non-black, is also a non-raven adds nothing to the confirmation of the hypothesis "All non-black things are non-ravens." Here (though not throughout his article) Hempel seems to be espousing an increase-in-probability concept of confirmation. He concludes that if the non-black thing observed is not yet known to be a non-raven, then its being observed to be a non-raven *would* confirm the hypothesis that all non-black things are non-ravens.

I agree with Hempel that we are often in a situation in which we know more about an item than that it is non-black. However, my point is not that our additional knowledge about the item makes the fact that the item is a non-raven irrelevant since it adds nothing to the evidence. (As seen in Chapter 10, I do not accept increase in probability as either necessary or sufficient for evidence.) My point is that in choosing non-black objects to observe we are (and know we are) following a selection procedure which (even if we do not know it) may be biased. If we are following S_3—which seems to be the one Hempel has in mind, since we will select non-black things known in advance to be non-ravens—the procedure is biased in favor of non-ravens. A procedure can be biased even if we don't know that it is. S_4—selecting non-black things from inside your house—is a selection procedure biased in favor of non-ravens, even if you don't realize this when you use it. Given your background information, it may be the case that

p(all observed non-black things are non-ravens, non-black things are selected for observation in accordance with S_4) ≈ 1 >>

p(all observed non-black things are non-ravens)

although you fail to recognize that this is so.

16. *Aspects of Scientific Explanation*, p. 19.

5. VARIETY OF INSTANCES

Our discussion of the ravens should provide illumination for the present topic. Investigators are warned not to choose evidence all of one type but to vary the instances observed. Why is this so?

Consider the following selection procedure for choosing ravens to observe:

S_m: Select male ravens for observation.

Since p(all observed ravens are male, ravens are selected for observation in accordance with S_m) = 1 >> p(all observed ravens are male), S_m is (strongly) biased in favor of males. If S_m is the selection procedure for ravens, then the observed ravens will not be varied with respect to sex.

More generally, let

h = all observed F's have property P

S = select F's to observe that have property P,

and assume that p(h) << 1. Then, since

p(h, S is employed) = 1,

S is biased in favor of P. If P (e.g., being male) is a property of a general type T (e.g., having a sex), then if S is the selection procedure for the observed F's, and if S is biased in favor of P, we may conclude that the observed F's (e.g., the observed ravens) are not varied with respect to T (e.g., sex). In short, the notion of a biased selection procedure introduced in the previous section can be used to understand a situation of the sort above in which the observed instances are not varied.

Returning to our example in which ravens are chosen for observation, let us suppose that S_m is used, so that the observed ravens are not varied with respect to sex. Under what conditions does this become a dangerous procedure? We are testing the hypothesis that all ravens are black. Suppose that given the background information about color uniformity among opposite sexes in other bird species we conclude

(1) p(female ravens have the same color as male ravens, all the observed male ravens are black) = r.

(The background information on which we base this judgment does not indicate anything about the particular color of female or male ravens.) From (1), assuming that S_m is the selection procedure used in observing ravens, we may infer that

(2) p(all ravens are black, all the ravens that have been observed are black) ≤ r.[17]

Now if r is small—less than k—then, with S_m as the selection procedure for ravens, we may conclude that

p(all ravens are black, all the ravens that have been observed are black) < k.

On the theory of evidence I advocate, this would mean that with S_m as our selection procedure the fact that all observed ravens are black is not evidence that all ravens are black. So what we have is the following situation. S_m is a selection procedure for ravens that is (strongly) biased in favor of males. Hence the observed ravens are not varied with respect to sex. In the light of background information we possess about other species, if the probability that female ravens have the same color as males, given that the observed males are black, is not high (if the probability in (1) < k), then the lack of sex variation in the observed ravens has a profound effect on what we count as evidence. It precludes taking the fact that all observed ravens are black as evidence that all ravens are black.

By contrast, suppose that in the light of background information we take r in (1) to be quite high (>>k). Then even if S_m is the selection procedure for ravens (which is biased in favor of males), we still may be able to conclude that the probability in (2) > k. So bias, and hence non-variation, with respect to sex does not necessarily preclude the observed black ravens from being evidence that all ravens are black. If we have background information suggesting that female ravens probably have the same color as male ravens, then sex bias in the observations will not be damaging. If, however, our background information is very meager, then it may be reasonable to assume that the probability in (1) is not high; again, if background information contains many instances of differences in color between opposite sexes in other species, then the probability in (1) is not high. In either of these cases we may conclude that the probability in (2) < k, and hence that the information about the color of the observed ravens is not evidence about the color of all ravens.

Now substitute the following for S_m:

S: Select both male and female ravens for observation.

---

17. The probability in (2) = p(all female ravens have the same color as male ravens and all male ravens are black). Let x = all female ravens have the same color as male ravens; y = all male ravens are black; z = all male ravens observed are black. Then the probability in (2) = p(x and y, z), since in (2) we are supposing that S_m is used for the observed ravens. Now p(x and y, z) = p(x, z) × p(y, x and z). But according to (1), p(x, z) = r. So p(x and y, z) = p(x, z) × p(y, x and z) ≤ r.

S is not biased in favor of males. If S is the selection procedure for ravens, then the observed ravens will be varied with respect to sex. With S as our selection procedure we cannot infer (2) from (1). So S does not run the same risk as S_m with respect to obtaining evidence for the hypothesis that all ravens are black. This is why S is preferable to S_m.

In general, then, when we test a hypothesis of the form "All F's are G's" the reason we should try to obtain a variety of F's is that we want *evidence*. If the observed F's are not varied—if they are biased in favor of some property P—then the fact that all observed F's are G's may not be evidence that all F's are G's.

6. THEORY AND OBSERVATION

How and when do observations provide evidence for a scientific theory? I shall suppose that a scientific theory T consists of a set of hypotheses. Let b be background assumptions made by users of T. (They will comprise facts that are, or could be, assumed by theorists whether or not they accept T.) Let e be a report of observations. I shall take e to be *evidence for theory T* if and only if e is evidence that h_T, where h_T is some (central) hypothesis in T or a conjunction of hypotheses in T. And "e is evidence that h_T" is to be understood in accordance with the definition of potential evidence given in Chapter 10. Thus evidence for a theory need not be evidence that each hypothesis in the theory is true; it is evidence that some (central) part of that theory is true.

Given that e is an observation report which is evidence that h_T, typically h_T will purport to explain e rather than conversely (or rather than h_T and e both being explained by something else). So instead of the broader "There is an explanatory connection between e and h_T" employed in Chapter 10, we can use "The reason that e is that h_T." Accordingly, e is evidence that h_T, given b, if

(1) e and b are true;

(2) e does not entail h_T;

(3) $p(h_T, e \& b) > k$;

(4) $p(\text{the reason that e is that } h_T, e \& h_T \& b) > k$.

In view of conditions (3) and (4), if h_T and h_T' are hypotheses in T, and e is evidence that h_T, it does not necessarily follow that e is evidence that h_T'.

Consider as an example Rutherford's nuclear theory of the atom (which Bohr used as a basis for his own quantized version). One central hypothesis in that theory is that (h_1) an atom contains a positive charge that is not evenly distributed but is concentrated in a nucleus whose volume is small compared to that of the atom. Another hypothesis is that (h_2) the positively charged nucleus is surrounded by a compensating charge of moving electrons.[18] There were observations which Rutherford took to be evidence for his theory in virtue of their being evidence that h_1. These observations were taken from scattering experiments in which alpha particles (doubly positively charged helium atoms) from radioactive sources are directed at a very thin metal foil (e.g., a gold foil 4×10^{-7} meters thick). A scintillation screen is placed behind the foil. Each alpha particle travels through the foil and hits the screen producing a flash of light. The number of these scintillations is observed with a microscope and their scattering angle is determined. Most of the alpha particles travel through the foil without being scattered; but some are scattered at wide angles. A good deal of background information is being assumed. One of the most important assumptions is that alpha particles are positively charged particles much more massive than electrons; another is that positive charges repel each other according to Coulomb's law.

Given the results of the scattering experiments and the background assumptions, it is probable that, since most alpha particles go through the foil without being scattered, the foil atoms are mostly empty of matter. Those alpha particles that are scattered are not scattered by the much less massive electrons. So they are probably scattered by a positive charge concentrated in one small portion of the atom. Therefore,

p(an atom contains a positive charge that is not evenly distributed but is concentrated in a small nucleus (h_1), results of scattering experiments and background information) > k.

Furthermore,

p(the reason that the scattering results are what they are (e) is that the atom contains a positive charge that is not evenly distributed but is concentrated in a small nucleus (h_1), e & h_1 & b) > k.

If the results reported in the scattering experiments are indeed correct, and the background assumptions are true, then, since the observational results do not entail h_1, we may conclude that Rutherford's observation reports of the scattering experiments constitute evidence that h_1. (All four conditions above are satisfied.) Since h_1 is a central part of Rutherford's theory, the observation reports constitute evidence for Rutherford's nuclear theory.

18. See E. Rutherford, "The Scattering of Alpha and Beta Particles by Matter and the Structure of the Atom," reprinted in *Foundations of Nuclear Physics*, ed. Robert T. Beyer (New York, 1949), pp. 111–30.

Notice that in the present case, unlike the simple examples used in Chapter 10, we need to present a theoretical argument to show that $p(h_1, e \& b) > k$. (In the lottery cases this probability can simply be "read off" from the descriptions of the cases; indeed, this is why I chose them.) But the fact that in many cases some non-trivial argument is needed to establish that $p(h, e \& b) > k$ does not vitiate this as a condition for evidence. Notice also that evidence for a theory T need not be evidence that each hypothesis in T is true. Although the results of the scattering experiments are evidence that h_1 they are not evidence that h_2 (that the positively charged nucleus is surrounded by a compensating charge of moving electrons). Given the results of the scattering experiments, h_2, and b, it is not probable that the reason that the alpha particles are scattered the way they are is that the positively charged nucleus is surrounded by a compensating charge of moving electrons (or that the reason for the latter is the former, or that some hypothesis explains both facts).

Now let's take the two most important conditions for evidence—(3) and (4)—and see what happens when one or the other is violated. Let T be a theological theory which contains h as a central assumption:

h = some 10,000 years ago God created the earth and He continues to sustain it;

e = the earth exists;

b = scientific background information now accepted.

Conditions (1), (2), and (4) for evidence are satisfied. Thus in the case of (4), given that the earth exists and that some 10,000 years ago God created the earth and continues to sustain it, it is highly probable that the reason that the earth exists is that some 10,000 years ago God created the earth and continues to sustain it. However, given our scientific background information b, we are very reluctant to say that the fact that the earth exists is evidence that some 10,000 years ago God created it and continues to sustain it. On my view, this is because the high probability requirement (3) is not satisfied in this case. Indeed, given our scientific background information which includes the results of carbon dating, $p(h, e \& b)$ is low.

This type of example illustrates a fundamental defect in the simple hypothetico-deductive method. According to the latter, if a hypothesis entails some observational conclusion, then the fact that the observational conclusion is true is evidence that the hypothesis is true. What this fails to require is that the hypothesis be probable in the light of the observational conclusion (plus background assumptions). Hypothesis h above entails e. But since h is not probable, given e and b, e is not evidence that h. Using the hypothetico-deductive method many "wild" hypotheses can be invented which entail the observed data. One reason we regard these hypotheses as "wild," which is also a reason we do not take the data to be evidence that they are true, is that their probability is so low, given the data and background assumptions.

To see what happens when the explanation condition (4) is violated, consider once again Rutherford's theory. Let

h_1 = an atom contains a positive charge that is not evenly distributed but is concentrated in a small nucleus;

b = the alpha particle scattering experiments have the results they do; plus additional background information about alpha particles, repulsion of like charges, etc.;

e = the scattering experiments were conducted in the years 1909–1913.

Is e evidence that h_1, given b? Intuitively, not. Yet conditions (1) through (3) are satisfied: (1) e and b are true; (2) e does not entail h_1; (3) $p(h_1, e \& b) > k$. The last obtains since $p(h_1, b) > k$, and e does not reduce h_1's probability, given b. The condition that is violated is (4). We cannot say that

p(the reason that the scattering experiments were conducted in the years 1909–1913 is that an atom contains a positive charge that is not evenly distributed, etc., $e \& b \& h_1) > k$.

More generally, given the truth of $e \& b \& h_1$, it is not probable that there is an explanatory connection between h_1 and e.

To obtain a case of evidence we might replace h_1 with

h_3: it was during the years 1909–1913 that Rutherford and his co-workers came to believe that an atom contains a positive charge that is not evenly distributed but is concentrated in a small nucleus.

Keep b and e the same. e is evidence that h_3, given b. (I shall assume that $p(h_3, e \& b) > k$.) Condition (4) is now satisfied since

p(the reason that h_3 is that e, $e \& h_3 \& b) > k$.

But $h_3 \neq h_1$.

Finally, let me briefly mention the mediating role that can be played by hypotheses in theories in certain cases of evidence. An example involves the use of Gay-Lussac's law (that gases combine in simple ratios by volume) and of certain theoretical background information available to Gay-Lussac in 1809. Let

e_1 = the gases hydrogen and oxygen combine in a simple ratio by volume;

e_2 = the gases nitrogen and oxygen combine in a simple ratio by volume.

The background information b includes the fact that in gases, by contrast to solids and liquids, the force of cohesion between molecules is slight; therefore, gases, by contrast to solids and liquids, should "obey simple and regular laws."[19] e_1 can be taken to be evidence that e_2, given b. How is this possible? Even if the high probability condition is satisfied, how can e_1 and e_2 satisfy the explanation condition? The latter must be taken to be not the restricted (4) above but the more general

p(there is an explanatory connection between e_1 and e_2, e_1 & e_2 & b) > k.

Now, recall that a sufficient condition for there being an explanatory connection between e_1 and e_2 is that there is some hypothesis h such that the explanations (The reason that e_1 is that h; explaining why e_1) and (The reason that e_2 is that h; explaining why e_2) are both correct. Indeed, given the fact that the gases hydrogen and oxygen combine in a simple ratio by volume, as do the gases nitrogen and oxygen, and given the background information b, the probability is high that the reason that the gases hydrogen and oxygen combine in a simple ratio by volume is that (it is a law) that all gases do; and the same for nitrogen and oxygen. Given e_1 and e_2 and b, the probability is high that Gay-Lussac's law mediates between e_1 and e_2 by explaining both via a special-case-of-a-law explanation.

19. Joseph Louis Gay-Lussac, "Memoir on the Combination of Gaseous Substances with Each Other," reprinted in *The World of the Atom*, eds. H. A. Boorse and L. Motz (New York, 1966), pp. 160–70. See p. 161.

Index

A priori requirement, 162-64, 257ff.
Achinstein, Jonathan, 365n.
Adams, Frederick R., 265
Alternation-State, 62
Appropriate instructions, 56, 112-16
Argument: α, 124, α', 125, β, 126
Aristotle, 5-7, 13, 68, 93, 121, 259-60, 313-14
Arthritic bike rider case, 344-45
Association (of proposition with explaining act), 86
Austin, J. L., 16, 83
Ayala, Francisco, 266

Background information and evidence, 339-42
Bayes's theorem, 333, 352
Becquerel, Henri, 228-29
Belnap, Nuel D., 29n.
Bennett, Jonathan, 266
Body of a question, 299
Boër, Steven E., 215-17
Bohr, Niels, 109, 110, 119ff., 238
Boorse, Christopher, 265, 272n.
Bootstrap theory, 355-62
Boyle's law, 145-47
Brody, Baruch, 11-13, 121, 127, 176-78, 185-86, 218
Bromberger, Sylvain, 15-16, 29n., 62, 66-68

Canfield, John, 263
Carnap, Rudolf, 323, 328n., 331n., 364-65
Cartwright, Nancy, 219, 247n.
Causal: explanation, Ch. 7; definition (A), 223, definition (B), 227; model (Brody), 11-12, 127, 185-86; -motivational model, 186; principle C_1, 232; C_2, 235; C_3, 239; principle of inclusion, 232
Causation: as a relation, Ch. 6; Aristotle's doctrine of, 5-7; See Causal
Causey, Robert, 219, 291, 306-09.
Classification explanation, 233-35
Collins, Arthur, 252n.
Complete: answer form, 30; content-giving proposition, 40; knowledge-state, 55; presupposition of a question, 29
Complex derivation explanation, 239-43
Confirmation: see Evidence
Consequence condition for evidence, 362-64
Consistency condition for evidence, 362, 364-65
Content: -giving proposition, 39; -giving sentence, 37; -noun, 31-39; -question, 49
Converse consequence condition for evidence, 362, 365
Correct: answer to a content-question, 105; explanation, 103-6; minimal scientific correctness, 158, 189ff.
Cotes, Roger, 232, 291, 295, 301
Counterpart what-causal explanation, 225-26
Creath, Richard, 171n.
Cummins, Robert, 288, 310-11

Davidson, Donald, 16, 27n., 186n., 187n., 205, 208
De re vs. de dicto: knowledge, 24; intentions, 64
Deductive-Nomological (D-N) model, 7-9, 70-71, 93, 101-2, 164; D-N dispositional model, 165, 168; D-N motivational model, 165-66, 169
Derivation explanation, 237-39
Descartes, René, 230-31

383

384 Index

Distributive sentence, 34
Dretske, Fred, 78n., 199-201
Ducasse, C. J., 186n., 188n.

e-sentence, 80
Ellis, Brian, 314-18
Emphasis, 78-81, 89-91, Ch. 6 *passim;* paraphrase views of, 203-9; -selecting words, 196-97; semantical vs. non-semantical uses of, 195-97, 213-17
Entailment condition for evidence, 362, 364
Equivalence condition for evidence, 362-64
Essential property model, 12, 176-78
Evaluation of explanations: correctness, 103-6; good explanations, 116-17; illocutionary evaluations, 107-8; in science, 117-55; role of general methodological values, 148-50
Even number case, 347-49
Evidence: bootstrap theory, 355-62; conditions of adequacy for, 362-67; definition combining probability and explanation, 336, 340; explanation definition, 334-36; potential, 323; probability definitions, 328-34, 351-55; role of background information, 339-42; subjective, 326; for theories, 377-81; veridical, 324
Explainability, 292-96
Explaining (act), 3, 15-19, 52
Explanandum, 158
Explanans, 158
Explanation (product): argument view, 81-83; new proposition view, 94-98; no-product view, 98-100; ordered pair view, 85-88, 91-92, 99-100; sentence and proposition views, 74-76
Explanatorily ultimate question, 296-302
Explanatory: connection, 324-25; limits, Ch. 9

Feigl, Herbert, 60n.
Feinberg, Gerald, 139
Fodor, Jerry, 25n.
Frankfurt, Harry, 268
Friedman, William, 347n.
Functional: explanation, 284-90; interdependence model, 166-67, 169
Functions: design-functions, 272-74; and ends, 280-83; explanation doctrine of, 266, 269-71; goal doctrine of, 264, 268-69; good consequence doctrine of, 264, 266-68; service-functions, 276-78; use-functions, 274-75
Fundamental facts, 305

Galileo, 142-44, 153
Gay-Lussac's law, 380-81

Geach, Peter, 39n., 281, 282n.
Ginet, Carl, 24-25
Glymour, Clark, 306n., 355-62
Goldman, Alvin I., 186n., 188n.
Good explanation, 116-17; *see* Evaluation of explanations
Grice, H. P., 19n., 215-17

Hanson, N. R., 334
Harman, Gilbert, 335n.
Heilbron, John, 128n.
Hempel, Carl G., 7-9, 93, 101-102, 108n., 118, 121, 123, 157-58, 159n., 164-66, 218, 243n., 253, 263, 264, 267, 291, 295-96, 323, 362-63, 365, 367, 373-74
Hesse, Mary, 328n., 331n.
Homogeneity, 10, 108n.
Hull, David, 263

Identity: explaining identities, 306-10; explanations, 235-37
Illocutionary: act, 3, 16; evaluations, 107-8, 116-17; force problem, 76-78, 88-89; product, 3, Ch. 3
Instructions, 53-56; appropriate, 56, 112-16; universal scientific, 117-25; B, 120; Cn, 133; H, 121, 124-25; Mi, 131; QL, 153; Un, 132
Interrogative transform, 298

Jammer, Max, 127n., 129n.
Jobe, Evan K., 178-81

Kim, Jaegwon, 171-72, 198, 203-4, 209-13
Knowledge: by acquaintance, 24-25; state, 24
Kuhn, Thomas, 128n.

Laslie, Adele, 345n.
Lees, Robert B., 33n., 34n.
Lehman, Hugh, 263
Levin, Michael, 204, 266
Lewis, David, 219
Loch Ness monster case, 342-43
Lottery cases: first, 328-29; second, 330; third, 334

McCarthy, Timothy, 187n.
Mackie, J. L., 351-53
Martin, Jane, 57
Matthews, Robert J., 20-21
Meixner, John B., 176n.
Methodological values, 137-38, 148-50
Models of explanation, 158, 257-61; causal model (Brody), 11-12, 127, 185-86; causal-motivational model, 186; deductive-nomological (D-N) model, 7-

9, 70-71, 93, 101-2, 164-66; essential property model, 12, 176-78; functional interdependence model, 166-67, 169; and the illocutionary theory, 188-92; priority model (Jobe), 178-81; question model (van Fraassen), 181-85; statistical relevance (S-R) model, 9-11, 125-26, 173-76
Moravcsik, Julius, M. E., 5-6

n-state, 61-63
Nagel, Ernest, 264, 265
Natural behavior, 310-18
NES requirement, 159-62, 257ff.
Newton, Isaac, 153-54, 246, 303
Nicod's criterion, 362, 365; Modified Nicod, 363, 366
Nominal, 33
No-product view, 98-100
Nozick, Robert, 353-55

Oppenheim, Paul, 157
Ordered pair view, 85-88, 91-92, 99-100

ϕ-existential term, 30

Paradox: of ideal evidence, 329; of the ravens, 367-74
Poole, B., 268
Popper, Karl, 329n., 353
Presupposition of a question, 29; complete presupposition, 29
Principle of reasonable belief, 327
Priority model of explanation, 178-81

Quark model, 139-42
Question: direct and indirect, 16-17; model of explanation, 181-85; related to a science, 318; sound question, 61-62

Ravens paradox, 367-74
Reason-giving explanation, 237, 254-57
Referential: opacity, 194, 197-98; transparency, 78, 194
Restructured sentence, 92, 98
Ruse, Michael, 263, 264
Russell, Bertrand, 24
Rutherford's theory of the atom, 119, 377-78

Salmon, Wesley C., 9-11, 93, 108n., 121, 125-26, 173-76, 186n., 218, 253, 331n., 332n., 333
Schiffer, Stephen R., 20n
Semantical aspect-selectivity, 195
Scriven, Michael, 23n.
Searle, John, 83
Selection procedure, 368; bias in, 369
Severed head case, 346
Shoemaker, Sydney, 232n.
Skyrms, Brian, 346n.
Sorabji, Richard, 263, 276
Special-case-of-a law explanation, 230-33
Stalled car case, 335
Statistical explanation, 251-53
Statistical relevance (S-R) model, 9-11, 125-26, 173-76
Steel, Thomas B., 29n.
Swinburne, Richard, 328n.

Toulmin, Stephen, 60n., 291, 302
Truncated version of a content-giving sentence, 105

u-restriction, 49
Ultimate question of its kind, 297-98
Understanding, Ch. 2 *passim;* and content, 42-45; and explaining, 63-66; and instructions, 53-56; understanding q_I, 54
Unger, Peter, 214-15, 249
Universal scientific instructions, 117-25
Universalists, 122

van Fraassen, Bas, 181-85
Variety of instances, 375-77
Vendler, Zeno, 15, 33n., 193n.

Walker, Herbert, 53n.
Wh-question, 29
What-causal: equivalent, 223; explanation, 225; interrogative, 220; question, 221; transform, 222
Wheaties case, 332
Woodfield, Andrew, 263, 264, 282n.
Woodward, James, 166-67, 170, 186
Wright, Larry, 266, 267, 269-71, 278-80

Lightning Source UK Ltd.
Milton Keynes UK
UKHW010823150620
364934UK00001B/2

9 780195 0374